DJORDJE I. NIKOLIĆ
ALEKSANDAR M. OGNJEVIĆ

DORNIER
THE YUGOSLAV SAGA
1926 - 2007

First Edition
© by KAGERO Publishing

AUTHOR
Djordje I. Nikolić, Aleksandar M. Ognjević

PHOTO CREDITS
Djordje Nikolić, Aleksandar Ognjević, Dejan Milojević, Aleksandar Smiljanić, Aleksandar Kolo, Ognjan Petrović, Mario Hrelja, Milan Micevski, Bojan Dimitrijević, Nebojša Milovanović, Miloš Milosavljević, Momir Milinović, Aleksandar Milošević, Mileta Protić, Igor Černiševski, Marko Babić, Petar Bosnić, Željko Marković, Vojislav Stankov, Spomenko Marković, Djordjević Family, Aleksandar Aleksić, Predrag Grandić, Boris Ciglić, Šime Oštrić, Predrag Stamenković, Dragan Drašković, Josip Novak, Robert Čopec, Mario Raguž, Tomislav Aralica, Danijel Frka, Dinko Predoević, Tomaž Perme, Dalibor Jovanović, Giancarlo Garrelo, Gregory Alegi, Roberto Gentilli, Oliver Fisher, Peter Petrick, Michel Ledet, Peter Simeon, Jan van den Heuvel, Andrew Stamatopoulos, Denés Bernád, Gyorgy Punka, Ian Carter, Andrew Crawford, Andrew Thomas, Guenter Frost, Herald Schiess, Aviation Museum - Belgrade, Airbus Corporate Heritage, Dornier Museum Friedrichshafen, HPMS, Stato Maggiore Aeronautica, Luftfahrt-Archiv Hafner, The Museum of Flight, ETH Zürich, Staatsarchiv St. Gallen, Swissair00

DRAWINGS
Vojislav S. Stankov, Oleksandr Boiko

COLOUR PROFILES
Vojislav Carević

TRANSLATION/PROOFREADING
Djordje I. Nikolić

DTP
KAGERO STUDIO – Łukasz Maj

LUBLIN 2021

ISBN 978-83-66673-61-8

KAGERO Publishing
Akacjowa 100, os. Borek, Turka
20-258 Lublin 62, Poland
phone/fax +48 81 501-21-05
e-mail: kagero@kagero.pl, marketing@kagero.pl, shop@kagero.pl
www.kagero.pl, shop.kagero.pl

Table of contents

Acknowledgments

This book would not be possible without the generous help from numerous friends and fellow aviation enthusiasts around the world. We would like to use this opportunity to most sincerely thank: Josip Novak, Robert Čopec, Mario Raguž, Danijel Frka, Dinko Predoević and Tomislav Aralica from Croatia, Michel Ledet from France, Edelgard Piroth, Udo Hafner, Ingo Weidig, Günter Frost, Oliver Fisher, Volker Koos, Peter Petrick and Hans-Ulrich Wilbold from Germany, Andrew Stamatopoulos from Greece, Gyorgy Punka from Hungary, Giancarlo Garello, Gregory Alegi and Roberto Gentilli from Italy, Jan van den Heuvel from Netherlands, Denés Bernád from Romania, Dejan Milojević, Aleksandar Smiljanić, Nenad Jovanović, Predrag Miladinović, Dragan Drašković, Aleksandar Kolo, Ognjan Petrović, Šime Oštrić, Mario Hrelja, Milan Micevski, Bojan Dimitrijević, Nebojša Milovanović, Miloš Milosavljević, Momir Milinović, Aleksandar Milošević, Vladeta Vojinović, Igor Černiševski, Marko Babić, Petar Bosnić, Željko Marković, Vojislav Stankov, Dejan Beda, Spomenko Marković, Djordjević Family, Aleksandar Aleksić, Slavko Cirkvenec, Predrag Grandić, Nenad Arizanović, Boris Ciglić, and Predrag Stamenković from Serbia, Harald Schiess from Switzerland, Tomaž Perme, Dalibor Jovanović and Marko Ličina from Slovenia, Tod Rathbone from USA, Ian Carter, Andrew Crawford, Andrew Thomas, Jeff Jefford, Kris Hendrix and Ken Stewart from UK.

With the assistance from the following archives and institutions, we were able gain access to valuable documents and records which aided our research and helped shed light on many historical facts: Airbus Group Archives, Luftarchiv Hafner, Dornier Museum, Militararhiv, Bundesarchiv, Vojni Arhiv, Muzej vazduhoplovstva - Beograd, Arhiv Vojnoistorijskog Instituta, Arhiv Jugoslavije, The Museum of Flight, USAF Museum, National Air and Space Museum, Imperial War Museum, RAF Museum, The National Archives, 230 Squadron Archives, Stato Maggiore Aeronautica, Staatsarchiv St. Gallen, Hrvatski Pomorski Muzej and Hrvatski Povijesni muzej.

Margot Nistl from Airbus Corporate Archives helped us immensely in answering all of our questions, helping with all inquires and emails, searching through archive documents day after day. The authors can say freely that without her help many important aspects of this book would be incomplete. We would like to thank her immensely for working with us and we hope to cooperate on future project as well.

We would also like to express our gratitude to Vojislav Carević and Vojislav Stankov, who created superb color profiles and technical drawings respectively, adding a new dimension to our book.

Our thanks goes to Kagero Publishing and Mr. Damian Majsak, who recognized the importance of this project, decided to have it published and make it available for all readers interested in the history of the legendary Dornier company as well as Yugoslav and Serbian aviation. Thank you as well to Maciej Łacina and Łukasz Maj from Kagero for an excellent cooperation and support!

Most importantly, we are forever grateful for persistent support we received from our spouses and children, for their encouragement during this journey to leave a lasting legacy in the history of aviation in Yugoslavia and Serbia.

Authors

Introduction

Claude Dornier's contribution in the field of aviation started with the recognition by Count Zeppelin, who immediately following his employment at *Luftshiffbau Zeppelin*, discovered young engineer's potential and provided him with a team and tools to succeed. Design after design, Claude Dornier and his team proved their ingenuity and vision, designing airplanes and floatplanes unlike any seen before. His own words put this into perspective:

I was always endeavoring to pursue my design principles even when I was risking the possibility that my ideas would be regarded as out of date. The greatest part of our success can be attributed to the fact that, during the decades over which our company has been built up, I was assisted by a large number of excellent engineers who gave their best at all times. It is this loyalty which enables me to look into the future with great confidence.

It is not unusual that following World War I, when the need arose to renew its aging air force which consisted mostly of obsolete war relics, the Kingdom of Yugoslavia (KJ – *Kraljevina Jugoslavija*) recognized the qualities of Dornier's airplanes and approached the factory instead of going exclusively to its traditional war time ally, France. Of course, wartime reparation payments played a role in this decision, however this is not to detract from the fact that at the time airplanes which the Kingdom of Serbs, Croats and Slovenes (KSHS – *Kraljevina Srba, Hrvata i Slovenaca*) and later KJ expressed interest in and eventually purchased from Dornier, had no equivalents. From 1926 when Do D was first introduced in PV (*Pomorsko Vazduhoplovstvo* – Naval Air Force) KSHS, various *Dornier* airplanes and floatplanes served with distinction and outlasted numerous changes in politics, names of countries, borders, flags, kings, dictators and presidents. Three of those, Do 17K, Do 22 and Do Wal, took part in the brief 1941 April war opposing the German Reich who itself fielded Do 17 bombers in its ruthless attacks against the Yugoslav capital, Belgrade.

After 94 years of distinguished service, the eventual sale of Dornier Do 28D-2 to a foreign buyer will bring about the closure to the maginficent and without a precedent Dornier Saga in the western Balkans. Even 50 years after the passing of great Claude Dornier, his brand of airplanes is still leaving a lasting mark in the world as well as Yugoslav and Serbian aviation history.

A spectacular photograph taken at Divulje with Do D in the center, the first of the Dornier types in Yugoslavia. Several other floatplanes then in service are visible, of which two license built Zmaj Hanriot 41 are sitting on Dornier's wings, clearly showing the robustness of its design. [Josip Novak]

Dornier-Werke GmbH history

The Dorniers trace their origins to a French family from the department of Isere. In 1862, Dauphin Dornier, a language teacher, came to Kempten in Bavaria. Dornier settled there for good, following the 1870-1871 Franco-Prussian War and married a daughter of a local family. On 14 May 1884 their first son, Claude Dornier, was born in Kempten. Claude grew up in his parent's home and attended a local school, with science being his prime interest. He attended *Technische Hochschule* (University of Applied Sciences) in Münich and in 1907 Claude Dornier earned his degree in engineering. Shortly after graduating, junior engineer Dornier was employed at *Machinenfabrik Nagel* (Machine factory Nagel) in Karlsruhe where he worked on strength calculations. After leaving *Machinenfabrik Nagel*, Dornier was briefly employed at *Eisenwerkes Kaiserslautern* (Kaiserslautern Iron works) in Kaiserslautern.

Claude Dornier joined *Luftshiffbau Zeppelin*, makers of the famous all-metal rigid airships, in 1910 where his abilities soon attracted Count Zeppelin's attention. In 1911, he began fundamental research to improve the strength of light sections and metal profiles. In May 1911, he succeeded in proving that the flanges increase the rigidity of aluminum angle sections by performing tests, which considerably influenced the profile of thin, stressed components. At the same time, Dornier was conducting numerous studies into the possibilities of further developing rigid airships.

By 1913, Count Zeppelin had gained such confidence in Claude Dornier's capabilities that he appointed him as his personal scientific advisor. In close cooperation with the Count, Claude Dornier began preliminary design work on a giant steel structure airship for transatlantic service. The Count soon came to realize that in order to fully exploit Claude Dornier's potential, he would need to provide him with better facilities.

Early in 1914, the "Do" Department was established within the *Luftschiffbau Zeppelin*. The so nicknamed "Carbonium" facility, a small gasworks located on the

Claude Dornier, the founder of Dornier-Werke. [Airbus Corporate Heritage]

At the edge of the existing Zeppelin airship factory, the so called "Carbonium" building housed the first "Do" department where Claude Dornier continued his research into the design of rigid airships. [Dornier Museum Friedrichshafen (Airbus Group)]

Seemos facility near Manzell where Count Zeppelin gave Claude Dornier the opportunity to practice his airplane construction skills. [Dornier Museum Friedrichshafen (Airbus Group)]

Following closure of Seemos yard, Claude Dornier decided to resume activities on the Swiss side of Lake Constance, he leased a small wooden building with a ramp leading into the lake. [ETH Zürich]

edge of the existing airship factories, provided Claude Dornier with space for two offices, a small workshop and a test area where he worked with an assistant engineer, several technicians and draftsmen. At his new facility, Claude Dornier continued research into the designs of rigid airships however this did not last long. Motivated by the emergence of a completely new means of lighter than air flight technology, his interests shifted towards airplane engineering.

Almost immediately following the outbreak of World War I on 28 July 1914, Count Zeppelin decided to build airplanes and established Seemoos facility near

Dornier Metallbauten GmbH Friedrichshafen Werk Manzell following the expansion. Note Do Wal in front of one of the hangars. [ETH Zürich]

Manzell, Germany, for this purpose. The Count gave Claude Dornier the opportunity to use his own design ideas for airplane construction. As a result of increasing airplane construction, new and large facilities for their time were constructed at this location.

Instead of relying on contemporary methods used to date, the Count confident in Claude Dornier's ingenuity entrusted him with the task of building giant metal flying boats at the new dockyard. This leap of faith laid the cornerstone in the evolution of metal airplanes. Considering the new task Claude Dornier began to evaluate and apply his research. The new problems however, could be solved only through a professional engineering approach based upon strict scientific criteria. The principles he proposed were to guide the entire development of the airplane industry in future decades and implied that all stressed-structure components should be of metal, steel or aluminium, depending upon the stress loads involved. Similarly, sections rolled from sheets and formed in accordance with the performance requirements should be of light metal construction and must carry loads. Finally, components must be attached by rivets or screws.

During World War I, Dornier worked on Zeppelin-Lindau Rs. I flying boat, which was ready for the first flight on 12 October 1915. This was the first Dornier designed airplane and was the first German airplane

Dornier logo of the time period. [Airbus Group]

to use duraluminium in its construction. Sadly, this airplane was destroyed when strong wind in the early dawn of 22 December broke its moorings and ran it aground where it was broken up by the waves. Claude Dornier did not allow this to set back his ambitions and by 30 June the following year the next Dornier designed flying boat, *Zeppelin-Lindau* Rs. II, took off. During the war years, metallic structures were further developed and applied to series of smaller airplanes. On

Dornier Altenrhein facility in Switzerland. [ETH Zürich]

3 November 1917 the stressed skin design was born with the first flight of the *Zeppelin-Lindau* CL. I, a two-seat escort and ground attack biplane. Structural frames and smooth sheet skin designed by Claude Dornier were slowly becoming the design standard of the future. The next flying boat to be successfully flown was Rs. III which proved to have excellent characteristics in heavy North and Baltic Sea weather conditions. The last airplane to fly before the end of World War I was D I, a single-seat fighter, which flew on 4 June 1918. Both the fuselage and the cantilever wing were stressed skin designs and it had jettisonable fuel tank beneath the fuselage. These features were ahead of their time. The end of the war brought other ambitious plans to an abrupt end however Dornier's amassed flying boat design and building experience would guide him in the years to come.

Shortly after the end of World War I, German aviation industry all but came to an end. *Zeppelin* factories at Reutin and Zech near Lindau were closed and almost all the staff was discharged. At the Seemoos facility, some 100 employees were producing buckets and wash boilers barely keeping the doors open. Claude Dornier however, continued his work against the dim prospects for the future, working on the two engine Gs. I flying boat, a predecessor to the famous Wal. In 1922, airplane manufacturing was prohibited by law

and the Seemoos yard had to be closed. Claude Dornier decided to resume activities outside of Germany where at Rorschach, on the Swiss side of Lake Constance, he leased a small wooden building with a ramp leading into the lake.

Following a suggestion by Dr. Hugo Eckener, the inter-war manager of *Zeppelin Werke GmbH Lindau*, in 1922 the factory was renamed to *Dornier Mettalbauten GmbH* and the company offices were moved from Lindau to Friedrichshafen. In 1923 the company purchased the nearby facilities of *Flugbau Lindau* in Manzell and the small dockyard at Seemoos was finally closed.

In the fall of 1924, the *Technische Universität* (Technical University) in Stuttgart presented an honorary engineering doctorate to Claude Dornier in "recognition of his merits in advancing airplane engineering".

As the company continued to prosper in the second half of the 1920s more and more airlines were using Dornier airplanes and asked for a quicker delivery. With the distance between the design offices at Friedrichshafen and the production facility at Marina di Pisa, precious time was being wasted in constant back and forth motion by the staff. Claude Dornier discovered a suitable site at Alternhein on the Swiss side of Lake Constance where he founded the *Aktiengesellschaft für Dornier-Flugzeuge* in the summer of 1926, which resulted in the Marina di Pisa facility being sold.

Marina di Pisa facility in Italy. Do Wals, such as the two newly completed I-AZAA and I-AZDC visible in the photograph, were manufactured here until 1932. [Airbus Corporate Heritage]

In 1929, design emphasis began to switch from commercial to military airplanes, particularly bombers. In the early 1930s the world economic crisis was casting its shadow on the aviation industry. *Luftshiffbau-Zeppelin* was losing interest in airplanes. Claude Dornier used this opportunity to acquire the remaining shares of *Dornier Metallbauten GmbH*, thus opening the way to new projects. When orders started to pick up again in 1933, a new branch was established; *Norddeutsche Dornier-Werke GmbH* at Wismar on the Baltic coast. Not long thereafter factories in Lubeck, Münich-Neuaubing and Münich-Oberpfaffenhofen were opened.

In 1931 the three engine Do Y bombers were manufactured at Altenrhein for KJ and in the same quest for export orders, tactical and reconnaissance airplanes and floatplanes followed amongst which were Do 10, Do C3, Do C2A and Do 22.

In July 1935 *Dornier Metallbauten GmbH* had a total of 7,080 employees. In 1937 factory changed is name from *Dornier Metallbauten GmbH* to *Dornier-Werke GmbH*. On 1 October 1938 *Dornier* factory at Friedrichshafen had 10,375 employees, which made a total of 7.1% of the entire German airplane industry. This number increased by 25% in the previous 9 months of the same year due to the high demand. Late in 1939, a total of 17,980 employees were working at the Dornier factories.

It is important to note that the *Dornier* factory at Friedrichshafen was a development firm, the opposite of the *Dornier* factory at Wismar, which was a typical licensed serial production facility. 6% of the total employees at Friedrichshafen were involved with new designs and construction of prototypes, 15% were employed with general duties and 54% with series production.

With the outbreak of World War II, virtually all of *Dornier* capacity had to be devoted to military needs. The famous flying boats were redesigned or rebuilt for reconnaissance, transport and emergency sea rescue roles and pressed into service with the Luftwaffe until the very end of the war.

Almost all *Dornier* factories were destroyed by allied bombing campaign and those which remained were dismantled in the process of German disarmament. The airplane production in Germany was prohibited while that in East Germany was lost to the Soviets. In order to make ends meet, with help of his sons and loyal associates, *Dornier* established in 1950 *Lindauer Dornier Gesellschaft GmbH* which produced textile machinery. This plant was built from ground up and with the engineering expertise its production was fully automated and efficient.

Similar to the condition following World War I, an overseas office called *Oficinas Téchnicas Dornier* was opened in February 1951 in Madrid, Spain,

Do Wal flying over Marina di Pisa facility. Three newly completed Do Wals awaiting delivery are visible on the ground. [Dornier Museum Friedrichshafen (Airbus Group)]

and was under management of Dornier's eldest son. Shortly thereafter at the request of the Spanish Ministry of Aviation a bid was presented for a liaison airplane with short take off capabilities. The airplane submitted in the bid was the Do 25. With design completed in six months, following the production of prototype, the first flight took place on 23 June 1954. Already October 1956 a new Do 27 appeared and met all the requirement of STOL (Short Take Off and Landing) set forth. In 1955 the ban on airplane production in Germany was lifted and *Bundeswehr* (Federal Defense) placed a contract for Do 27, which was however manufactured by CASA (*Construcciones Aeronáuticas SA*) in Spain. Do 27 was followed by Do 28 Skyservant which was manufactured both for

domestic use as well as foreign customers. Numerous designs for both military and civilian use followed and *Dornier* even expanded into the sphere of aerospace with work on satellites, Ariane rocket components and life support system for the American Spacelab space station.

Claude Dornier, the founder and the owner of *Dornier* companies died on 5 December 1969 at the age of 85. With his passing came the end of an era, marked with aviation milestones until his very last day. Claude Dornier said to his son Silvius, a year prior to his passing:

You know, I have lived a good life, I had a lot to do and design and experienced many things. And now I am simply tired.

Dornier Do D

Development

The story of Dornier Do D began on 6 February 1924 when *Dornier Metallbauten GmbH* signed a contract with *Kawasaki Dockyard & Co. Ltd* from Kobe for the development and construction of a total of eight airplanes. This included the development and construction of a single engine military torpedo and reconnaissance floatplane as well as single engine land airplanes. To speed up the development process, Dornier decided to base the new floatplane on the projects already under development, the Do B, which became the civilian airplane Komet III, and Do C, a light bomber and reconnaissance airplane. A sequential designation Do D was assigned, and the project development continued at a rapid pace. Due to its military role, the Do D project relied more on the Do C type. It borrowed its fuselage design and landing gear was replaced with floats which had to be placed at such an angle to allow for the weapons to be carried below the fuselage.

Construction of the first Do D for the Japanese licensee, under W.Nr. (*Werke Nummer* - Construction number) 57, began in March/April 1924. It was powered by a 375 hp (horsepower) 12 cylinder *Rolls-Royce* Eagle IX engine coupled with a four blade wooden propeller and radiator in the nose. The cockpit was located below the wing. According to the *Dornier* factory documents, this floatplane was 90% complete as of 27 September 1924 and was fully complete in the early October 1924

but the exact date of the first flight remains unknown. Following several test flights, the official hand off to the Japanese delagation took place on 29 October at Manzell. In 1925 it took part in the official Japanese Navy bid for

One of the Do Ds for PV KSHS under construction at Manzell. [Airbus Corporate Heritage via G. Frost-ADL]

The first prototype Do D, W.Nr. 57, completed in early October 1924 is seen here taking off from Bodensee during factory trials. [Airbus Corporate Heritage via G. Frost-ADL]

W.Nr. 131, the first PV Do D, prior to a factory test flight at Bodensee. [Airbus Corporate Heritage via G. Frost-ADL]

Do D on landing approach following another test flight. [Airbus Corporate Heritage, via G. Frost-ADL]

This original period plaque which lists all of the world records attained by W.Nr. 139 belonged to PV pilot Albin Pirc. [Nebojša Milovanović]

Between 16 July and 10 August 1927 W.Nr. 139 piloted by Georg Zinsmaier, Richard Wagner and Egon Fath attained numerous world records. [Airbus Corporate Heritage]

torpedo bomber and was the only one to satisfy the strict reqirements. Despite this, *Kawasaki* was unable to secure a production order. It was registered with the (IAACC) *Inter-Allied Aeronautical Commission of Control* officially as a civilian airplane since Germany was prohibited from developing and constructing military airplanes and this registration was officially received on 10 November 1924.

In 1926 a new 600 hp 12-cylinder BMW model VI engine became available. As soon as the type ap-

proval was granted, series production began and many airplane manufacturers considered it for their designs. Accordingly, in 1926/1927 the Do D series development continued with BMW VI engine without reduction gear, now under the designation Do D bis. Different versions were tested, with enclosed and open float struts, cockpit forward below the wing or behind the wing, belly or side radiator, different tail designs and wing arrangements.

In the early 1920, PV KSHS did not posses floatplanes which could be used as torpedo carriers or bombers which significantly limited its ability to protect the vast coastline along the Adriatic Sea. The domestic industry lacked the technical expertise as well as the machinery to produce a modern floatplane which satisfied the demands set forth. Several state commissions visited European aviation manufacturers to familiarize

W.Nr. 135 crashed into the Bodensee on 24 February 1927. Pilot von Mitterwallner survived but a mechanic, Lehre, drowned. [Djordje Nikolić]

As a replacement for W.Nr. 135, W.Nr. 149 was constructed. This floatplane included several modifications, of which the most prominent was moving the cockpit further aft behind the wing. [Airbus Corporate Heritage via G. Frost-ADL]

themselves with the floatplanes available and evaluate which supplier shall be awarded the contract. As a result, in June 1926 *Dornier* received a contract from KSHS for the purchase of 10 Do D bis through a Zürich based *Aero Metall AG*. Another factor which influenced this decision was the fact that these floatplanes were to be paid from the war reparations fund. They were assigned W.Nr. 131 to 140 and were produced at Manzell. They differed from the base Do D also in that they had different radiators which were now located on each side

W.Nr. 149 during take off. [Djordje Nikolić]

W.Nr. 149 taking off from Bondensee. The Manzell assembly building is visible in the background. [Airbus Corporate Heritage via G. Frost-ADL]

W.Nr 193 and 194 had float struts made from two distinct parts instead of a single float carrier. Note that the floatplane is still unpainted. [Airbus Corporate Heritage via G. Frost-ADL]

Do D next to the 12 engine Do X. Since Do X had its first flight on 15 July 1929, this Do D as likely one of the second series PV machines. [Djordje Nikolić]

of the fuselage as well as a new square window also on the fuselage directly below the wing trailing edge. The first Do D was completed in November and the factory validated that all the requirements were met. The first three were delivered to PV KSHS in January 1927, the next three in March, three by the middle of the year with exception of W.Nr. 139, which was used to attain eight world records between 16 July and 10 August 1927. This floatplane was modified for the occasion by adding the transmission to the BMW VI engine and using a more efficient propeller. It was registered on 13 July in Swiss airplane register as CH 177 and on 15 September, after the record flights, it was duly removed. These flights were conducted by three *Dornier* pilots: Georg Zinsmaier, Richard Wagner and Egon Fath. Before the delivery to PV KSHS in mid September 1927, the engine was reverted to the original configuration. The following world records were secured:

16 July – Maximum altitude of 5,851 m with 1000 kg load

4 August – Maximum speed of 190.435 km/h over a distance of 100 km with a load of 2,000 kg

8 August – Distance covered of 1,600 km with a load of 1,000 kg

8 August – Maximum speed of 175.600 km/h over a distance of 1,000 km with a load of 1,000 kg

10 August – Distance covered of 2,100 km without load

10 August – Distance covered of 2,100 km with 500 kg load

10 August – Maximum speed of 172.000 km/h over a distance of 2,100 km with a load of 500 kg

10 August – Maximum speed of 172.000 km/h over a distance of 2,000 km without load.

One of the Do D intended for the delivery, W.Nr. 135, crashed into Bodensee during flight testing on 24 February 1927 with pilot von Mitterwallner at the controls. The pilot allegedly fainted due to exhaust gas poisoning and despite hitting the water hard he was rescued before the floatplane sank, while one of the assemblers who was inside the fuselage, Lehre, drowned.

Do D and Do X taking off from Bodensee in tandem. Do D was used as a chase plane during Do X test flights. [Staatsarchiv St. Gallen]

All but two Dorniers from the first series were assembled at Kumbor under the supervision and with the assistance of Dornier engineer Josef Götz and technician Balluff. [Airbus Corporate Heritage]

Do D was subsequently recovered on 5 March. As a replacement for W.Nr. 135, Dornier began work on W.Nr. 149 which was constructed with several modifications requested by PV KSHS. These included moving the cockpit further aft behind the wing, eliminating the machine gunner position, moving the internal storage space forward as well as running the rudder control cables on the outside of the fuselage. This Do D was delivered in mid 1927. Moving the cockpit further aft did not improve visibility and following a crash landing on 28 July 1928 these changes were reversed back to the original configuration with exception of the control cables.

Towards the end of 1927 KJ ordered another 14 floatplanes to be delivered in 1929, thus becoming the major user of this *Dornier* type. The delivery did not take place sooner because war reparation payments assigned to KJ for the year 1928 were already booked. As the work commenced, *Dornier* believed that the

floatplanes could be delivered ahead of schedule. The first flight of the second series Do D took place on 15 September 1928. In the end these were delivered as contracted in 1929, with four in July, six in August and the last four in September. Pilots kbb Većeslav Dujšin and pk Danilo Hubmajer were assigned to receive the newly arrived Do Ds into PV KSHS inventory.

The second series incorporated several changes and improvements over the first series:

The fuselage and the float struts were strengthened

The fuselage underside was not recessed and as a result there were no provisions to carry torpedoes, only bombs on external racks

The underside fuselage contour was straight towards the tail instead of having a gentle curve

The rudder control cables ran on the outside of the fuselage

The ruder was covered with fabric instead of duraluminium sheets

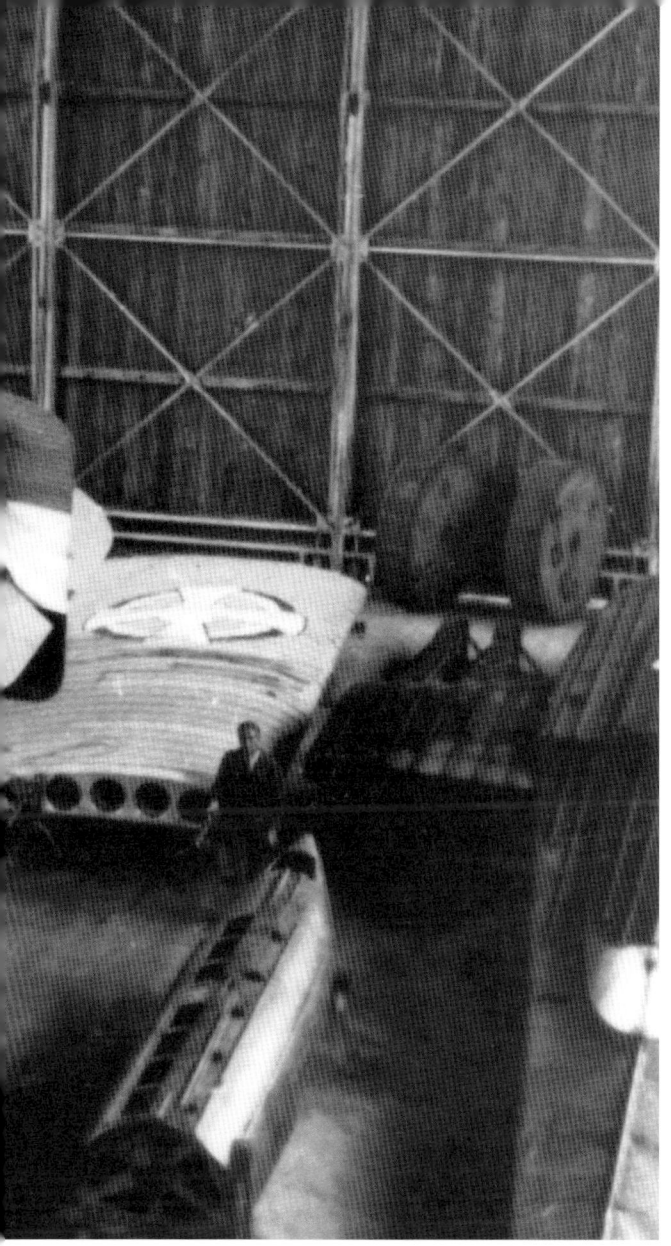

Two of the first floatplanes, W.Nr. 193 and 194, included additional modifications to the float to fuselage support struts, which were made from two distinct parts instead of a single float carrier. This change led to improved handling while turning but at the cost of decreased speed.

Apart from the KJ, RVM (*Reichsverkehrsministerium* - German Ministry of Transport) from Berlin purchased three floatplanes which were used for coastal flights by *Lufthansa* as well as by DVS (*Deutsche Verkehrsfliegerschule* – German commercial pilot school).

With the total of 29 floatplanes produced, Do D was not a major export success for *Dornier*, however it did help PV KSHS fill the gap in offensive capability which gradually began to degrade and would not be resolved until Do 22 arrived in 1938.

Introduction into Service

The first two Do D flew over to Kumbor with Dornier pilots at the controls while the remainder were delivered by rail and were assembled with the assistance of Dornier engineer Götz and technician Balluff who was sent by the factory to KJ to provide maintenance and training of PV mechanics. Following the formal introduction into service, Do D joined 20.HE (*Hidro Eskadrila* – Hydroescadrille). They represented the first modern and all metal floatplanes in PV service. Götz

Engineer Götz, who is sitting next to the PV Commander in the background, and technician Balluff, who is in the foreground, are enjoying a meal with the PV pilots to celebrate the hand over of Do D to the PV. [Airbus Corporate Heritage via G. Frost-ADL]

Do D 201 not long after delivery to PV KSHS. [Aleksandar Ognjević]

This photo taken in 1926 depicts 1st class of non-commissioned officers in front of three Do D at Djenovići. The wooden float support on the left reads D201. [Djordje Nikolić]

20.HE crews at Kumbor with por bb lk Gogala in front of Do D 203. [Djordje Nikolić]

25.HE crews posing in front of Do D 220. [Maden via Tomaž Perme]

Pomorsko Vazduhoplovna Škola, Kumbor, 4 December 1927. [Aleksandar Ognjević]

Do D 213 with young cadets during a celebration event, 1927. [Aleksandar Ognjević]

Do D 201 in front of a hangar at Kumbor is getting washed off to remove the salt water after a flight. [Dalibor Jovanović]

Do D 202 in a flight near the Adriatic coastline. [HPMS]

Do D 201 seen here during a high ranking inspection prior to the torpedo trials. [Djordje Nikolić]

and Balluff also helped establish an overhaul facility at this location as well as at Divulje naval base.

In the spring and summer of 1927, Do D's intended role was put to test when torpedo drop trials began with borrowed 530 mm ship torpedoes, at first at the controls of the German pilot while later PV pilots took over. Pbb Miroslav Gogala conducted one such test on 15 July 1927. Tests showed that a fully loaded Do D with a torpedo could not climb to an altitude greater than 60 m, while torpedoes were dropped from 20 m. This performance was not practical, the floatplane was underpowered to carry such large torpedo and as a result after repeated trials their use in this role was definitely abandoned.

Another view of the Do D 201 with torpedo between its floats. [Miloš Milosavljević]

The moment a torpedo hit the coastline. Due to the underpowered engine and heavy construction, Do D was not suitable for a torpedo bomber role with the available ship borne torpedoes and trials were soon discontinued. [Dejan Milojević]

20. HE CO and pilot, pbb Miroslav Gogala in white outfit, proudly poses with his comrades next to Do D 201, which he piloted during the 11 hr 57 min and 44 sec flight on 29 May 1929. [Djordje Nikolić]

In addition to the torpedo trials, very early on during its use Do D showed other limitations. Pilots complained that despite its robustness, this floatplane was very heavy and its engine underpowered which limited its ability to carry offensive weapons and be maneuverable enough to escape modern fighters which it may have encountered during an eventual war. Additionally, it was very tricky to land, and a number of accidents occurred, which further strengthened its dislike amongst the pilots. By the time the order for the second series was placed, only seven floatplanes of the original 10 remained, while the rest were destroyed in various accidents.

Do D 202 landing at Boka Kotorska. [HPMS]

Do D 205 is stuck on the beach as a result of a low tide at Crikvenica on 7 August 1927. [Djordje Nikolić]

Another view of Do D 205 with crew disembarking following a flight. [HPMS]

On 29 May 1929 Do D 201 flew at an altitude of 2,500 m for a total of 11 h 57 min and 44 sec back and forth between Kumbor and Sušak, for which the 20.HE CO and pilot, pbb Miroslav Gogala, received an award from the *Dornier* factory. During this flight, he was accompanied by mechanic pf Rafael Perhauc, pilot nar Špiro Knešević and radio operator nar Jovan Bek.

The second series of 14 airplanes arrived to KJ in several batches in 1929 where they joined the newly formed 25. and 26.HE while two were sent to the training HE . All three HE were based at the time at Kumbor. Along with the second series, sizeable quantities of spare parts, including floats and wings, were ordered in order to keep all floatplanes in flying condition.

Do D 204 has attracted a lot of attention at one of Dubrovnik's beaches. [Aleksandar Smiljanić]

Do D 212 with immobilized rudder behind remains of another floatplane. [Djordje Nikolić]

Naval fleet show of force at Boka Kotorska with six Do D visible. [Djordje Nikolić]

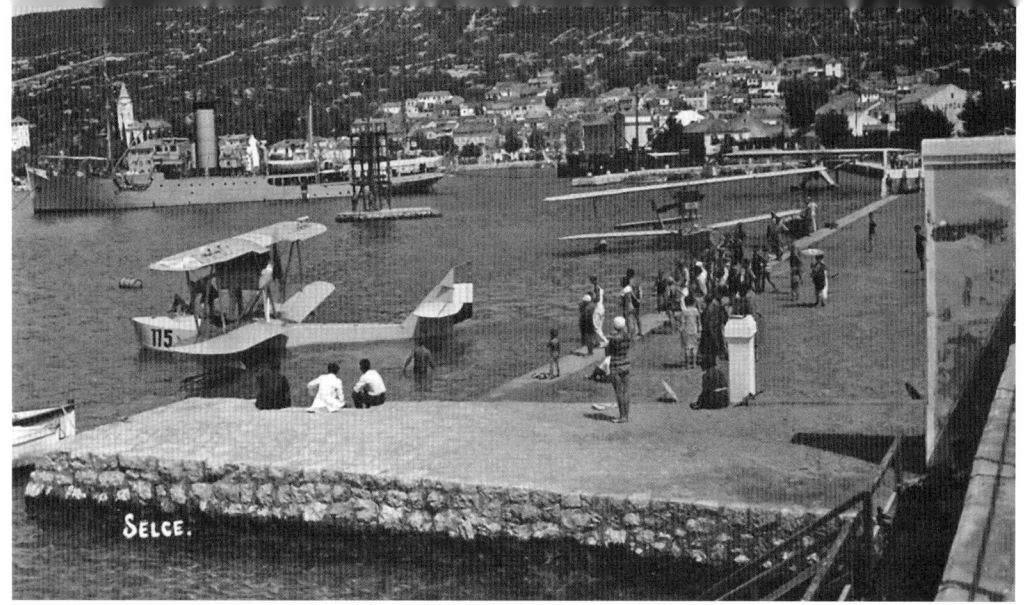

Unknown Do D and IO floatplanes 115 and 121 at Selce harbor during the 1930s. [Djordje Nikolić]

A formation of three Do Ds: 206, 201 and 210 taken on 5 July 1936. [Maden via Tomaž Perme]

A fly over by another trio of Do Ds, this time 206, 201 and 210. [Aleksandar Ognjević]

An interesting Christmas card showing a montage of three Do D in flight over the coast line. [Aleksandar Ognjević]

A lovely lady posing with her sun umbrella on a float belonging to an unknown Do D. [Aleksandar Smiljanić]

Do D 208 is hoisted on to the floatplane tender Zmaj with another (202) already on board. Photograph was taken on 20 June 1933. [Bojan Dimitrijević]

Do D on board Zmaj floatplane tender. Due to the lack of catapults, floatplanes could not be launched from the ship, hence they had to be lowered onto the water. [HPMS]

Each time naval aviators came ashore, they attracted significant attention from the curious ladies. [Aleksandar Smiljanić]

Do D 212 at Divulje. [HPMS]

1st series		2nd series	
W.Nr.	Escadrille No.	W.Nr.	Escadrille No.
131	201	193	211
132	202	194	212
133	203	195	213
134	204	196	214
135	(Crashed at Manzell, replaced by W.Nr. 149)	197	215
136	205	198	216
137	206	199	217
138	207	200	218
139	208	201	219
140	209	202	220
149	210 (replacement for W.Nr. 135)	203	221
		204	222
		205	223
		206	224

As of 1 May 1932, seven Do D were on strength from the first series and 13 Do Ds from the second series.

By 1934, seven Do Ds were struck off charge, leaving 17 in service. PVKJ retrospective review from 1937 lists a total of four Do D were struck off charge: 202, 208, 213 and 221 along with two BMW VI engines. In 1938, four Do D were struck off charge: 211, 216, 219 and 224 along with six BMW VI engines.

Do D periodically operated from the sole floatplane tender ship, *Zmaj*, which was ordered in 1929. Dorniers were unable to take off from the ship due to a lack of a catapult, hence they were lowered with the crane prior to and picked up the same way following flight.

According to the plan formed during the 4 February 1936 KM (*Komanda Mornarice* - Naval Command) meeting, by 1936 obsolete floatplanes which included Do D were supposed to be replaced. This task was accomplished with the purchase of Do 22 and their introduction in 1938. In 1936 there were a total of seven Do D within 2.HK 25.HE, another seven within 3.HK 26.HE and three within PVŠ (*Pomorsko-vazduhoplovna škola* – Naval Aviation School) at Divulje. Of the 17 Do D, seven were inoperable. 25.HE and 26.HE were tasked to with reconnaissance within 30 nm from the KJ territorial waters. It

Do D 217 used as a center piece in this group photograph from 1932. [Aleksandar Smiljanić]

Do D 216 and 222 at Crikvenica. [Aleksandar Smiljanić]

A line up of five Do Ds including 218 and 224 at Divulje naval base. [HPMS]

Another view of the line up at the same location this time with Do D 208 visible. [Maden via Tomaž Perme]

Four Dorniers, 219, 218, 216 and 221 at Divulje naval base. [HPMS]

Do D 213 and 221 at Kumbor. [Aviation Museum - Belgrade]

Do D 220 preparing for a flight. [Mario Raguž]

Do D 220 in flight. [Aviation Museum - Belgrade]

Do D 221 seen here with PV officers in front of a hangar. [HPMS]

Do D 223 at a pier at Šibenik. [Aleksandar Smiljanić]

Do D 222 resting in the calm and sunny waters. [Aleksandar Smiljanić]

Do D 222 at Crikvenica. Note that it lacks the machine gun turret. [Aleksandar Smiljanić]

Do D 224 from 26.HE at Divulje. [Dejan Milojević]

Rogožarski PVT-H 51 head on view with Do Wal/H 260 immediately behind and Do D 214 on the right. Do 214 crashed near Boka Kotorska on 25 April 1939. Three crew members were killed. [Djordje Nikolić]

Do D 205 crashed on 15 September 1927, with 5 crew members killed. [Djordje Nikolić]

The remains of the crew are carried to their final resting place by their comrades. [Djordje Nikolić]

is known that as of July 1936, 201, 206 and 210 were serviceable and in flying condition, while 215 was serviceable in 1937.

These floatplanes served in PVKJ until 25 August 1939 when an official order No.17 issued by the Naval Commander prohibited their further use, since new Do 22s were available in numbers to fully replace them. Do D were however allowed to be only used in sea rescue roles while speeding across the water. They were intended for removal from service in accordance with the report submitted to MViM (*Ministarstvo Vojske i Mornarice* - Ministry of Army and Navy) on 10 May 1940.

Do D 207 crashed on 26 January 1928 near Zelenika. [Djordje Nikolić]

Due to the damage to the front of the 207 floatplane, the crew is presumed dead. [Djordje Nikolić]

Accidents

Year after year a number of accidents with Do D occurred as a result of frequent use resulting in constant attrition and decrease in available number of serviceable floatplanes. Until 1939, one to two Do Ds were lost each year, for a total of 17 floatplanes. The following reconstruction was made based on available documents and photographs.

Do D 210 (W.Nr. 149) crashed near Gruž on 25 July 1928. [Djordje Nikolić]

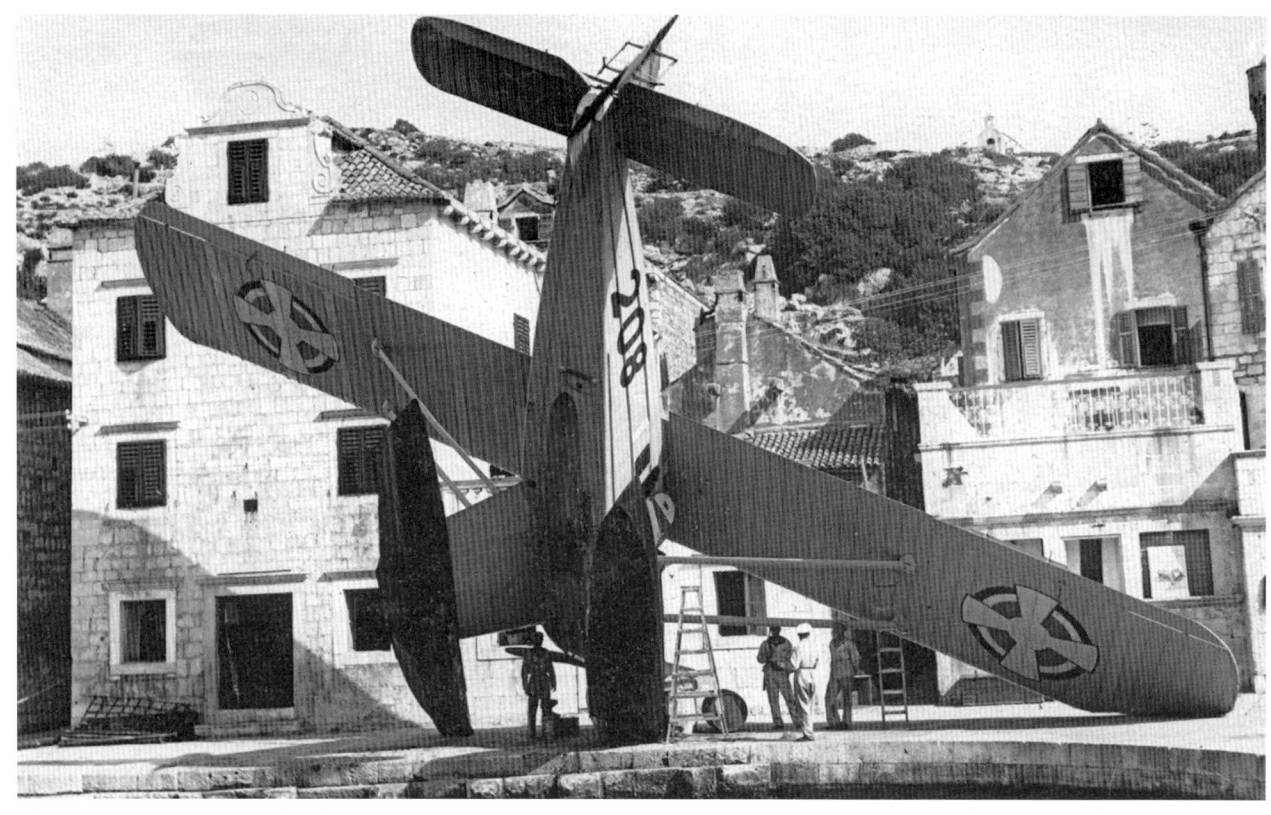

While trying to impress a lady friend, Pbb Lenčič crashed his Do D 208 at a Hvar island pier on 1 August 1929. [Djordje Nikolić]

Do D 201 hoisted aboard *Zmaj* floatplane tender following the accident on 13 August 1931. [HPMS]

The remains of Do D 222 are brought to shore. [Dejan Milojević]

Funeral proceedings for the crew of Do D 222. [Dejan Milojević]

Severely damaged remains of Do D 222 show how violent the crash was on 21 October 1931 when the entire crew perished. [HPMS]

Do D 202 struck a buoy during a night landing. [HPMS]

On 14 September 1937 Do D 208 stalled and fell on land near Boka Kotorska. [HPMS]

Sad remains of Do D 214 which crashed on 25 April 1939 with all hands lost. [HPMS]

Apparent hard landing resulted in a port float carrier and starboard wing tip damage on Do D 218 on an unknown date. [HPMS]

One of the not so severe accidents took place when the Do D 201 ran aground at Boka Kotorska. [HPMS]

View of the damaged float carrier on Do D 218. [HPMS]

Do D 204 which crashed under unknown circumstances is seen here being hoisted to shore by an auxiliary boat crane. [HPMS]

Sailors posing next to the fuselage of Do 215 in a hangar. [HPMS]

Do D 219 suffered a broken fuselage, likely the result of a hard landing. [HPMS]

The crash of Do D 220 in June 1936 near Trogir. [HPMS]

Escadrille No.	Date	Notes
205	15 September 1927	Crashed during training flight near Herceg Novi due to pilot error. The entire crew perished
207	26 January 1928	Crashed into water near Zelenika under unknown circumstances. The entire crew perished
210	25 July 1928	Crashed into water near Gruž under unknown circumstances
208	1 August 1929	Pbb Lenčič crashed into the pier at Hvar island, tipping the floatplane on its nose. Minor damage
201	13 August 1931	Crashed into water with pilot Bernuš at the controls. The remainder of the crew was likely injured. The floatplane was hoisted on *Zmaj* following the accident
222	21 October 1931	Crashed in Trogir bay with its entire crew led by pilot pf Šoštarić and co-pilot pf Pisačić. The entire crew perished
220	June 1936	Crashed into ground near Trogir bay with all crew presumed dead
202	27 July 1937	Force landed in the night at Gruška harbor striking a buoy, which destroyed the floatplane and its engine, but fortunately the crew remained unharmed
208	14 September 1937	Stalled and fell on land near Boka Kotorska. As a result, the flight mechanic was severely wounded, and the floatplane and its engine were completely destroyed
214	25 April 1939	Crashed near Boka Kotorska. Three crew members killed
201	Unknown exact date. After 5 July 1936	Crashed into a structure on the shore in Boka Kotorska. Light damage
204	Unknown	Unknown circumstances
209	Unknown	Unknown circumstances
215	Unknown	Unknown circumstances
218	Unknown	Port float and starboard wing damaged. Unknown circumstances
219	Unknown	Apparent hard landing. Damaged fuselage
224	Unknown	Unknown circumstances. Judging by the available photographs, the crew likely did not survive

The Italian forces captured one of three PV Do Ds at Divulje naval base. They proudly displayed the Italian flag. [Manca collection via Aleksandar Ognjević]

April War

When the April war began, only three Do D remained. On paper Do D 212 was assigned to 1.HK (*Hidro Komanda* – Hydro Command) 15.HE, 203 to 3.HK 20.HE and 204 to 3.HK 21.HE.

203 was listed as unserviceable as of 31 January 1941 and was to remain out of service until 1 April but its engine was listed as serviceable, so the reason why it was grounded and remained listed as unserviceable as late as 5 April remains unknown. Until that time, the register shows that this floatplane flew a total of 1,224 times since 1927, underwent six overhauls and its engine S/N 16700 had 650 hours and 25 minutes working time and was overhauled eight times. It was captured by the Italians at Divulje and was shortly thereafter adorned with an Italian flag for propaganda purposes.

204 was listed as serviceable also as of 31 January 1941 but unserviceable as of 5 April. It had a total of

Another view of captured Do D 203. Allegedly this photograph was taken in 1943, which begs to question why the Italians kept such an obsolete floatplane for two years after it was captured. [Airbus Corporate Heritage]

A photograph of hangar II taken by the Italian forces at Kumbor shows silhouettes of Do D on the hangar door. [Stato Maggiore Aeronautica]

790 hours and 38 minutes flight time during a total of 735 flights and was overhauled a total of seven times. Its engine S/N 16139 was serviceable and had a total of 319 hours and 56 minutes working time and was overhauled six times.

212 during the April war remained at 2.HK workshop at Divulje since it was unserviceable and was intended to be struck off charge and scrapped. Due to their obsolescence, the Italians did not press these floatplanes into service and they were eventually scrapped.

Do D did not take part in combat operations as it lacked any combat value due to its obsolescence. It was only suitable for support roles and would have been struck off charge eventually and scrapped in the coming year.

Construction Features

Do D was a braced high wing monoplane of mostly metal construction. Heavy duty parts were made from steel while all others were made from duraluminium in order to improve corrosion properties due to exposure to saltwater as well as to reduce weight. duraluminium was further protected from corrosion with the application of a special coating. Both materials were used in the form of profiled sheets which were joined together with rivets, which allowed for their simple alignment, installation and replacement.

The massive rectangular wings consisted of a center section and two outer halves which had rounded tips. The wing center section was connected to the fuselage

Do D in this factory photo shows its overall metal construction with sturdy fuselage, float carriers, wings and floats. Note the attached depth charges between the floats. [Airbus Corporate Heritage via G. Frost-ADL]

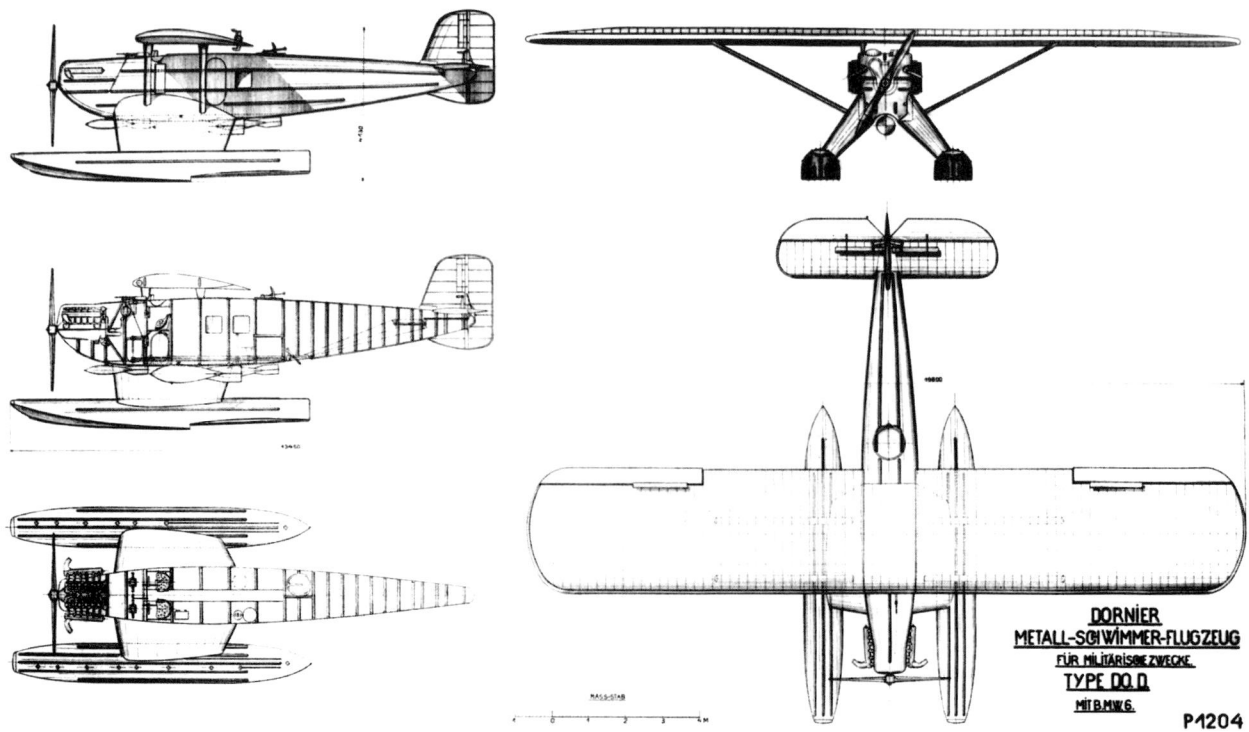

Original factory drawing showing Do D with various weapons configurations as well as the fuselage cutaway. [Airbus Corporate Heritage via G. Frost-ADL]

with two short, rigid pairs of stubs. The outer wing halves were held in place with two struts each, which were in turn attached to the float carriers. The inside of the wings consisted of spars made from steel profiles and box ribs which stiffened the construction. The corrugated duraluminium sheets which attached to the box ribs further strengthened the wing. The raised edges of the sheets were riveted together at the joints covered

with a narrow, U shaped, profiles. The ailerons were covered with fabric and had trimmers installed on top.

Fuselage was of rectangular cross section and was subdivided into engine in the nose, two seat cockpit for pilot and co-pilot and mechanic and radio operator area which also housed machine gun positions. Fuselage tapered from nose to tail, and it was stiffened with two external longerons. On the interior, the outer skin

The massive wing was supported by two pairs of struts on each side and its center section sat on two short and rigid pairs of stubs. [Airbus Corporate Heritage via G. Frost-ADL]

Do D 201 and Do 202 under construction at Manzell. Note the longerons which run along the fuselage and serve the purpose to stiffen its construction. [Airbus Corporate Heritage]

A view of Do D inner wing structure. This wing came from the destroyed Do D 207. [Djordje Nikolić]

The cockpit of W.Nr. 149 clearly shows it was moved further aft. The pilot and the copilot had basic flight instruments, of which some were mounted on the trailing edge of the wing cutout. Note the simple windshields. [Airbus Corporate Heritage]

Do D cockpit interior was very basic. Pilot and co-pilot used a yoke wheel to fly the floatplane and sat on rudimentary metal seats which were leather lined and adjustable in height and position. [Airbus Corporate Heritage]

Do D 221 shows here the float carrier mounted machine gun mounts which did not prove very practical. [HPMS]

Only the first series Do Ds were envisioned as torpedo carriers. The second series was modified to remove the recess below the fuselage. [HPMS]

A nice view of the floats with stiffening ribs on the top as well as the port side door, window and the foot steps. [Dejan Milojević]

strengthening profiles which were positioned perpendicular to the direction of the flight were attached to the bend resistant frames. The resulting construction allowed for a spacious inside area which permitted inspection of the riveting until the very back of the fuselage. The engine was bolted to two supports which were attached to the two strongest front U-shaped fuselage frames. The

fuselage below the floats on the first series floatplanes had a recess which enabled Do D to carry torpedoes. On the second series Do D, this feature was removed and instead external bomb cradles were used. The cockpit was located on top of the fuselage below the wing, apart from the W.Nr. 149 which temporarily had the cockpit relocated further aft. Neither solution offered

12 cylinder water cooled BMW VI engine during maintenance by PV mechanics. Easy to remove panels simplified engine maintenance. [Dejan Milojević]

good visibility for the crew. Crew was protected from the wind by a small windshield on either side. Pilot and co-pilot used a yoke wheel to fly the floatplane and sat on rudimentary metal seats which were leather lined and adjustable in height and position. All instruments were mounted on a control panel and in the case of W.Nr. 149 some were even mounted in the wing structure. The crew entered the floatplane through an oval door in the port side of the fuselage, just below the trailing edge of the wing. The fuselage also had a large square window on either side, directly behind the cockpit. The crew did not carry parachutes.

The tail assembly was attached with four bolts on top of the very end of the fuselage. The horizontal and vertical stabilizers were both lined with duraluminium sheets while on the first series rudder was also made from duraluminium and on the second series it was fabric lined. Both the ailerons and the rudder were provided with trimmers installed on horizontal and vertical stabilizers respectively. Horizontal stabilizer angle was adjustable on the ground.

Do D rested on two watertight and compartmentalized floats which had longitudinal stiffening ribs. Access panels on the top allowed for inspection and repair while screws on the bottom allowed for drainage while on land. Floats were prone to leaks, which was a design flaw. Floats were connected to the fuselage with large carriers which *Dornier* designed purposely to look like a hydrofoil and these were set at an angle of 50º to the direction of flight. Due to their size they actually provided additional lift. Only two floatplanes, 211 and 212, had four individual float carriers which did not improve performance and resulted in decrease in speed. Hand and footrests were mounted on the float carriers for easier access and maintenance.

Do D was powered by a 600 hp BMW VI water cooled engine, which was the first 12-cylinder engine built by BMW. The engine was easily accessible for maintenance due to foldable access panels. It was coupled to a two-blade fixed pitch wooden propeller. The engine was isolated from the rest of the cockpit with a fire wall. The radiators were mounted on both sides of the fuselage, just below the wing leading edge.

73 octane fuel was carried in two wing tanks in the center section, each with 285 l capacity. Fuel was

Do D 206 undergoing engine maintenance in a hangar. Mechanics pose proudly with the floatplane under their care. [HPMS]

An unknown second series Do D showing its unique radiator arrangement. [Mario Raguž]

PV Do Ds were painted in overall DKH (Dr. Kurt Herberts) L40/52 Light Grey color and had their escadrille numbers applied in Black color on the fuselage. [Djordje Nikolić]

gravity fed to the engine. Do D could also be equipped with additional fuel tanks below pilot's and co-pilot's seats with 150 l volume each. Two oil tanks with 50 l volume each were installed directly behind the engine.

The first series floatplanes were intended to be torpedo carriers and for that purpose they were armed with a single 530 mm torpedo carried in the fuselage recess between the floats. Factory photographs show that depth charges could also carried on the same mounts. The second series did not have the recess for the torpedo, instead one 250 kg bomb or two 100 kg depth charges could be carried on external cradles. Bomb sight was manufactured by *Goertz* company.

Defensive armament consisted of twinned 7.7 mm *Darne* machine guns in a turret on top of the fuselage and a single 7.7 mm *Darne* machine gun protruding through the hatch in the fuselage belly. Some PVKJ floatplanes were modified with machine gun mounts installed on the float supports which would require one of the crew to get out and operate the gun in flight while exposed to the wind and risking a free fall.

Do D was equipped with a *Telefunken* radio with a 200 m antenna which was stored in a drum and which was typically extended in flight from the rear fuselage. A reconnaissance camera could also be carried, depending on a mission.

Technical Specifications

The technical specifications declared in the *Dornier* documentation differed significantly from the performances obtained during service in the PV. The speed, range and maximum service ceiling were all tested to be significantly lower than those declared by the factory and are shown in below table.

Technical specifications Dornier Do D (1st series)	
Quantity used:	10
Crew:	3
Years of Service:	1927-1941
Span:	19.6 m
Length:	13.5 m
Height:	4.1 m
Wing area:	62.0 m²
Engine:	One 500[1]/600[2] hp BMW VI
Empty weight:	2,600 kg
Maximum weight:	3,900 kg
Maximum speed:	170 km/h
Cruise speed:	150 km/h
Service ceiling:	3,300 m
Maximum range:	750 km
Armament:	One 530mm anti-ship torpedo, one 250 kg bomb or two 100 kg depth charges. Three 7.7 mm Darne machine guns for defensive purposes, two in twinned turret on top and one in the belly

Technical specifications Dornier Do D (2nd series)	
Quantity used:	14
Crew:	3
Years of Service:	1927-1941
Span:	19.6 m
Length:	13.5 m
Height:	4.1 m
Wing area:	62.0 m²
Engine:	One 500[1]/600[2] hp BMW VI
Empty weight:	2,800 kg
Maximum weight:	3,900 kg
Maximum speed:	160 km/h
Cruise speed:	130 km/h
Service ceiling:	3,000 m
Maximum range:	637 km
Armament:	One 530mm anti-ship torpedo, one 250 kg bomb or two 100 kg depth charges. Three 7.7 mm Darne machine guns for defensive purposes, two in twinned turret on top and one in the belly

[1] Nominal power
[2] Takeoff power

Do D were always kept in pristine condition as seen here. [Airbus Corporate Heritage]

Camouflage and Markings

During the initial testing in Germany Do Ds remained unpainted. Prior to delivery to KSHS, floatplanes were painted in overall *DKH* (*Dr. Kurt Herberts*) L40/52 Light Grey color and had their escadrille numbers applied in Black color on the fuselage. "Kosovo cross" insignia were applied both on top and below the wings while the rudder was painted in the colours of the Yugoslav flag. The floats were painted entirely Black. Throughout their service, Do D kept this camouflage scheme. It is likely that the they were touched up with paint during service and overhaul. It is unknown if paint was sourced locally or was purchased from abroad. As seen in the photographs, they were always kept in pristine condition.

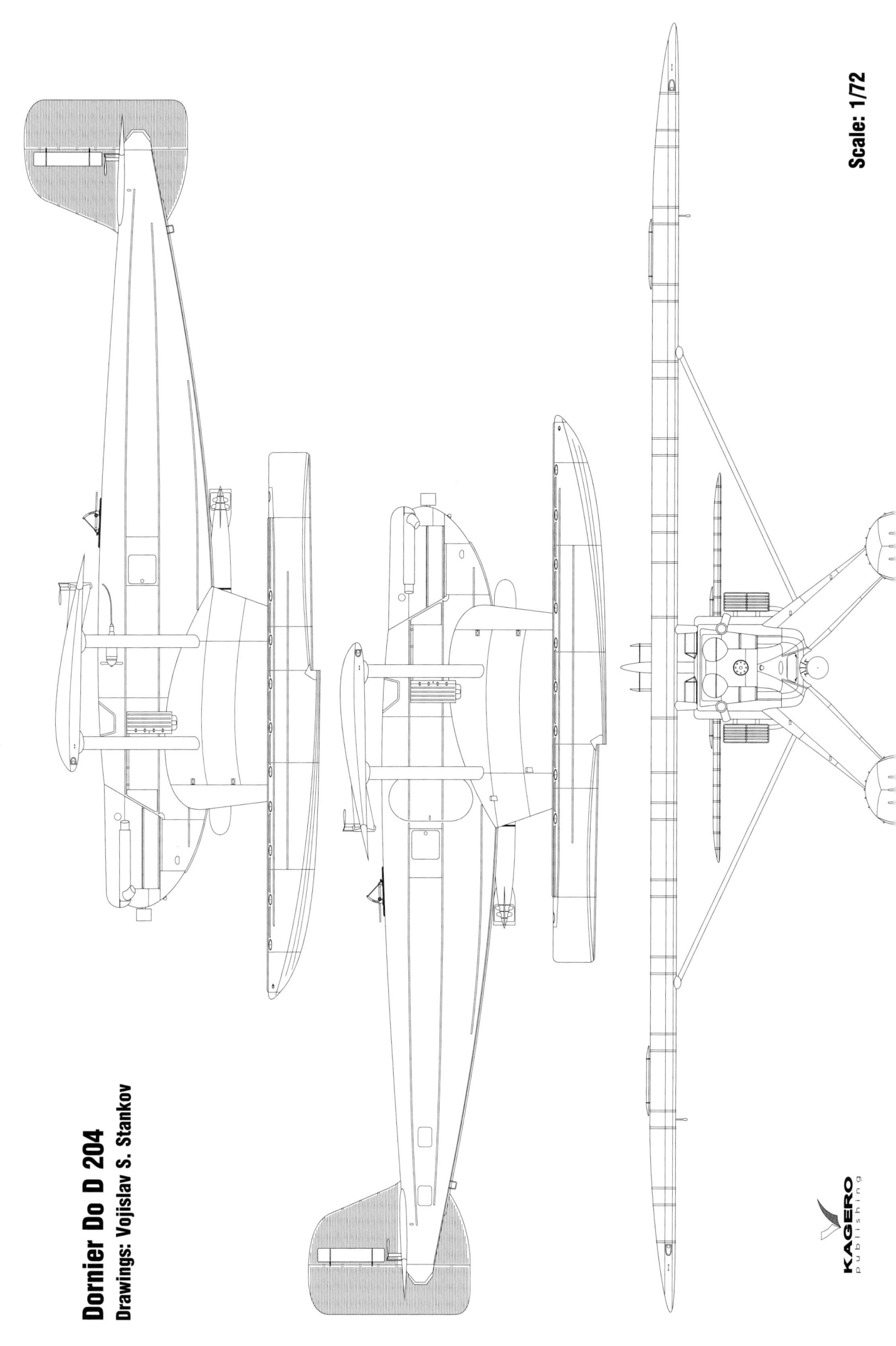

Dornier Do D 204
Drawings: Vojislav S. Stankov

KAGERO publishing

Scale: 1/72

Dornier Do D 204

Drawings: Vojislav S. Stankov

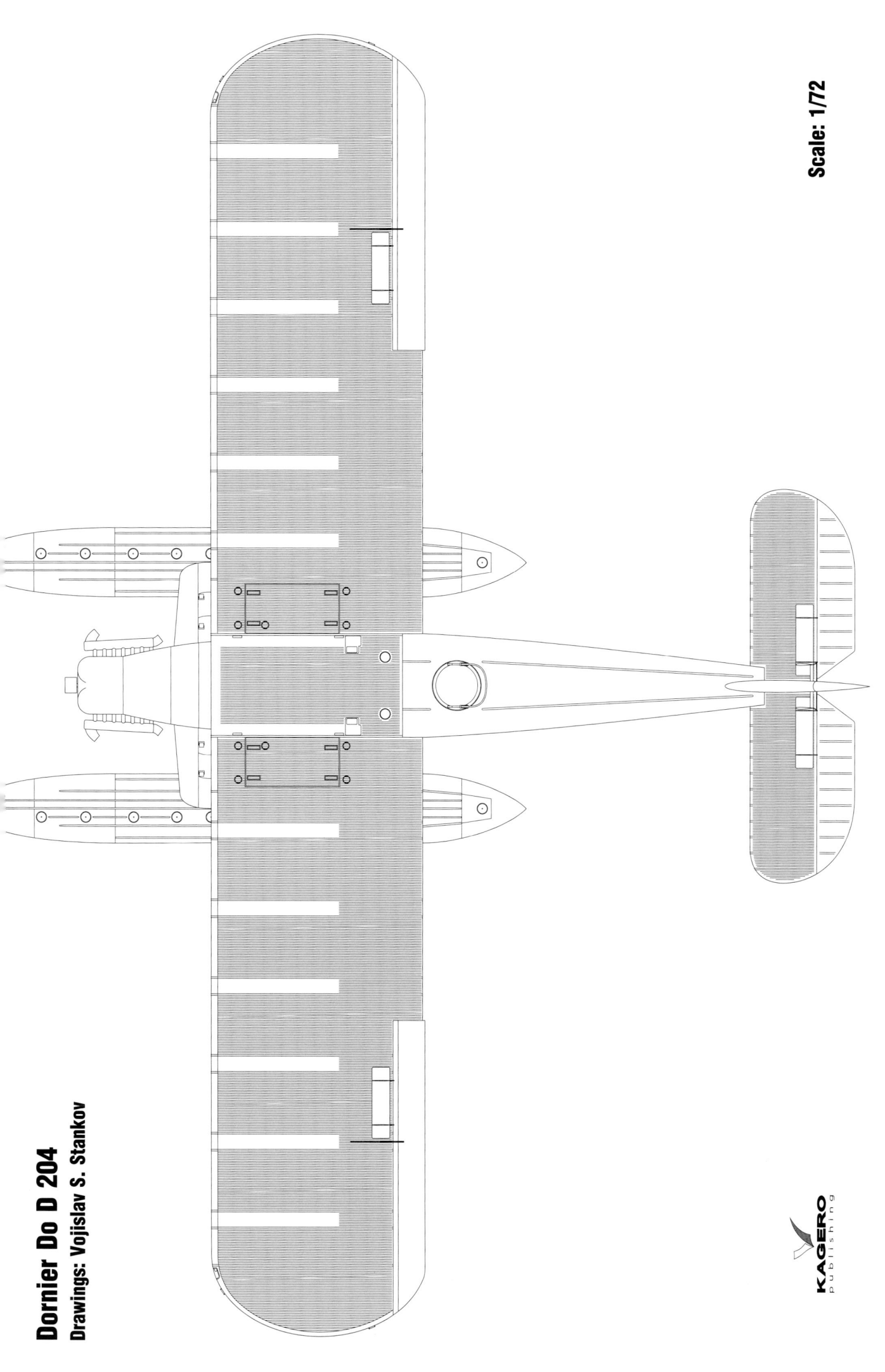

KAGERO
publishing

Scale: 1/72

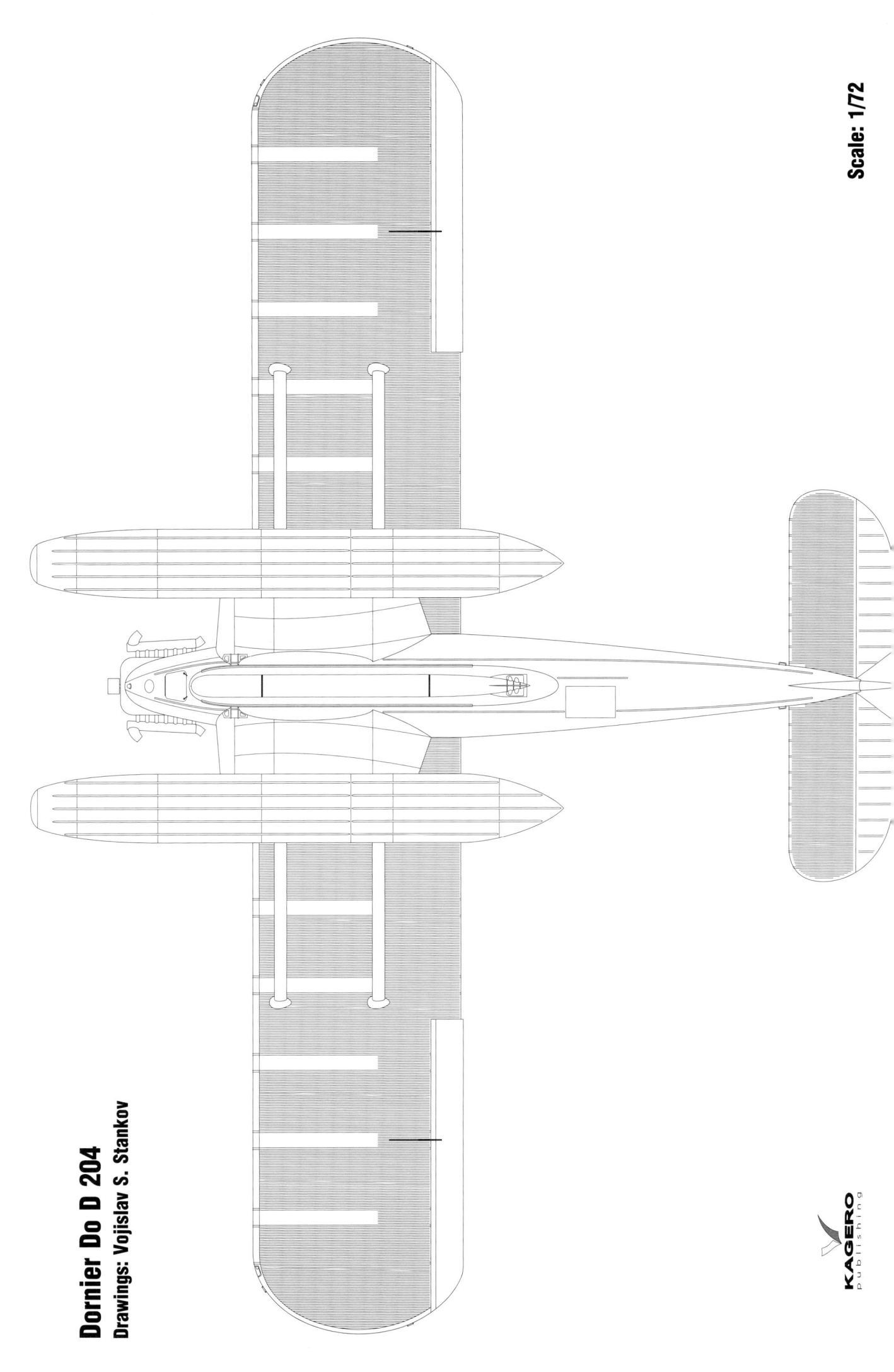

Dornier Do D 204
Drawings: Vojislav S. Stankov

Scale: 1/72

KAGERO
publishing

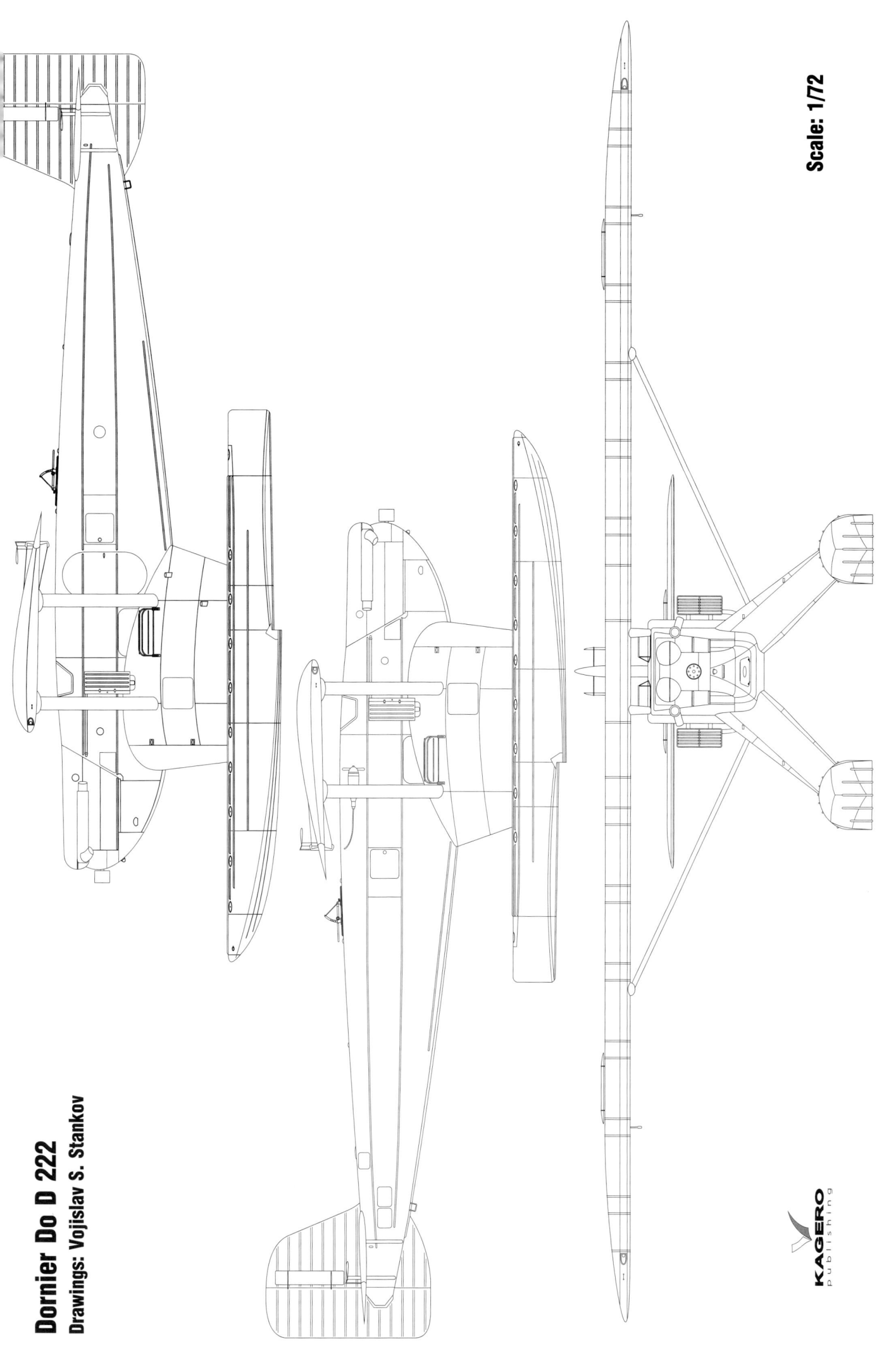

Dornier Do D 222

Drawings: Vojislav S. Stankov

Scale: 1/72

KAGERO
publishing

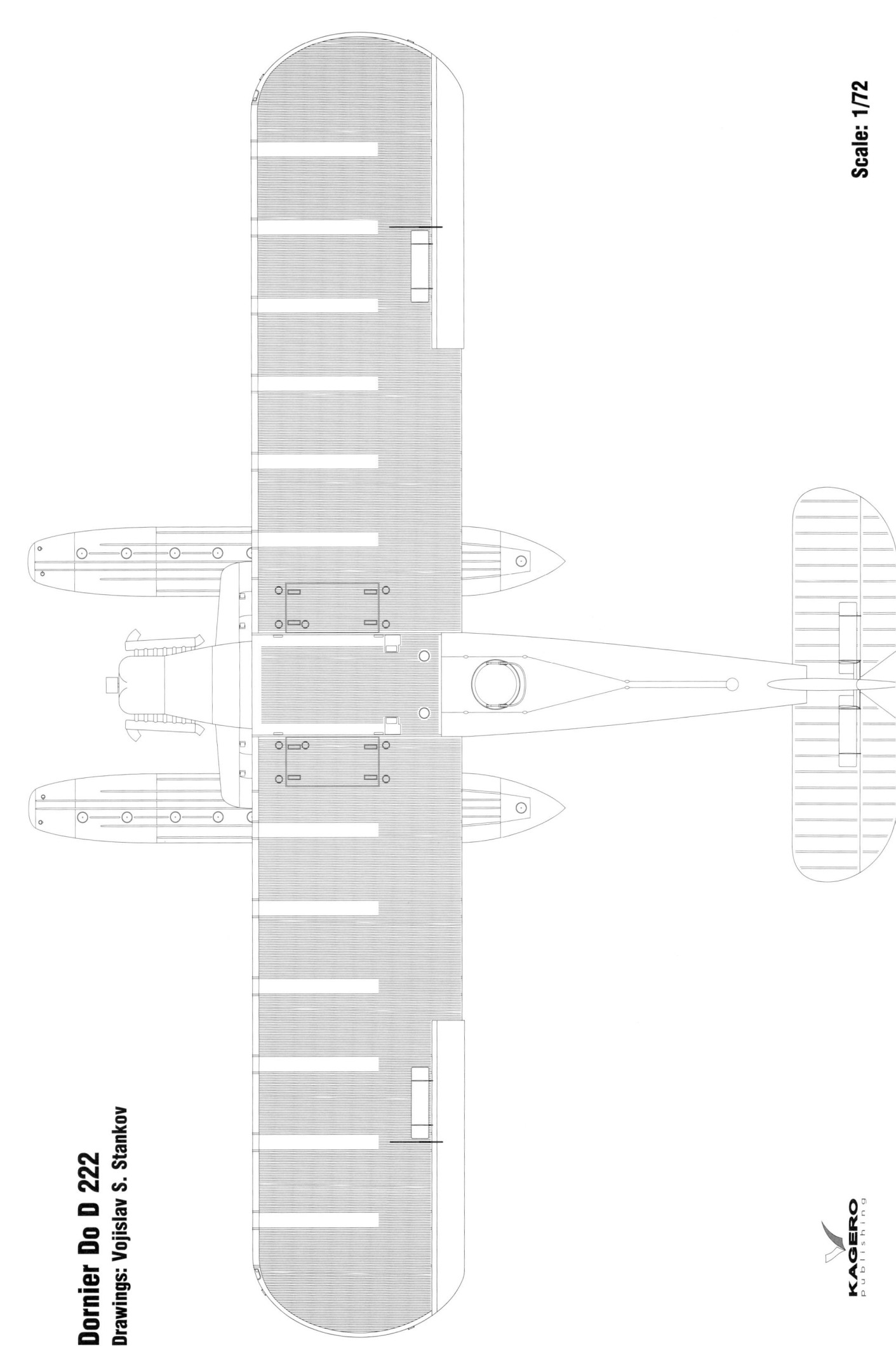

Dornier Do D 222
Drawings: Vojislav S. Stankov

Scale: 1/72

KAGERO
publishing

Dornier Do D 222

Drawings: Vojislav S. Stankov

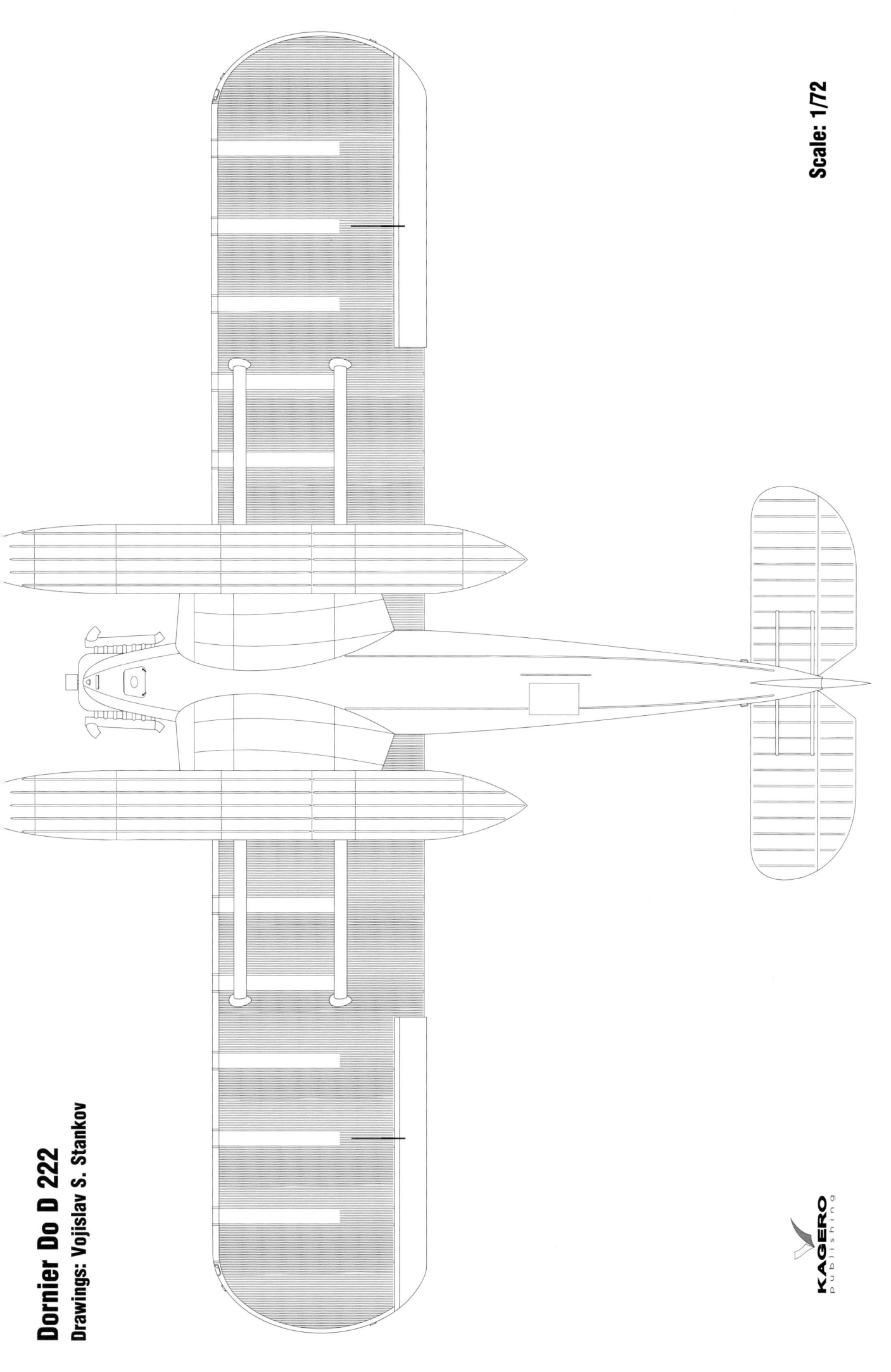

Scale: 1/72

KAGERO
publishing

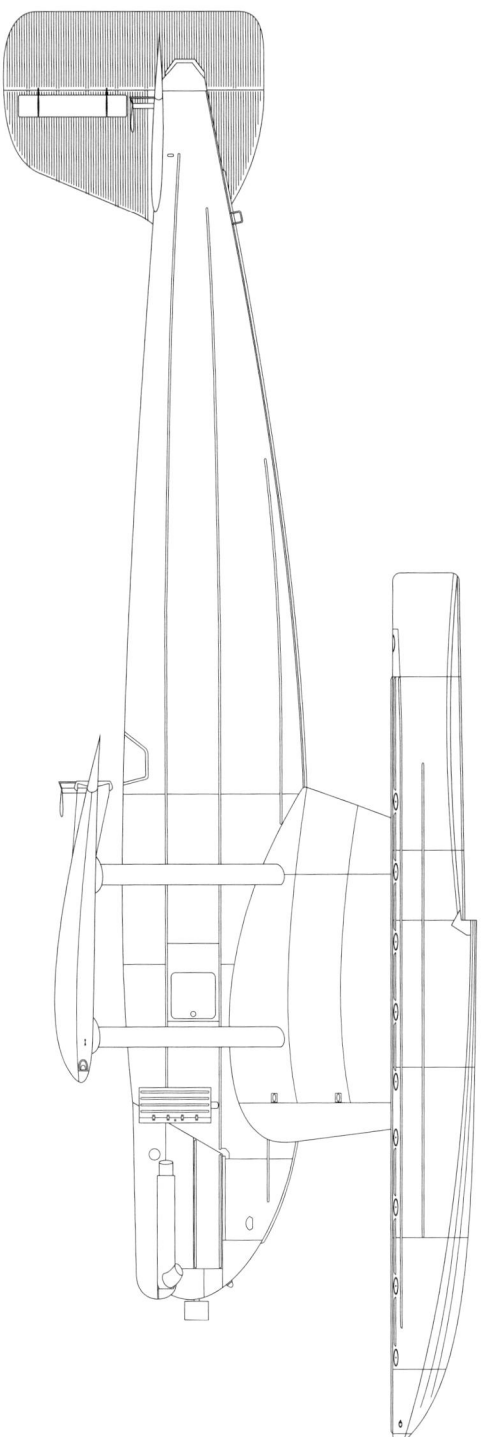

Dornier Do D 210
Drawings: Vojislav S. Stankov

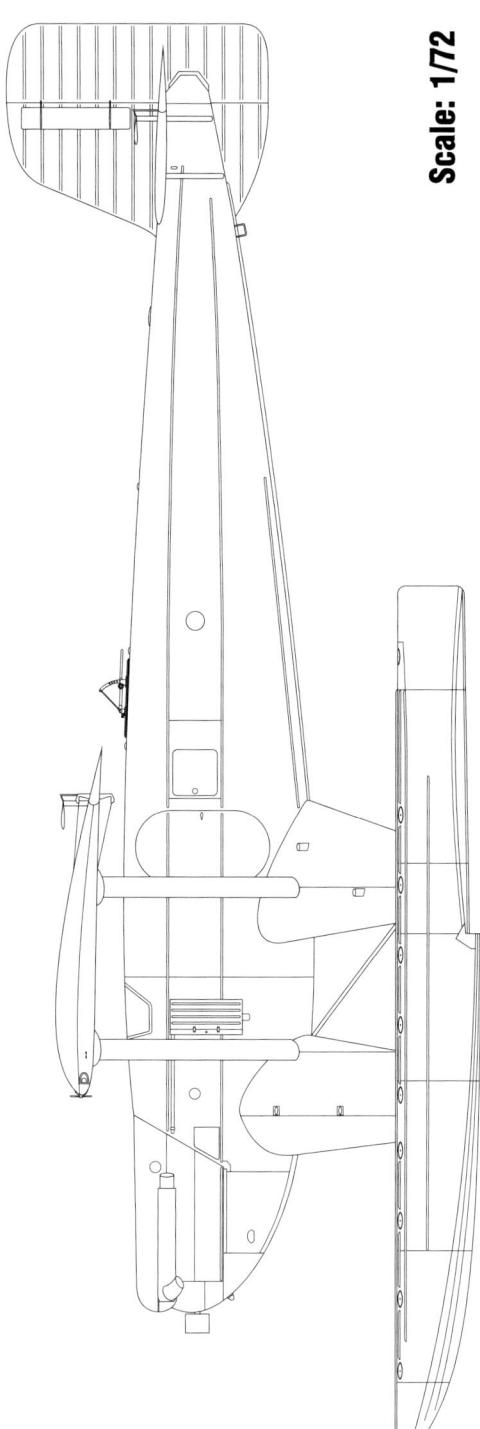

Dornier Do D 211
Drawings: Vojislav S. Stankov

Scale: 1/72

KAGERO
publishing

Dornier Do D 204

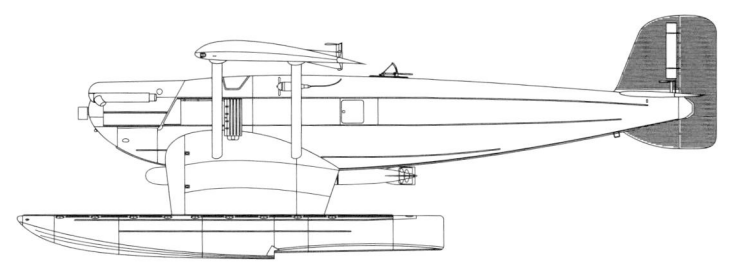

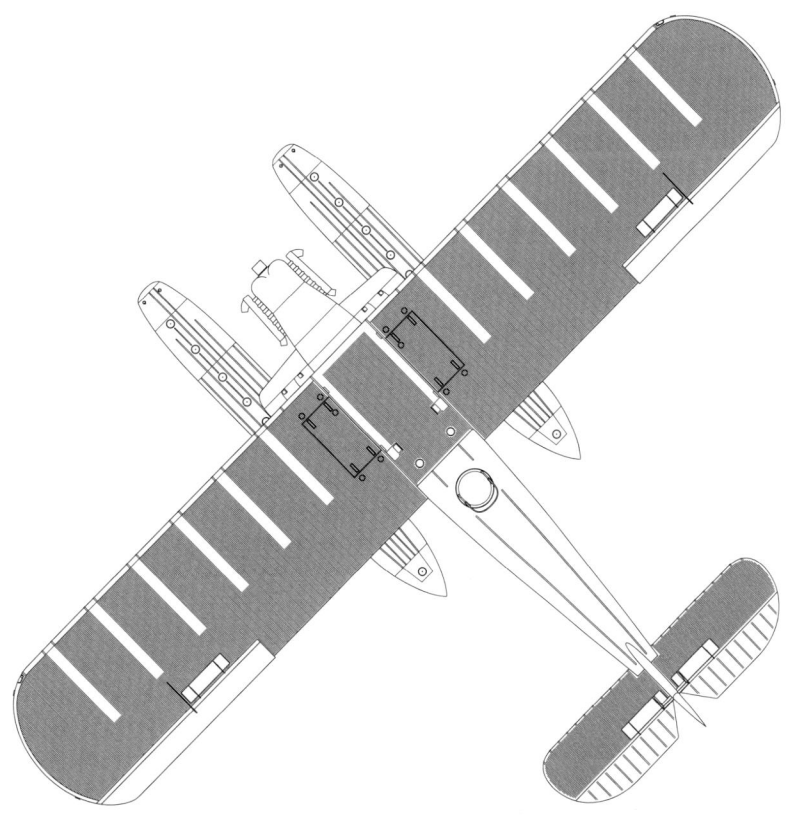

Dornier Do D 210

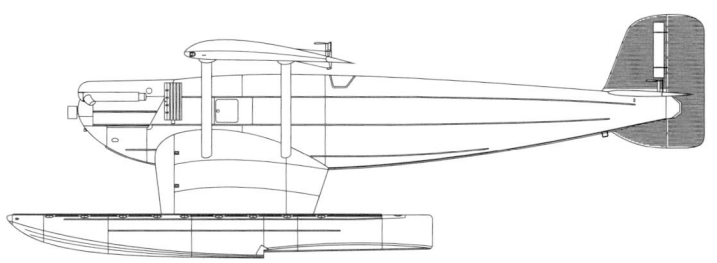

KAGERO
publishing

Drawings: Vojislav S. Stankov

Scale: 1/144

Dornier Do D 204

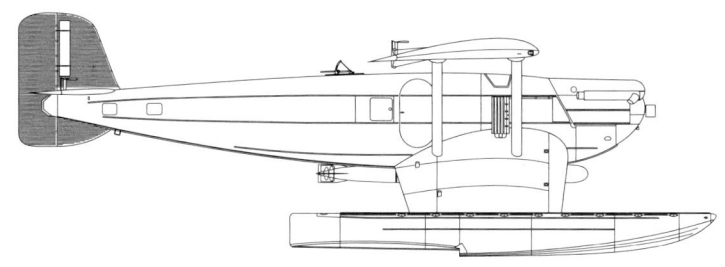

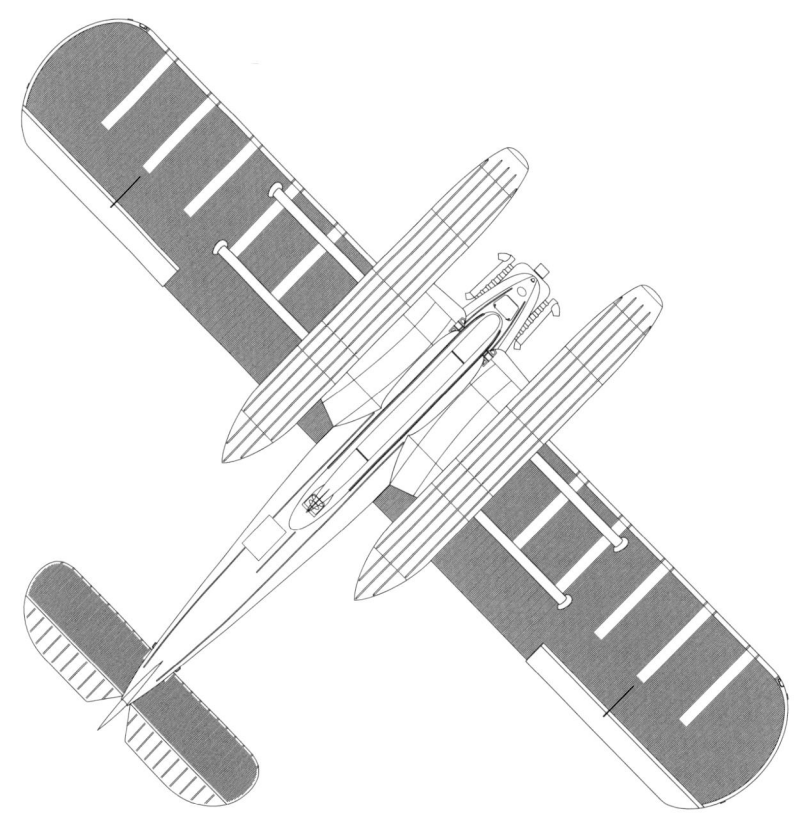

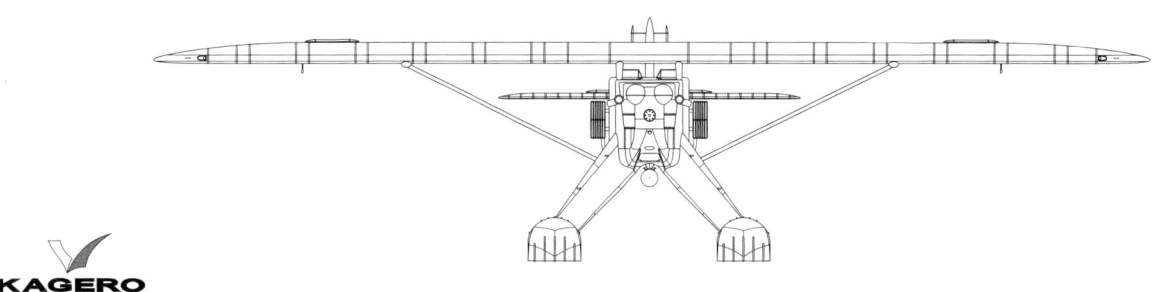

KAGERO
publishing

Drawings: Vojislav S. Stankov

Scale: 1/144

Dornier Do D 222

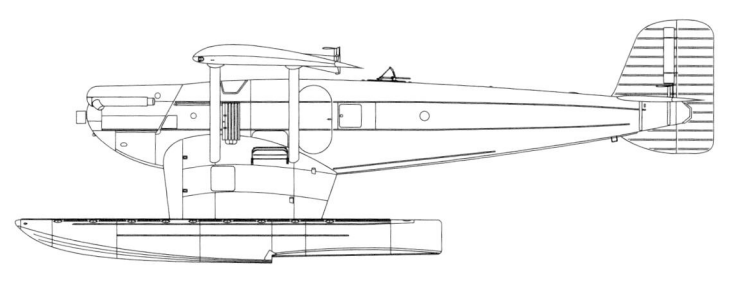

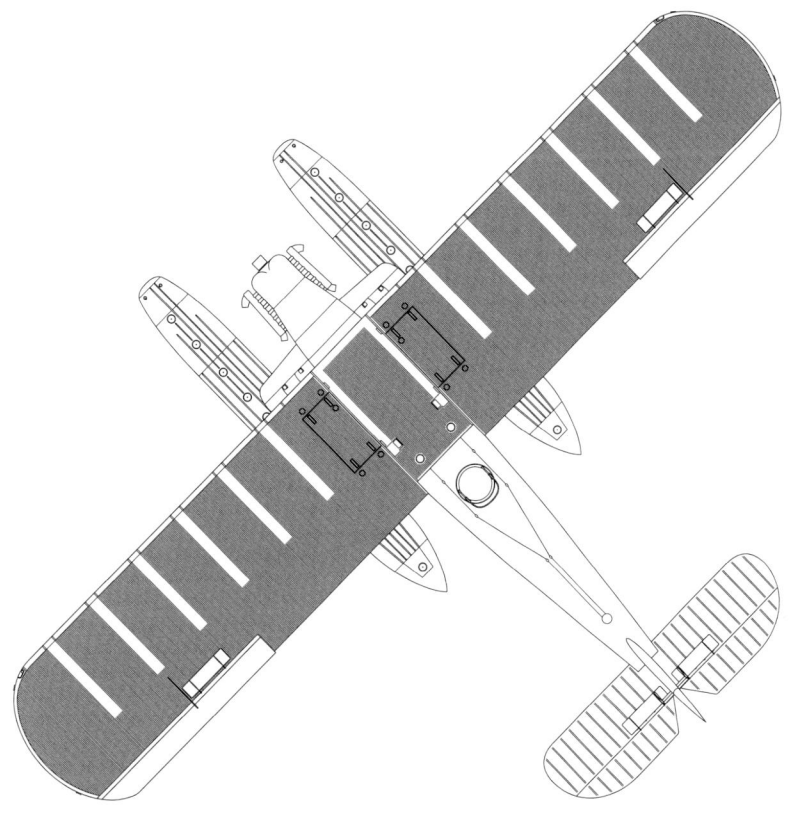

Dornier Do D 211

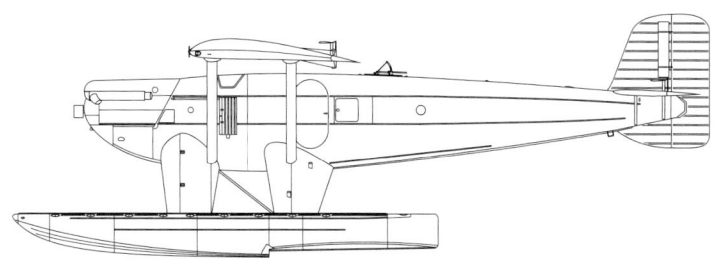

KAGERO
publishing

Drawings: Vojislav S. Stankov

Scale: 1/144

Dornier Do D 222

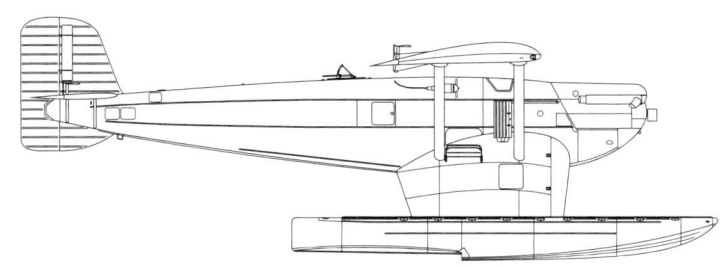

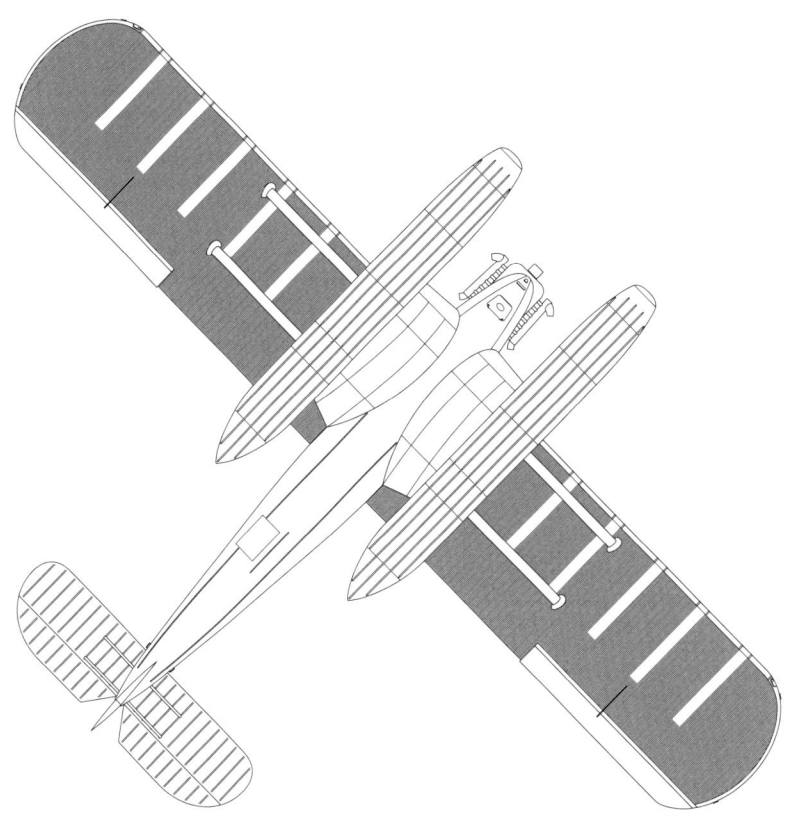

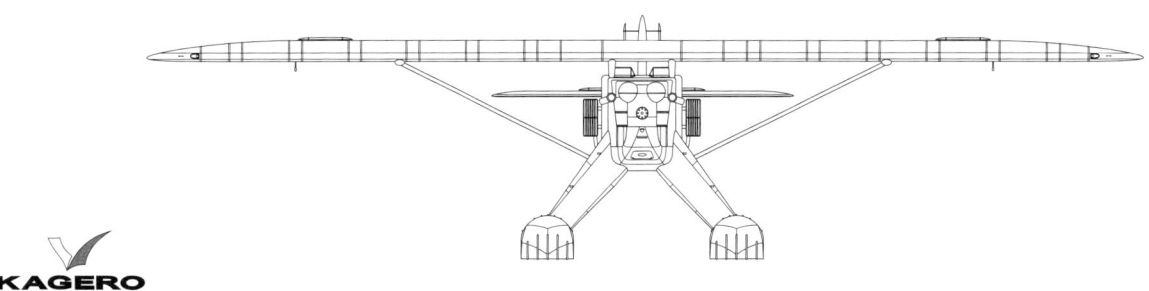

Scale: 1/144

Drawings: Vojislav S. Stankov

Dornier Do Wal

Development

The story of Do Wal began towards the end of World War I with the start of construction of the first of three ordered model Gs I flying boats in August 1918. These flying boats, marked with Navy registration numbers 8805/18 to 8807/18, had a military role which could be provisionally changed to the civilian one after the war. The first flight of Gs I took place on 31 July 1919 from Bodensee, while the test flights with Swiss airline *AD ASTREA* continued between October and December of the same year. Netherlands and Sweden showed interest and Gs I was flown from Friedrichshafen to Netherlands via Potsdam and Norderney in five hours to take part in demonstrations. In order to prevent handing over this flying boat to the Allies, Gs I was sunk by its crew in the Baltic Sea on 25 April 1920.

At the time of the airplane construction ban in Germany Dornier and his designers reworked the unfinished Gs II type in 1921/22 to bring it up to date with the latest aviation technology trends of the time period. They produced a military ocean-going flying boat suitable for long range reconnaissance and bombing missions. It was natural to resort to the fuselage design of the precursor model Gs I, which was developed for this purpose on behalf of the *Kaiserliche Marine* (Imperial Navy). The design

for Gs II was trialed first with a 1:1 scale mockup following which on 20 July 1922 the construction began at Marina di Pisa. The prototype construction progressed slowly due to financial difficulties and was completed by 30 October. Eventually Gs II was renamed "Wal 1922" and received Hersteller-Nr. 1 and WerkNr. 1 as well as fuselage code M-MWAA. The first flight took place on 6 November 1922 from Marina di Pisa with factory pilot Ulrich Niemeyer at the controls.

Due to the construction ban which was in effect until May 1922, Dornier Wal could not be built in Friedrichshafen. Main production site was the *Dornier Metallbauten GmbH* branch SAICM (*S.A. Italiana Costruzioni Meccaniche*) at the Italian coastal city Marina di Pisa, which acquired the Zeppelin parent company in December 1921. From November 1925, the factory was renamed under the new name CMASA (*Costruzioni Meccaniche Aeronautiche S.A.*). The new flying boat was known in the first years of construction simply under the type designation Dornier Wal, with engine types used initially not specifically marked. The distinction between military and civilian use happened only through additions of "*Militare*" or "*Cabina*" designations. Finally, in 1926/27 Wal was assigned type designation Do J. At the DLV's (*Deutscher Luftsportverband* – German Air Sports Organization) request, these flying

The first flight of Gs I took place on 31 July 1919 from Bodensee. [Airbus Corporate Heritage]

Dornier Gs II was in essence the first Do Wal. It was renamed "Wal 1922" and received Hersteller-Nr. 1 and WerkNr. 1. [Airbus Corporate Heritage]

The first Do Wal was purchased by the Spanish Military and received code M-MWAA. [Airbus Corporate Heritage]

boats registered in Germany were also marked with sub-versions and built-in engines (such as Do J Bas and Do J Gas).

As a standard powerplant for the first Do Wal, *Dornier* selected the English 12-cylinder *Rolls-Royce Eagle* engine, which had a good reputation worldwide and was considered very reliable. In the preferred version, IX, it had an output of 360/395 hp. However, it was also possible to install many other engines in the power range of 300-500 hp because the fuselage had sufficient strength and because the engine arrangement close to the center of gravity resulted in virtually no stability problems due to the different installed weights.

In the early January 1922 Dornier negotiated with the Spanish military administration the purchase of six military Wals, with the official order following in March. The first prototype Do Wal was delivered along with other five newly built machines fulfilling this order. Dornier later commented on these beginnings:

We had acquired the work without having enough resources. There was also no order when we started. At the moment when the situation started to get critical, the Spanish order came.

The military version of Do Wal was used in total of 12 countries, with the most numerous user being the USSR. The next followed Netherlands which used most at its Indochina colonies, Germany, Spain, Chile,

The first Do Wal arrived to KJ on 16 June 1927. It was originally equipped with four blade propellers. [HPMS]

Not long after the arrival, Do Wal/B 200 is seen here in flight above the Adriatic Sea. Note the dark color of the fuselage compared to the wings and horizontal stabilizers. [HPMS]

Italy, Argentina, Norway, Japan, Kingdom of Yugoslavia, Portugal and Uruguay. Apart from the factories in Italy (CMASA) and Manzell, Do Wal was additionally manufactured under license in USSR (*Zavod No.45*), Netherlands (*Aviolanda*), Spain (CASA) and Japan (*Kawasaki*). The production ended with over 250 produced "whales".

The only modifications and alterations which were found to be necessary were those to the engines and equipment, with the structure remaining largely unchanged. The following engines were installed: *Hispano-Suiza*, *Rolls-Royce*, *Gnome-Rhône*, *Bristol-Jupiter*, *Isotta-Fraschini*, BMW, *Napier-Lion*, *Lorraine-Dietrich* and *Farman*. Pilots Richard Wagner and Tulio Crosio flying a *Rolls-Royce* Eagle IV powered Do Wal attained no fewer than 20 world records during flights on 4, 9, 10 and 11 February 1925 while carrying loads of 250 kg, 500 kg, 1,000 kg, 1,500 kg and 2,000 kg. In January 1924 Major Ramon Franco conducted a round trip flight across the distance of 4,500 km from Spain to the Canaries attracting a lot of attention. Roald Amundsen used two Do Wals in May 1925 to fly from Spitzbergen to deep into the Arctic. In 1926 Ramon Franco flew to Argentina and in 1927 Major Sermento de Beires flew from Lisbon to Rio de Janeiro. Between 1933 and 1937 Dornier Wal crossed the Atlantic 328 times, totaling 999,400 km distance!

It is safe to say that the name Dornier became well known around the world because of Do Wal. In its almost unchanged form this machine flew continuously for more than 20 years, which was at the time a remarkable technical and aviation history achievement.

Introduction into Service

Following the entry of Do D in service, PV KSHS began a search for a long range reconnaissance and bomber flying boat to supplement this type in service

Chief Dornier pilot Richard Wagner along with pbb lk Miroslav Gogala flew the first Yugoslav Do Wal from Marina di Pisa to Kumbor. [Dornier Museum Friedrichshafen -Airbus Group]

and to finally replace all obsolete trophy flying boats which remained following the end of World War I. Multiple designs were evaluated from the offering of the traditional ally, France, but none met the necessary technical requirements set forth by the PV Commission. According to Miloš Crnjanski, PV Commission "*visited almost all flying boat factories in Europe and following extensive studies and tests, it selected a famous world known machine, German Dornier Wal, so far the most perfect type of*

Minister of Army And Navy disembarking from the first Do Wal, escadrille number 200 at Kumbor on 12 July 1927, several weeks after its arrival. [Djordje Nikolić]

Aft view of Do Wal 200 with curious looks from high ranking Navy officials. [Djordje Nikolić]

Do Wal and Do D in the calm waters of Rab harbor. Note the darker camouflage of the Do Wal compared to Do D. [Dejan Milojević]

Do Wal at the beach has gathered significant attention. [Dejan Milojević]

Do Wal coasting in one of the Adriatic Sea bays. Note two men in a dingy observing the passing Wal. [HPMS]

flying boat in the world." It is certain that another factor played a decisive role and that is the perceived ability to pay for these machines from the war reparation funds.

The first order was placed in 1926 for a single BMW VI equipped Do J Wal (W.Nr. 78) which was first flown on 10 June 1927 and just six days later, on 16 June 1927, it flew from the Italian factory at Marina di Pisa to Kumbor by PV pilot pbb Ik Miroslav Gogala and *Dornier* chief test pilot Richard Wagner along with two *Dornier* mechanics. The route was Marina di Pisa

Do Wal 200 seen here during the survey of the accident by Do D 208 at Hvar on 1 August 1929. Note that two blade propellers are now installed.
[Djordje Nikolić]

Do Wal 200 resting in calm water. The font of the number 200 has changed from the time it arrived in KJ. [HPMS]

Crew members posing in the cockpit. [Aviation Museum - Belgrade]

A rendezvous at sea. Do Wal 200. [HPMS]

– Messina-Spartivento – Santa Maria di Leuce – Boka Kotorska and the flight lasted 7 h and 50 min, across the distance of 1,470 km at an average speed of 193.3 km/h. Following the arrival, Do Wal received escadrille number 200, which was ahead of those assigned to the previously arrived Do Ds, since Do Wal was considered a "Command" or "Admiral" flying boat. Do Wal joined 20.HE at Boka Kotorska. Sometime later it was as-

Curious beachgoers using the opportunity to pose with the "admiral" Do Wal. [Aleksandar Smiljanić]

Do Wal 200 tied to a buoy. [HPMS]

Do Wal at Divulje behind the wreck of a crashed Do D. [HPMS]

signed to 26.HE at Divulje, then to 25.HE while its final posting was back again with 26.HE.

Based on the 1930 *Dornier Metallbauten GmbH Friedrichshafen a.B. Jahresbericht 1930* (Annual Report) , following long negotiations a contract was signed in the middle 1930 for the delivery of 12 Do Wals and a number of spare parts. This contract was valued at 4 Million RM including the engines. The machines were to be delivered in the period of 1 March 1931 to March 1933. Six of these were to be furnished from

With its engines off Do Wal is slowly coasting towards a ship. [HPMS]

Towards the later part of its service, the escadrille number was moved to the aft fuselage. [HPMS]

One of Do Wal/J for KJ is seen during production. [Oliver Fischer]

During ferry flights Do Wal/J carried Swiss code on the vertical stabilizers as well as black fuselage numbers. [Aleksandar Smiljanić]

Do Wal/J 254 taxiing in water. [Djordje Nikolić]

Do Wal/J 254 in shallow water with curious swimmers around. [Djordje Nikolić]

Do Wal/J 253 in flight. [Josip Novak]

Do Wal/J 254 in flight. [HPMS]

the war reparation funds. The Annual Report interestingly points out that "based on the experiences to date, it can be counted on that the contract will be executed smoothly".

PV placed an order for six additional Do J II Gis (8 t – Wal) machines in 1930, which were equipped with the *Gnome-Rhône* Jupiter VI engines and aerodynamic improvements which included rounded wing tips, horizontal stabilizers and the rudder. Despite aerodynamic improvements, their performance was not as impressive as that of the 200, which is mainly

Do Wal/J on take off. [Josip Novak]

Do Wal/J 256, behind it is one of four Yugoslav submarines. [Aleksandar Ognjević]

A poor quality photograph of three Do Wal/J. Far right machine is 253. [HPMS]

Do Wal/J 252 at Divulje. [HPMS]

View of the rear fuselage turret and bomb rack on Do Wal/J 252. [HPMS]

3.HK base at Kumbor with three Do Wals tied to buoys. [Djordje Nikolić]

A duo of Do Wal/Js on take off. [Aleksandar Smiljanić]

Do Wal/J 251 and Do Wal/J 255 tied to a buoy. [Josip Novak]

A feast in the honor of foreign guests. Note Do Wal in the back. [Aleksandar Smiljanić]

Do Wal/J 255 at Rab island. [Aleksandar Smiljanić]

An unknown Do Wal J at Hvar island. [Djordje Nikolić]

A celebration in a hangar with Do Wal/J, Do D and DH.60. [Djordje Nikolić]

Two Do Wal Js at Hvar island. [Djordje Nikolić]

the result of less powerful engines. These machines completed at Manzell, W.Nr. 211 through 216, were marked for the ferry flight with fuselage numbers 303, 304, 313, 314, 316 and 319 and had Red painted tails with White CH letters on vertical stabilizers as well as Red painted wing bands with White CH letters and the same numbers as on the fuselage. In KJ they received escadrille numbers 251 through 256 respectively. Of these only four were paid through the war reparation funds since Germany cancelled all the payments following the so called Hoover plan which came into effect on 1 July 1931. At the time of the war reparation payment moratorium, 251 was entirely complete while 252, 253 and 254 still lacked engines. The remaining two machines, 255 and 256, were paid by KJ to ensure delivery. Following the delivery of 251, 252, 253 and 254 in 1931 and 255 and 256 in 1932, all Do Wals joined the newly formed 21.HE based at Kumbor. At the same time plans called for more Wals

An unknown Do Wal J at Kumbor with a naval aviator in the front of the photograph. [Djordje Nikolić]

Two Do Wal/Js, of which one is 254, at Rab island. [Djordje Nikolić]

Do Wal/J 256 at Boka Kotorska. Note the interesting triangle with black number 6. The reason for such atypical marking is not known. [Andrew Crawford via Djordje Nikolić]

Sailor Dušan L. Šeferović posing on Do Wal/J aboard Zmaj seaplane tender, 4 August 1937. [Momir Milinović]

Three Do Wal/J, 254, 255 and an unknown one, tied to buoys. [Djordje Nikolić]

A momento with Do Wal/J. [Josip Novak]

A very nice fly over shot of Do Wal/J 251, dated 1934. Note that barely visible on the vertical stabilizer there is a triangle marking, similar to the one seen on 256. [Aleksandar Ognjević]

Do Wal/J 251 this time flying head on towards the photographer. [Aleksandar Ognjević]

to equip 27.HE but despite an existing order, they were never delivered due to the German reluctance to supply them free of charge.

In 1934 additional four Do Wals were purchased with payment in full and these arrived with no equipment. This time, three Do Wals (W.Nr. 245, 246 and 260) were equipped with *Hispano Suiza* Ydrs engines while one (W.Nr. 240) had a BMW VI engine which was more powerful than the one from Do Wal 200 and was housed in a more streamline nacelle resulting in higher speed. Following their arrival to KJ in 1935 and 1936 respectively, the new flying boats received escadrille numbers 257 through 260 and along with the 200 joined the ranks of 26.HE which until then used Do D floatplanes solely.

Do Wals were used in numerous exercises along the Adriatic coast, especially for long range patrols. They trialed laying smoke screen in low level flight from 50 m altitude by using *Dominko* brand smoke generator hung on bomb racks. Smoke screen was laid in the duration of 60 to 70 seconds. Interestingly, Do Wal 257 was used in 1930s during filming of a movie "The Coral Princess" in Dalmatia, Croatia. During this event, a film camera was mounted in front of the nose turret.

The contemporary newspaper *Novo doba* wrote that on 24 August 1936 Do Wal 257 and 258 flew the Duke

A drawing by cadet Mišić, which he hand sketched during his training. [Aleksandar Ognjević]

Do Wal/B 257 carried fuselage code D-7 and a large swastika during the delivery. Here it is shown at Divulje. [Josip Novak]

Do Wal/B 257 during tow by a motor boat. [HPMS]

Do Wal/H 259 coasting by the beach at Makarska, 1939. [Boris Ciglić]

A smiling lady posing at the beach with Do Wal/H in the background. [Aleksandar Smiljanić]

and the Duchess of Kent from Divulje to Cavtat, while on a visit to a local businessman. The Duchess flew in 257 piloted by nar Ik Keršić, with the observer C-in-C of PV commander Ivo Preradović. The Duke and his secretary flew in 258 with pilot nar Ik Kamnikar at the controls.

During a friendly visit by KM (*Kraljevska Mornarica* - Royal Yugoslav Navy) to Greece on 1 September 1937, apart from Navy ships, Do Wals 251, 252 and 254 flew for five hours from Boka Kotorska to Phaleron Naval base. Prior to this trip Do Wals thoroughly rehearsed long distance flights to ensure all equipment was operating as required. A KM ship *Spasilac* was in the Argostoli harbor in the Ionian Sea in case Do Wals experienced trouble during the flyover. Do Wals successfully returned four days later on 5 September.

From 2 February until 15 July 1938 engineer por Svetozar Bukalov was stationed at *Aerometal* factory in Zürich, Switzerland, to supervise engine pylon construction, likely for those machines which were designated for conversion to *Rolls-Royce* Kestrel XVI engines.

Families of pilot Josip Kauzlarić on the left and Josip Ferenc on the right, Divulje 31 October 1938. [Josip Novak]

High ranking PV officers visiting a naval base. Do Wal/H 260 and Do Wal/B 257 are visible in the background. [Dejan Milojević]

The British Crown Prince Edward Albert Christian and Wallis Simpson, later known as the Duchess of Windsor, preparing for a flying from Divulje To Cavtat, August 1936. [Aleksandar Ognjević]

Kbb Ivo Preradović helping Wallis Simpson board Do Wal/B 257. Do Wal/H 258 is visible in the background. [Aleksandar Ognjević]

Do Wal/B 257 was piloted by nar Keršić during this occasion. [Gustav Ajdič via Tomaž Perme]

Do Wal/H with protective canvas over the four blade propeller. [Josip Novak]

In the second half of 1940, the equipment on Do Wals was standardized to facilitate easier maintenance and training of the crews. The equipment varied due to the use of three different engine types at that time.

21.HE was used to test 100 kg domestic bombs made by *Stanković* company from Višegrad. These tests were conducted at the entrance to Boka Kotorska bay and the records show that every other bomb failed to detonate.

Of all the Do Wals, those with the *Gnome-Rhône* Jupiter VI engines exhibited the poorest performance, hence they were declared obsolete and to prolong their

Rogozarski Sim XIV in the foreground with Do Wal/H, likely 259 and Do Wal/B 200. Also note two Do 22s on the right, of which one is 303. [Aviation Museum – Belgrade via Aleksandar Ognjević]

service life, efforts were taken to re-engine them. As early as 1938, the Annual Navy report lists that two *Rolls-Royce* Kestrel XVI engines were acquired but they were actually received from the reserve stocks of 6.VB (*Vazduhoplovna Baza* - Aviation Base) at Zemun since these engines were primarily used for VVKJ (*Vazduhoplovstvo Vojske Kraljevine Jugoslavije* – Royal Yugoslav Army Air Force) Hawker Fury Mk. II biplane fighters which were at the time rendered obsolete. Do Wal 256 was the first to receive the new engines and was followed by Do Wal 253. This Do Wal was listed as out of service until November of 1941 however it is known that it was operational and conducted flights during the 1941 April War. As of 31 January 1941, another Do Wal, 254, was

at 2.HK workshop undergoing conversion. Since the Kestrel engine was equipped with the compressor, once obsolete Do Wal with Jupiter engines, received a new life with improved performance. Plans existed to eventually modify all of the remaining Jupiter machines to the new engine, but these did not materialize by the beginning of war in Yugoslavia on 6 April 1941. On 22 March 1941

KM (*Komanda Mornarice* – Naval Command) attempted to purchase spare parts for *Hispano Suiza* 12 Ydrs engines in inventory from *Avia A.D*, however due to attack against the KJ, these deliveries never materialized. This indicates that even with the imminent danger of war, PVKJ attempted to keep its Do Wals with the *Hispano Suiza* engines in active service for several more years.

Do Wal in PVKJ service

W.Nr.	Escadrille No.	Delivery year	Engine type at delivery
58	200	1926	BMW VI
211	251	1931	Gnome-Rhône Jupiter VI
212	252	1931	Gnome-Rhône Jupiter VI
213	253	1931	Gnome-Rhône Jupiter VI
214	254	1931	Gnome-Rhône Jupiter VI
215	255	1932	Gnome-Rhône Jupiter VI
216	256	1932	Gnome-Rhône Jupiter VI
240	257	1933	BMV VI
245	258	1933	Hispano Suiza Ydrs
246	259	1934	Hispano Suiza Ydrs
260	260	1934	Hispano Suiza Ydrs

Do Wal at 1.HK 3.HG 26.HE. Overview from 31 January 1941

Do Wal Type	Escadrille No.	Total flight time	Total flights	Status	Engine Type and S/N	Engine working hours
Wal/B	257	611 h 20 min	781	Inoperative (engine overhaul)	BMW VI 17206 17207	608 h 35 min 616 h 35 min
Wal/H	258	336 h 30 min	458	In service	Hispano Suiza Ydrs 485323 485639	232 h 35 min 383 h 30 min
Wal/H	259	306 h 10 min	425	In service	Hispano Suiza Ydrs 485324 485421	167 h 45 min 336 h 10 min
Wal/H	260	368h 40 min	462	In service	Hispano Suiza Ydrs 485420 485644	363 h 40 min 365 h 15 min

Do Wal at 2.HK Workshop. Overview from 31 January 1941

Do Wal Type	Escadrille No.	Total flight time	Total flights	Status	Engine Type and S/N	Engine working hours
Wal/J	254	450 h 15 min	521	Inoperative	Installation of Rolls-Royce Kestrel XVI 9115 9123	N/A

Do Wal at 3.HK 2.HG 21.HE. Overview from 31 January 1941

Do Wal Type	Escadrille No.	Total flight time	Total flights	Status	Engine Type and S/N	Engine working hours
Wal/B	200	826 h 10 min	890	Inoperative (airplane and engine major overhaul – out of service until 28 November 1941)	BMW VI 16065 16745	699 h 50 min 553 h 30 min
Wal/J	251	508 h 40 min	698	In service	Gnome-Rhône Jupiter VI 3536 3537	548 h 35 min 501 h 15 min
Wal/J	252	551 h 05 min	777	In service	Gnome-Rhône Jupiter VI 3525 3526	558 h 45 min 557 h 45 min
Wal/J	255	516 h 20 min	884	Inoperative (out of service until 28 November 1941)	Gnome-Rhône Jupiter VI 3535 3532	515 h 05 min 530 h 00 min
Wal/K	253	437 h 15 min	443	Inoperative (out of service until 28 November 1941)	Rolls-Royce Kestrel XVI 9109 9117	6 h 15 min 6 h 15 min
Wal/K	256	205 h 50 min	227	In service	Rolls-Royce Kestrel XVI 9163 9165	141 h 20 min 141 h 20 min

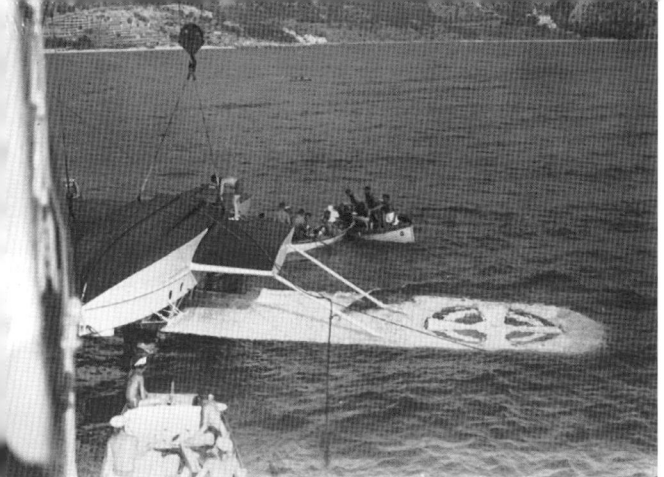

On 27 July 1934, while landing in rough seas in Boka Kotorska, Do Wal/J 256 flipped over. [HPMS]

Do Wal/J 256 was quickly raised the same day by seaplane tender Zmaj and after repairs would eventually become the first Rolls-Royce Kestrel powered Do Wal/K. [Gustav Ajdič via Tomaž Perme]

Accidents

During its entire service with PVKJ, Do Wal suffered only one accident. On 27 July 1934, while landing in rough seas in Boka Kotorska, 256 flipped over. It remained afloat and was raised the same day by seaplane tender *KB Zmaj*. It remains unknown if any of the crew were injured but 256 was later repaired and was converted to the *Rolls-Royce* Kestrel XVI engine.

April War

Two months prior to the war, 21.HE conducted daily reconnaissance of the Adriatic Sea between Boka Kotorska in the northwestern direction towards Korčula island and southeastern direction towards the border. Two missions per day were usually flown, one at sunrise and one at sunset, at a distance no further than 10 nautical miles from the coast. 21.HE was only supplied with six parachutes, which was only enough for one crew at the time, so these changed hands before each flight. Spare parts were very scarce as they were sourced from abroad and their sourcing at the time proved difficult. Each escadrille had only one spare engine and the only means of communication were radio stations installed on Do Wals themselves.

Immediately prior to the war 21.HE and 26.HE conducted several exercises. Their dislocation to wartime positions did not take place in a timely manner, hence Do Wals remained at their peace time bases in early April 1941. 21.HE dispersal location was intended to be at Dobrota, in Boka Kotorska bay, not too far from 3.HK home base at Kumbor while 26.HE relocated to 1.HK home base, Vodice, on 28 March.

All HEs belonging to 3.HK were activated between 31 March and 1 April 1941 due to the imminent war danger, even though they were already under constant

Aerial bombardment of Divulje naval base. Note the hits on PV hangar. [Ten. Vittorio Sanseverino via Gregory Alegi]

An aerial view of Kumbor taken from an Italian airplane shows Do Wal/H 260 with an interesting ad hoc camouflage along with three Rogožarski Sim XIV floatplanes. [Giancarlo Garrelo]

The same Do Wal/H 260 shown after the war with German soldiers posing on the nose. Note the ad hoc camouflage visible on the leading edge of the wings. [Oliver Fisher]

Do Wal/J 255 was eventually towed to a beach where it was attached to the beaching gear. [Oliver Fisher]

Late spring or early summer photographs shows Do Wal/J 255 which attracted significant attention from soldiers who were swimming nearby. [Oliver Fisher]

alert for a period of several months. Their crews always stayed near their machines and sometimes slept in them. Do Wals were at the time armed only for self defense and carried no bombs. Their intended war time use was limited due to slow speed, however their range still made them suitable for long range reconnaissance which, with the presence of danger from the faster Axis fighters would be very risky and was recommended only at nighttime and in favorable weather conditions. During this time, two exercises took place to practice dislocation in shortest time possible. Despite the re-location efforts, there was an acute lack of material to camouflage these machines at their dispersal locations.

On average, Wal/H and Wal/K consumed a total of 60 t of 87 octane fuel monthly with each Wal flying on average 30 h. Hence, the 87 octane fuel war reserves allowed for 85 flying hours for Wal/H and Wal/K. Do Wal/J and Wal/B consumed a total of 43 t of 73 octane fuel monthly with each Wal flying on average 30 h, and the war fuel reserves could last up to six months. In war time 21.HE was intended to sustain six flying days each week.

Do Wal order of battle on 6 April 1941				
Do Wal Type	1.HK	2.HK	3.HK	
	3.HG 26.HE	Workshop	2.HG 21.HE	Workshop
Wal/B	257	–		200
Wal/J	–	–	251, 252, 255	–
Wal/H	258, 259, 260	–	–	–
Wal/K	–	254	253, 256	–

On the eve of the war, 5 April, rear-admiral Emil Domainko arrived along with two officers to PVKJ HQ (Headquarters) in a Do Wal 253/K belonging to 21.HE. Following consultations, they left towards Šibenik but

Do Wal/J 255 did not take part in the war and was captured at Dobrota by the Italian forces. [Giancarlo Garrelo]

Hangar at Kumbor shows a partial view of an unknown Do Wal/H which was captured there with damaged port horizontal stabilizer. [Statto Magiore Aeronautica]

Another aerial view of Kumbor shows two Do 22, 301 and 304, along with two unknown Do Wal in the water and another one in front of the hangar. [R. Gentilli via Bojan Dimitrijević]

This poor quality aerial photograph shows two Do Wal tied to buoys at Kumbor. [Andrew Stamatopoulos]

An unknown Do Wal/J is visible to the right of the Italian soldier at Kumbor. On the left is Do 22 301. [R. Gentilli via Bojan Dimitrijević]

Do Wal/H 258 at a hangar at Divulje along with Rogožarski PVT-H and Sim XIV. This flying boat was the last Yugoslav Do Wal to fly during the April War. [Giancarlo Garrelo]

Do Wal/H, likely 258. [Ten. Vittorio Senseverino via Gregory Alegi]

Captured Do Wal/J, likely 254 at a hangar at Divulje along with Do D 203. [Manca Collection]

their flying boat broke down en route when the rear *Gnome-Rhône* Jupiter VI engine malfunctioned. Late in the evening of the same day, Navy commander informed the PVKJ HQ that the attack would commence in the early hours of 6 April. The entire 3.HK at Kumbor was on high alert, and the majority of aviators spent the night by their machines and slept in hangars. In the early morning hours of 6 April 1941 21.HE lowered its Do Wals in the water.

Observation post at Mljet island sent a notice at 06:40 that a large formation of unknown airplanes was flying above in the direction of Mostar airfield. Immediately thereafter, 21.HE was ordered to dislocate. The enemy bombers arrived at 07:05 bombing Tivat and Zelenika, inflicting first casualties. 21.HE dislocated its Do Wals to Dobrota and Kotor. During the afternoon attack at 15:30 by five German Ju-88 against Kotor, Do Wal/K 256 suffered bomb shrapnel damage to the rear *Rolls-Royce* Kestrel XVI engine. Do Wal/K 253, which

was tied to a buoy at Šibenik and whose rear engine was not operational, was ordered at 22:16 to relocate to Skradin for repairs and then be made available to 1.HK CO. Some time later, pilot pbb Ik Vladislav Kandare attempted to fly this Do Wal from Prokljansko lake but due to the faulty engine was unable to do so. As a last resort, pilot torched his Dornier near Skradin to prevent its capture.

At 1.HK in the early morning at 02:00 an alert status was raised and all 26.HE crews went to their machines which were moored at Vodice. The attack against 2.HK home base at Divulje took place at 06:50, when Fiat Br.20 bombers made an inaccurate bombing run causing no damage. Escadrille CO ordered his crews to start their engines and spread out along the bay, but due to the lack of return signal, most Do Wals maneuvered in water for almost two hours at random. By noon, all serviceable machines left the area due to difficult water conditions and to avoid destruction. All

Vertical stabilizer once belonging to the Rogožarski Sim XII and an unknown destroyed Do Wal. [Stato Maggiore Aeronautica]

Inside of the decimated hangar at Divulje showing completely destroyed set of Do Wal wings. [Stato Maggiore Aeronautica]

Another view of the destroyed wings which were neatly disassembled during servicing. [Stato Maggiore Aeronautica]

Remains of destroyed Do Wal at the back of the hangar. [Stato Maggiore Aeronautica]

but one coasted through Šibenik channel to Jadratovac at around 13:30, while the lone Do Wal flew to Zlarin. At Jadratovac there were three buoys available to which flying boats were tied off.

During most of the first and second day of the war, 6 and 7 April, all Do Wal HE worked hard to prepare their dispersal locations while avoiding detection and destruction which rendered their war time use early in the conflict nonexistent. On 7 April an order was issued by 3.HK Command to supply 26.HE with bombs and fuel as combat missions were expected for

the next day. On the same day, Do Wal from Zlarin joined the rest of the machines at Jadratovac despite the worsening weather.

The first enemy attacks against Do Wals were not successful but without proper camouflage equipment and adequate protection from air attacks, on 8 April at 13:00 four Do Wals from 26.HE were discovered at Jadratovac and strafed by three Macchi C.200. Following this attack, machine guns were removed from the two heavily damaged Do Wal/H 258 and 259 and were placed on shore and used as makeshift anti-aircraft defense, shooting

An overhead view of the destroyed Do Wal fuselage. [Stato Maggiore Aeronautica]

down one enemy airplane according to the reports from the ground. Do Wal/H 259 had its useable parts removed and installed on Do Wal/H 258, while Do Wal/B 257 and Do Wal/H 260, which were only lightly damaged, dislocated towards Prokljansko lake on 12 April in the dark. Do Wal/B 257 was damaged during this flight when it ran aground and was not repaired for the remainder of the conflict. Do Wal/H 258 remained at Jadratovac because pilot was unable to start the engines, likely due to damage sustained during the strafing attack.

Pbb IIk Ilija Goršić witnessed the events at Jadratovac firsthand:

During the attack by Italian bombers against Šibenik, it was clear that 26.HE airplanes at Jadratovac had been discovered, but the Commander forbade their fly over to Prokljansko Lake.

The same day around 13:00 Italian Fiat (Macchi – a.c.) fighters attacked the flying boats belonging to 26.HE at Jadratovac heavily damaging two of them while one sustained light damage, but they did not manage to sink

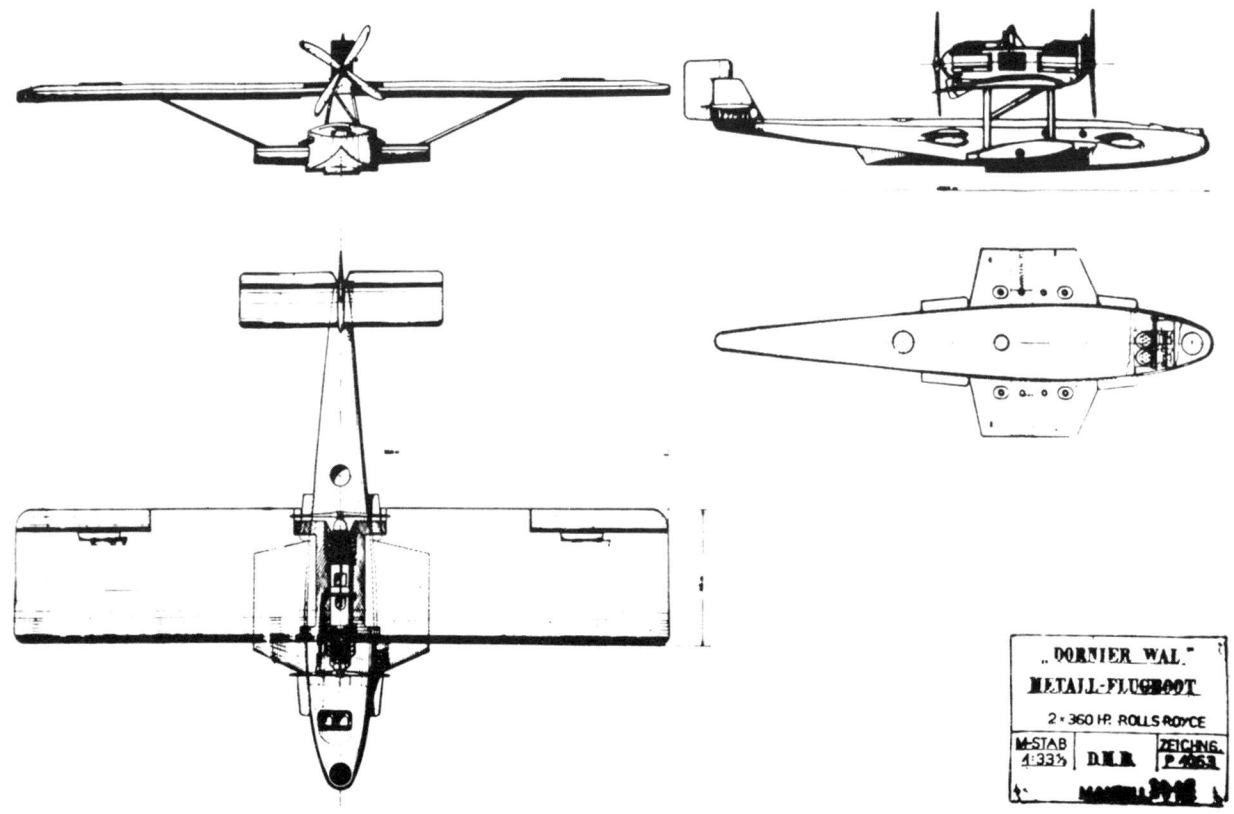

Original Dornier factory drawing showing the first prototype WerkNr. 1. The basic configuration has changed little between subsequent models, with exception of the engines and the tail assembly. [Airbus Corporate Archive]

88

either of them. 3.HG Command ordered them to fly over to Prokljansko lake. Only two flying boats flew over, and one of them ran aground at the coast by village Vrulje. It was damaged and the workshop did not manage to repair it.

In the evening hours of 9 April, PVKJ HQ issued a written order to relocate all floatplanes to Boka Kotorska, amongst which were three Do Wal. The dislocation continued on 10 April. Do Wal/H 260 flew from Prokljansko lake with pbb IIk Ilija Goršić at the controls but due to the lack of fuel it was forced to land in heavy winds near the islands Bobara and Mrkan. Its crew fired signal rockets which were observed and a motorboat arrived to tow them to Cavtat. There, fuel was delivered by a truck from Kumbor. This Do Wal finally arrived as late as 15 April to Kotor Bay. On 13 April Do/H Wal 258 which was finally repaired at Jadratovac managed to land at Kaštel Kimbelovac to refuel (after receiving temporary camouflage) and thereafter it flew to 21.HE at Dobrota in Boka Kotorska. Another Italian attack commenced as soon it arrived.

As of 14 April, the entire crew of 21.HE was still on strength. Mechanics made necessary repairs to all Do Wals which were at the time at Boka Kotorska, and their machine guns were removed and placed on land to act as makeshift anti-aircraft defense.

In the evening of 15 April, a decision was made to fly all serviceable machines to Greece. 21.HE CO pbb Ik Oskar Bizjak did not follow the order by pbb Ik Vladeta Petrović to fly over with his escadrille, with an excuse that *"his machines were old and unserviceable and that his men were not the best skilled or were not present at all"*. Reportedly fearing that pilots may actually attempt to fly over, pbb Bizjak even placed some of his pilots on house arrest. Pbb Petrović wrote later in his war diary:

The fact that the unit remained in country was largely also my mistake because I was unable at the time to evaluate how reliable pbb Bizjak is during our earlier service together – otherwise I would have personally flown over with 21.HE and I would handed over 20.HE to pbb Zobundžija (Ladislav – a.c.) until Corfu. It was duty of pbb Bizjak to inform me between 10 and 16 April that he does not wish to leave the country with his unit.

According to pbb Avgustin Grošelj at least one of the Do Wals was unable to make the long-distance flight for valid reasons:

Wal 258 which belonged to my group (21.HE), considering the state of its engine, was unable to fly over overseas to Corfu.

Kv Vicenc Grošelj gave a different account where he indicates it was the crews who declined to fly over:

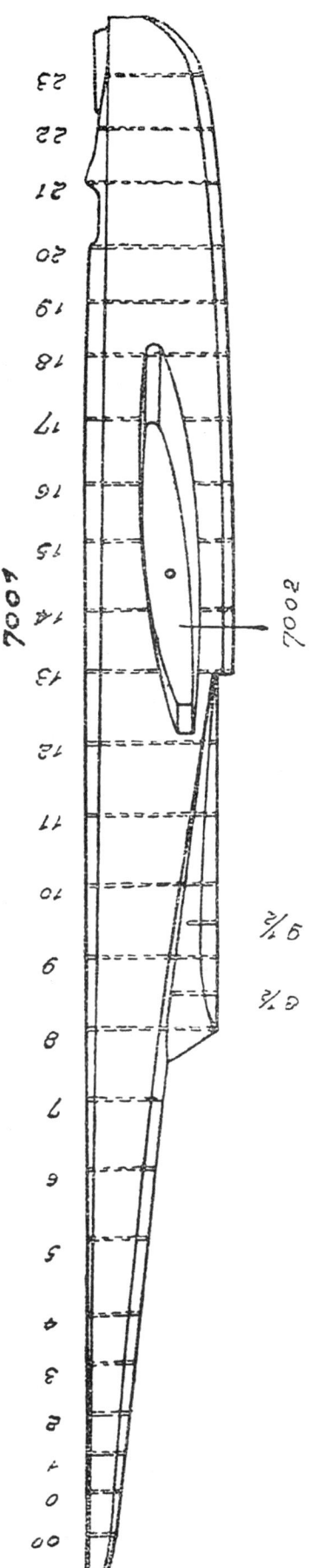

Do Wal fuselage section detail. [Catalogo nomenclatore per idrovolante Wal Militare]

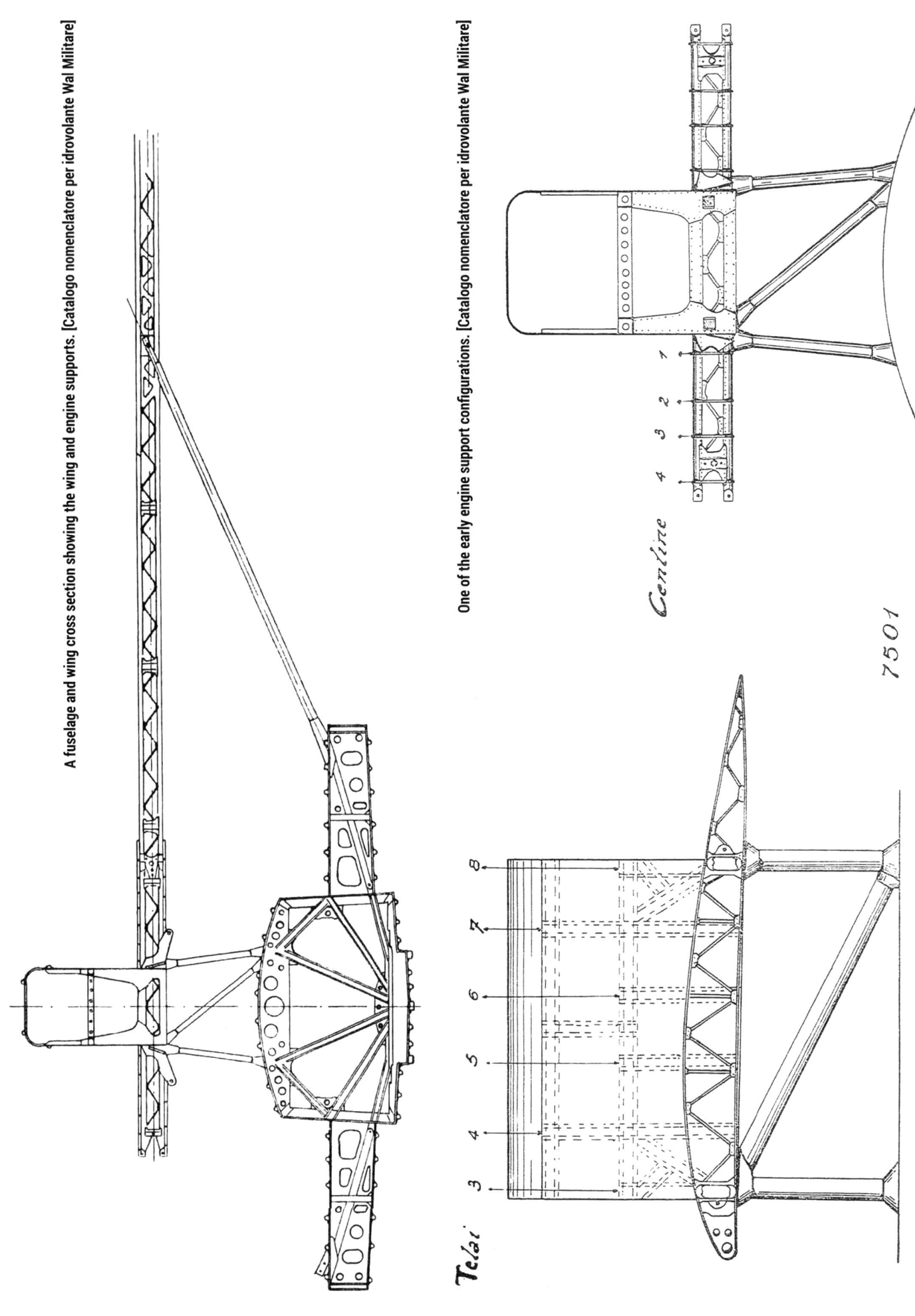

A fuselage and wing cross section showing the wing and engine supports. [Catalogo nomenclatore per idrovolante Wal Militare]

One of the early engine support configurations. [Catalogo nomenclatore per idrovolante Wal Militare]

Centine

Telai

7501

90

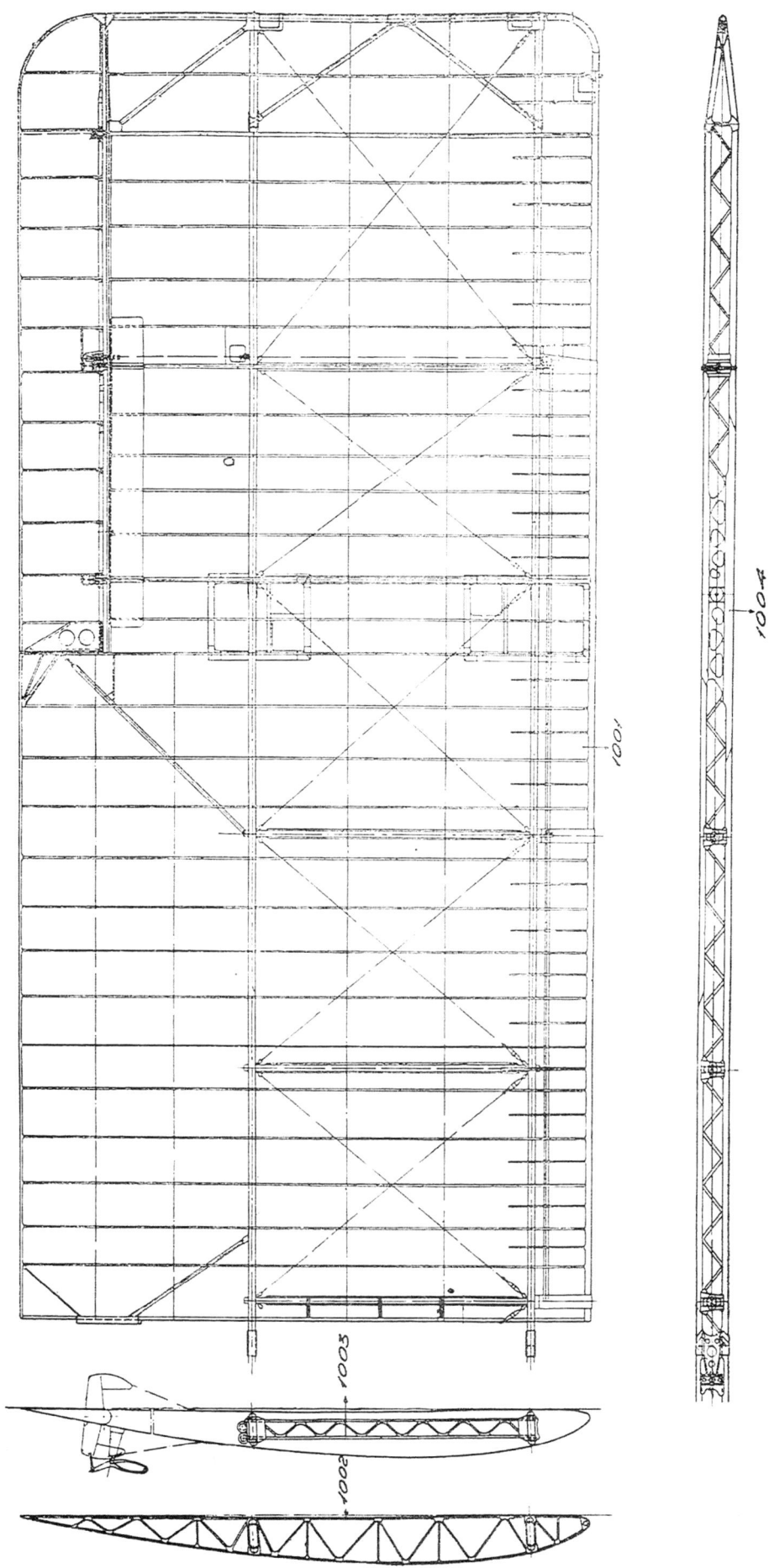

A wing detail showing the two spars, ribs and bracing cables. [Catalogo nomenclatore per idrovolante Wal Militare]

91

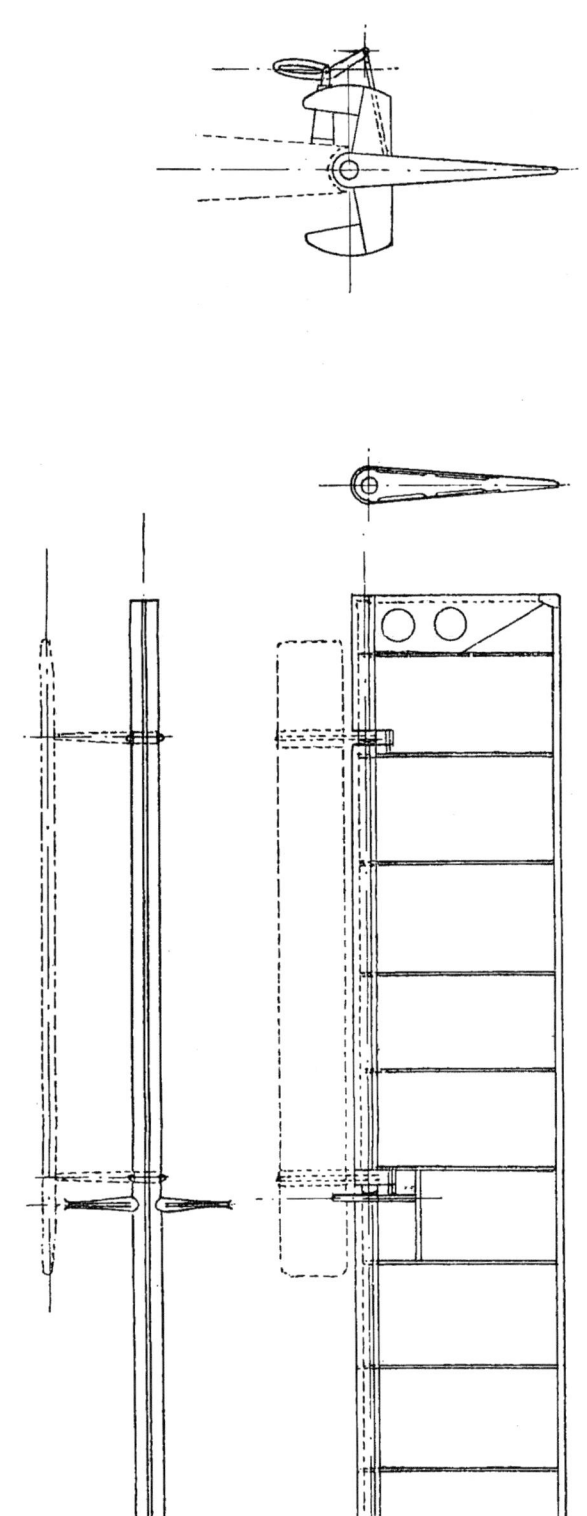

A detail of the aileron showing the trim tab arrangement. [Catalogo nomenclatore per idrovolante Wal Militare]

Amongst the 26.HE crews, the question was raised if we should fly over along with Do H (Do 22) to Corfu, but pilots declined this indicating that their flying boats were damaged or unable to take off and, that even if they were capable of flying, they could not make it since they were not equipped with night flying equipment.

On 17 April, floatplanes which relocated from their home bases began to return. Amongst them Do Wal/H 260 coasted in the water to Kumbor under own power and its crew used the opportunity to throw weapons and radio equipment overboard. Pilot Ilija Goršić then deliberately ran it against a sewage pipe on shore damaging the fuselage. Pnar Milivoj Boroša from 21.HE on own accord damaged and partially sunk an unknown Do Wal which was tied in front of 3.HK base.

On 18 April Do Wal/H 258, despite sustained damage during the attack at Jadratovac, was flown over from Boka Kotorska by pf Franjo Jagatić to Divulje, marking the last flight of PVKJ Do Wal during World War II. Maj Borivoj Šrajer described this event:

When the Italians entered Divulje one of our Do Wal flew in from Boka with pilot Jagatić thinking that PVKJ Command was still working. The Italians captured the machine and let the crew free.

Pf Jagatić described this event in his own words as well:

We received an order to transfer the airplanes to our bases. I took off with my crew and landed at Kumbor, where 21.HE was based. After we arrived, I asked escadrille CO for permission to fly over to Divulje, since I belonged to that base. He did not give me specific orders, but he said that I can do as I please. I took off and landed at Divulje and handed off my airplane to kbb Nikola Nardeli, who immediately thereafter sent me to my home. The remainder of my 26.HE was disbanded on 14 April and all airplanes were either destroyed or sunk by own crews. During the entire war, the escadrille did not conduct a single mission to defend the country.

21.HE did not receive nor undertake any combat missions and its combat value was significantly reduced with the lack of two of the most capable Do Wal/K 253 and 256. It was pbb Ik Vladeta Petrović's plan that 21.HE conducts night bombing of Albanian harbors, bomb eventual shore landing zones along the Yugoslav coast as well as transport aviators in case of organized retreat. During the entire course of the war, Do Wal/B 200 remained at 3.HK workshop at Kumbor due to the lack of spare parts while Do Wal/J 254 was undergoing conversion to the *Rolls-Royce* XVI engine at Divulje. Do Wal/J 254 was subsequently destroyed at a hangar during the *Regia Aeronautica* bombing.

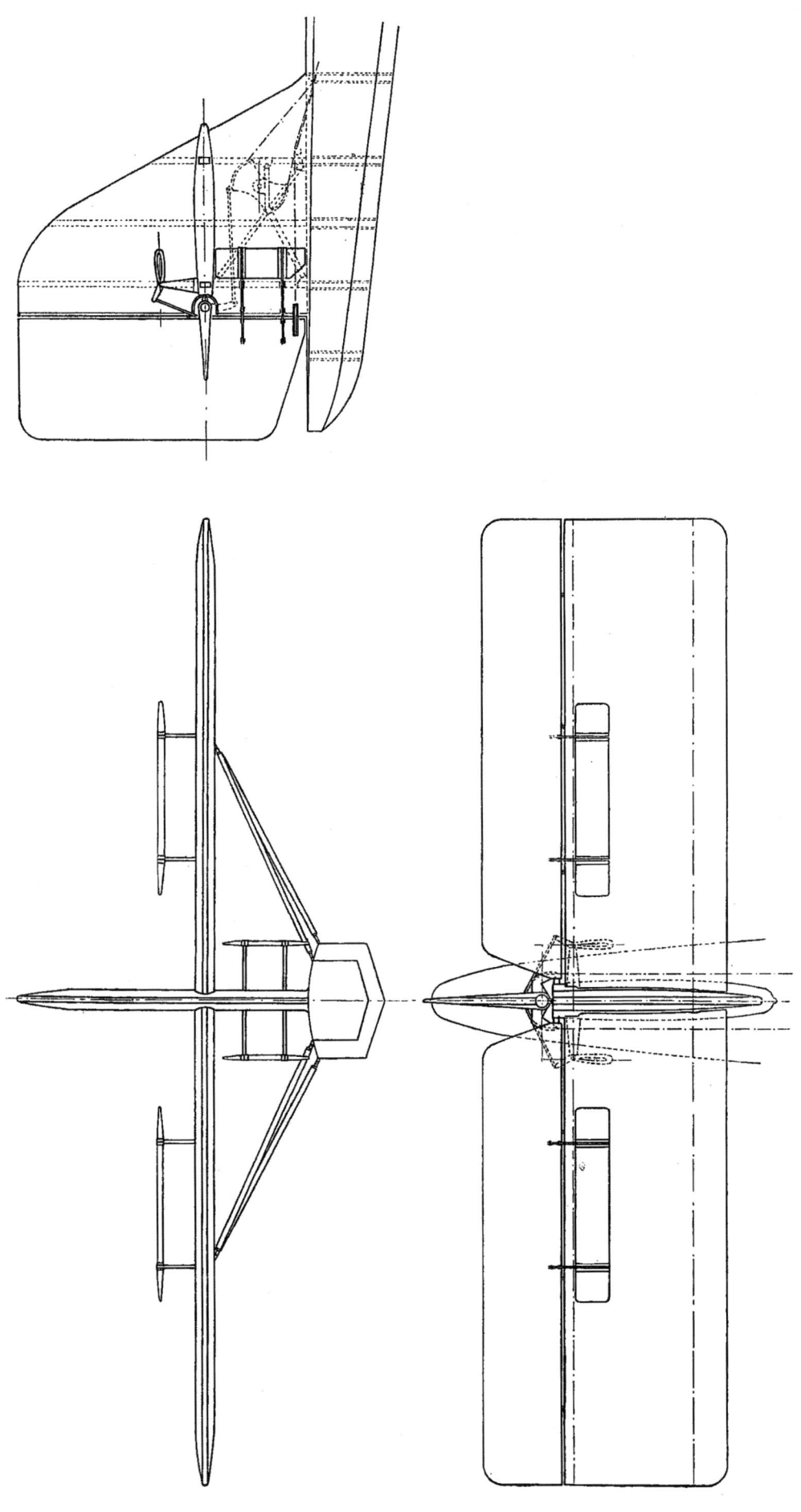

Do Wal tail assembly of the Do Wal/B 200 showing the vertical and horizontal stabilizer arrangement along with the support struts, trim tabs and control cables. [Catalogo nomenclatore per idrovolante Wal Militare]

A detail of the vertical stabilizer to rear fuselage attachment. [Catalogo nomenclatore per idrovolante Wal Militare]

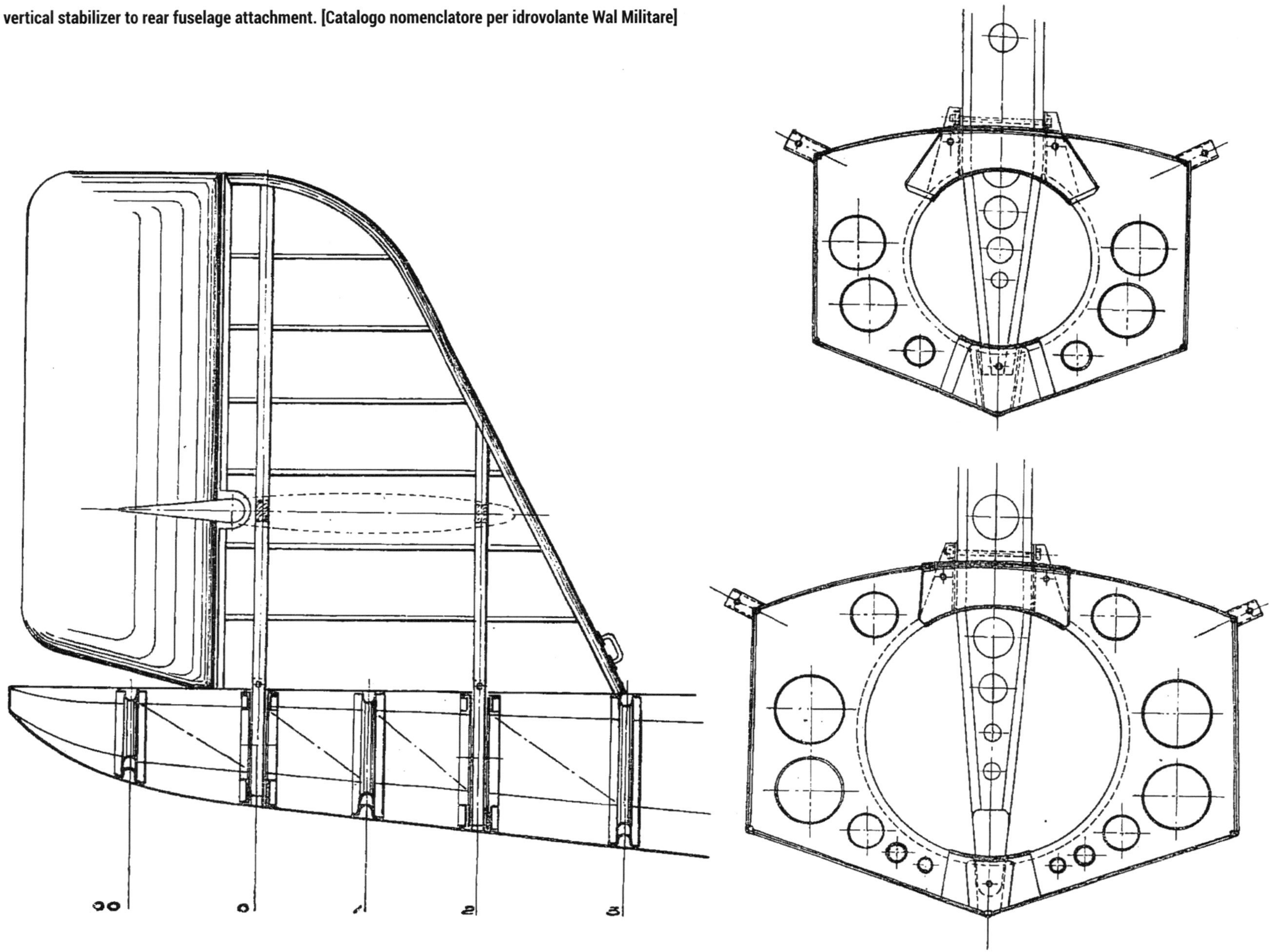

Do Wal/J 256 showing here with the beaching gear. Note the rear machine gun turret in the fuselage and the aileron trim tabs. [Josip Novak]

Italian documents indicate that a total of eight Do Wals were captured of which the known escadrille numbers are 200, 251, 252, 255, 256, 258 and 260. Do Wals were allegedly offered to Germans who declined to use the type due to its obsolescence and as a result these machines were likely scrapped.

Construction Features

Do Wal was a braced high wing monoplane of all metal design with fabric covered wing and control surfaces. The wing halves were connected to the center section, which formed the engine nacelle and were braced against the sponsons by a pair of struts each. The entire wing except for the step on areas was covered with fabric. The center section mounted to the fuselage by means of the so called cabane struts also server as the main support for the engine nacelle.

The intrinsically stable two step hull was equipped with Dornier designed sponsons which significantly improved stability on the water surface. Hull and sponsons had six bulkheads. Hull was subdivided to the observer's seat with machine gun in the nose, two seat flight deck with dual controls, fuel tanks and bomb dropping section and observer's seat with machine gun in the tail. Small windshields were installed ahead of the front and rear turrets as well as ahead of the pilot and copilot's seats. Many instruments in the cockpit were furnished by *Askania* company. Do Wal/B 200 during its early service had two additional lifting surfaces in front of the sponsons, these were removed later in the service as seen in the photographs.

The horizontal stabilizers, rudders and ailerons were balanced by small trim tabs. The pointed front keel gradually turned into a curved hull floor with a longitudinal step in the center. Between the hull floor and the rear displacement keep there was a cross step.

Large hull resulted in exceptional stability on water surface in even most severe sea conditions.

The tail assembly consisted of adjustable horizontal stabilizer, a fixed keel fin, rudder and elevator with automatic balancing, ailerons separated from the wing by a slot and installed on bearing brackets and fabric covering. The tail consisted of one tail fin and one horizontal stabilizer with attached rudders.

On PVKJ Do Wals, the tail assembly as well as the horizontal stabilizer supports differed between models.

Across the entire production of Do Wal different engines were installed in tandem. Paired with various engine iterations were different radiator arrangements as well as two or four blade wooden propellers. Wooden propellers were covered with special canvas covers when not in use to protect them from the elements. PVKJ Do Wals used BMW VI, *Gnome-Rhône* Jupiter VI, *Hispano Suiza* Ydrs and *Rolls-Royce* Kestrel XVI engines. Do Wal/B 200 was initially delivered with a four-blade propeller however at least as of mid 1929 it was replaced by a four blade one.

The offensive weapons of PVKJ Do Wals consisted of 100kg and 250kg domestically made *Sartid* bombs. Only Do Wal/K 253 and 256 were equipped with 250 kg bomb racks while all others used 100 kg bomb racks. Bomb racks were carried either on each side of the cockpit or each side of the rear fuselage turret. 7.7 mm *Saint-Étienne Darne* machine guns wee mounted on turrets with *Scarff* rings, one machine gun turret was at the very nose of the airplane and the other at the rear fuselage. Do Wal/J and Do Wal/B 200 used *Lofte* bomb sight manufactured by *Karl Zeiss* from Jena while Do Wal H and K used *Goerz-Boykow* bomb sights manufactured by *K.P. Goerz* from Bratislava.

Hispano Suiza powered Do Wals had an antenna which spanned from the engine nacelle to two rods close to each wing tip.

Do Wal cockpit with pilot and copilot yokes and basic flight instruments. [Staatsarchiv St. Gallen]

Aft view from the cockpit towards the bulkhead to the next compartment. [Staatsarchiv St. Gallen]

Depending on the model, one or two auxiliary wind driven electrical generators were used. [Staatsarchiv St. Gallen]

A detail of the aft Gnome-Rhône Jupiter VI engine cylinders. Note the four blade wooden propeller. [Staatsarchiv St. Gallen]

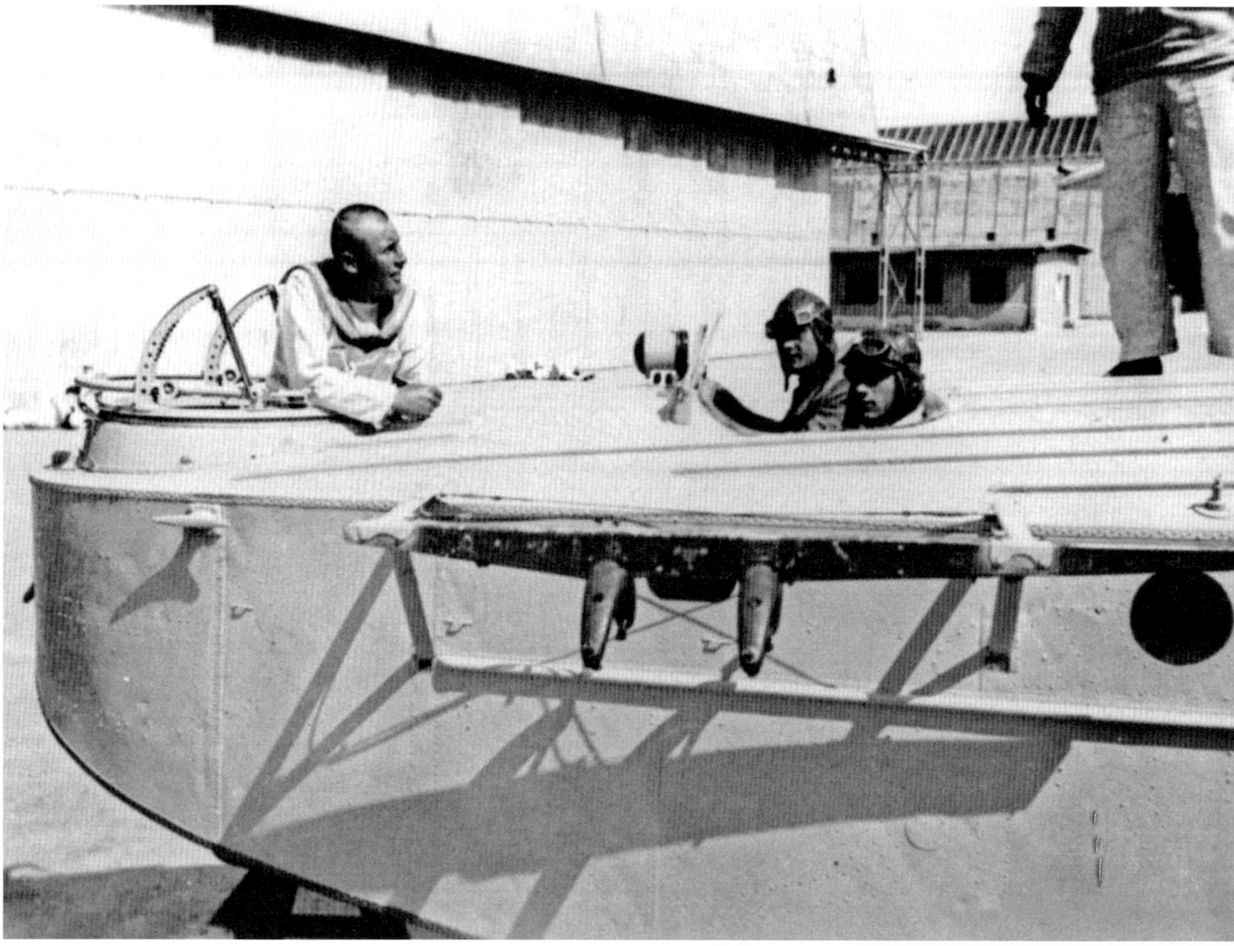

Do Wal/B 200 showing the front machine gun turret, old style bomb rack and cockpit which was offset to the starboard. [Josip Novak]

Abandoned PV bombs at Kumbor which were captured by the Italian forces. [Stato Maggiore Aeronautica]

Technical Specifications

Technical specifications Dornier Do Wal "BMW" (Do Wal/B)

Quantity used:	2
Crew:	6
Years of Service:	1927-1941
Span:	22.5 m
Length:	17.6 m
Height:	5.6 m
Wing area:	95.0 m²
Engine:	Two 500/600 hp BMW VI (200) Two 530/700 hp BMW VI (257)
Empty weight:	4,180 kg
Maximum weight:	6,700 kg
Maximum speed:	190.5 km/h
Cruise speed:	185 km/h
Service ceiling:	5,300 m
Maximum range:	1,180 km
Armament:	Four 250 kg bombs. Three 7.7 mm Darne machine guns for defensive purposes, two in a twinned turret in the nose and one in the rear fuselage turret.

Technical specifications Dornier Do Wal "Jupiter" (Do Wal/J)

Quantity used:	6
Crew:	6
Years of Service:	1931/32-1941
Span:	23.2 m
Length:	18.2 m
Height:	5.5 m
Wing area:	96.0 m²
Engine:	Two 480 hp Gnome-Rhône Jupiter VI
Empty weight:	3,950 kg
Maximum weight:	6,950 kg
Maximum speed:	196.5 km/h
Cruise speed:	150 km/h
Service ceiling:	3,000 m
Maximum range:	1,000 km
Armament:	Four 250 kg bombs. Three 7.7 mm Darne machine guns for defensive purposes, two in a twinned turret in the nose and one in the rear fuselage turret.

Technical specifications Dornier Do Wal "Hispano" (Do Wal/H)

Quantity used:	3
Crew:	6
Years of Service:	1935/36-1941
Span:	23.2 m
Length:	18.5 m
Height:	5.9 m
Wing area:	96.0 m²
Engine:	Two 760/835 hp Hispano Suiza Ydrs
Empty weight:	4,730 kg
Maximum weight:	7,100 kg
Maximum speed:	247 km/h
Cruise speed:	220 km/h
Service ceiling:	7,300 m
Maximum range:	1,500 km
Armament:	Six 250 kg bombs. Three 7.7 mm Darne machine guns for defensive purposes, two in a twinned turret in the nose and one in the rear fuselage turret.

Technical specifications Dornier Do Wal "Kestrel" (Do Wal/K)

Quantity used:	2
Crew:	6
Years of Service:	1940-1941
Span:	23.2 m
Length:	18.2 m
Height:	5.85 m
Wing area:	96.0 m²
Engine:	Two 670[1]/750[2] hp Rolls-Royce Kestrel XVI
Empty weight:	4,320 kg
Maximum weight:	7,000 kg
Maximum speed:	228 km/h
Cruise speed:	200 km/h
Service ceiling:	6,000 m
Maximum range:	1,140 km
Armament:	Four 250 kg bombs. Three 7.7 mm Darne machine guns for defensive purposes, two in a twinned turret in the nose and one in the rear fuselage turret.

[1] Nominal power
[2] Take off power

Camouflage and Markings

Prior to delivery to KSHS, Do Wal/B 200 was painted in a Medium Gray color on the top surfaces while the undersurfaces were painted in Black with a specially formulated paint which was used to prevent corrosion. The vertical stabilizer had a large Yugoslav state flag while large "Kosovo cross" insignias were applied across almost the entire wing chord both on top and below the wings. The escadrille number 200 was in Black color initially with a White outline and interesting cursive number 2. Just several weeks after its arrival to KJ, the White outline was removed. While in service Do Wal/B was painted, several times, with a Light Grey color which was most likely the DKH L40/52 which was applied on the Do Wal/J, Do Wal/B and Do Wal H delivered in the 1930s. Do Wal/B 200 had its escadrille number changed several times, at first the font changed and towards the end of its service the escadrille number was moved to the rear fuselage, to match all other Do Wals.

The first Do Wal/B 200 at Kumbor on 12 July 1927, several weeks after its arrival still carries the original darker Gray color paint, likely 2000/67 Grau. [Djordje Nikolić]

Several years after its arrival, the escadrille number font has changed, White outline was removed, and the machine received a new coat of paint, light Gray in color, which matches DKH L40/52 applied from 1935 onwards. [V.Koos – ADL]

During its final years in PV service Do Wal/B 200 had its escadrille number relocated to the aft fuselage. [Josip Novak]

Shown here prior to delivery, PV Do Wal/Js with Red vertical stabilizer and rudder with Swiss code CH in White. The identical code is applied to the port and starboard wing. The machine carries ferry code number 314. [G.Frost – ADL]

A tail view of this Do Wal/J shows a Blue-White-Red Yugoslav tricolor flag across the entire rudder. [HPMS]

Do Wal/B 257 received ferry registration D-7 and a large swastika on its vertical stabilizer and rudder which were quickly removed following its arrival. [Josip Novak]

On arrival, Do Wal/Js, had Red vertical stabilizer and rudder with Swiss code CH in White as well Black fuselage ferry code numbers 303, 304, 313, 314, 316 and 319. Wings had a wide Red band with Swiss Code CH on the port wing as well as Red band on with ferry code numbers on the starboard wing. After the arrival to KJ, tail and the wings were painted over, Yugoslav flag was added to the rudder escadrille numbers 251 through 256 respectively were painted on the rear fuselage and the same size "Kosovo cross" insignia as on Do Wal/B 200 were added to the wings. During their service, it is known that at least 251 and 256 received a Yellow triangle outlined with Red color containing number 1 or 6 respectively, although for unknown reason and duration.

Do Wal/J 256 shows two large Kosovo cross insignia on top of the wings as well as an interesting yellow triangle with a red border and large black number 6 inside. [Mario Raguž]

Do Wal/H 260 shown here after capture at Kumbor had hastily applied green paint on wing top surfaces as well as part of the fuselage topside behind the wings. [Oliver Fischer]

Do Wal/B 257 was delivered with Red vertical stabilizer and rudder with a large swastika inside of a White circle. The ferry code D-7 was applied on the rear fuselage. Soon after the arrival, this Do Wal was painted over in the same manner as the machines which arrived before it.

Do Wal/H 260 interestingly at some point during the April War received a hastily applied camouflage with Green patches along the entire upper wing surface and over the "Kosovo Cross" insignia, as well as on the rear fuselage. Interestingly, on Do Wals captured during the April War, it is clear that the Black anti-corrosion color beneath the fuselage belly was not applied, either its application was considered unnecessary or the paint itself could not be sourced at the time.

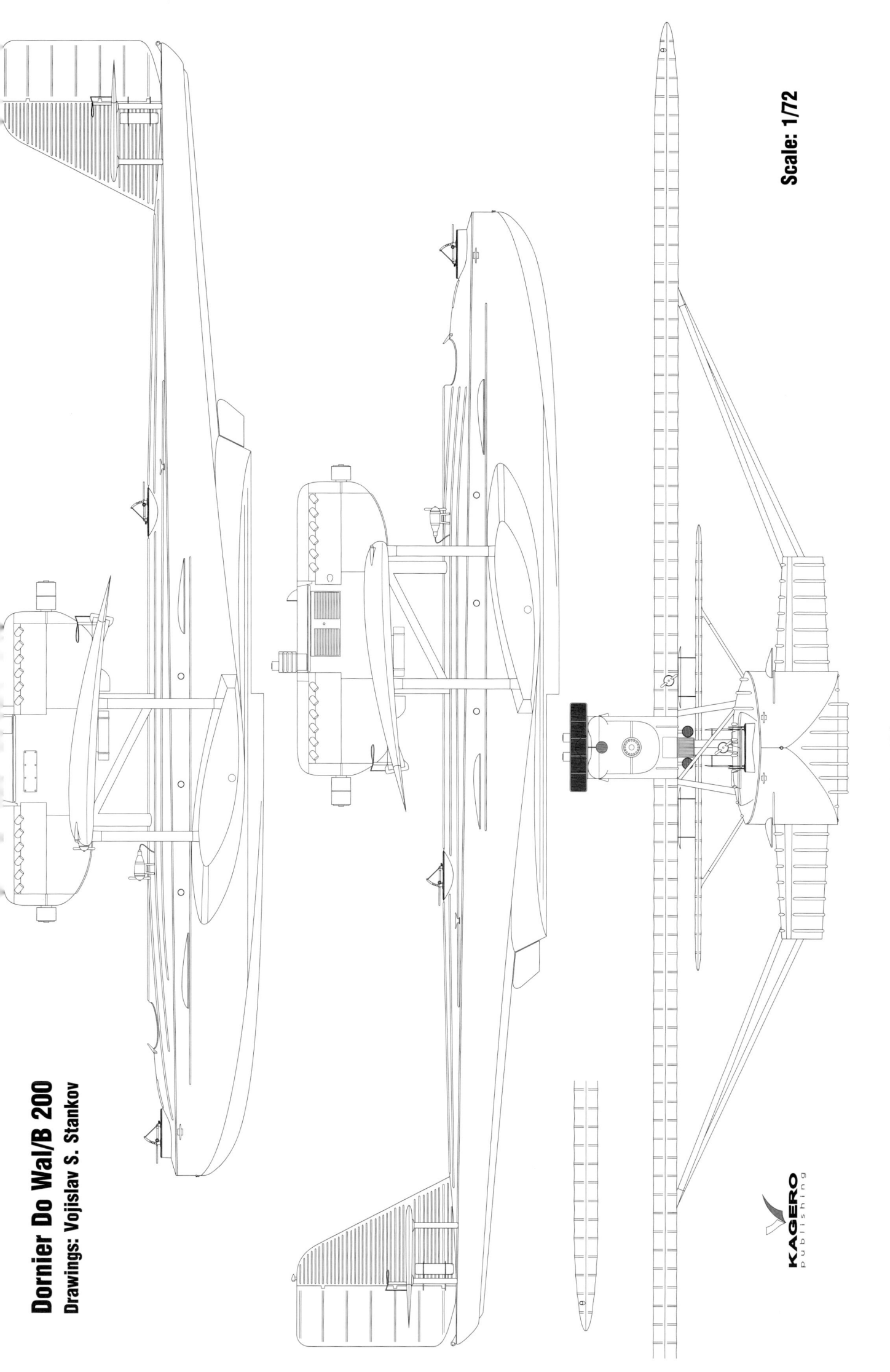

Dornier Do Wal/B 200
Drawings: Vojislav S. Stankov

Scale: 1/72

KAGERO
publishing

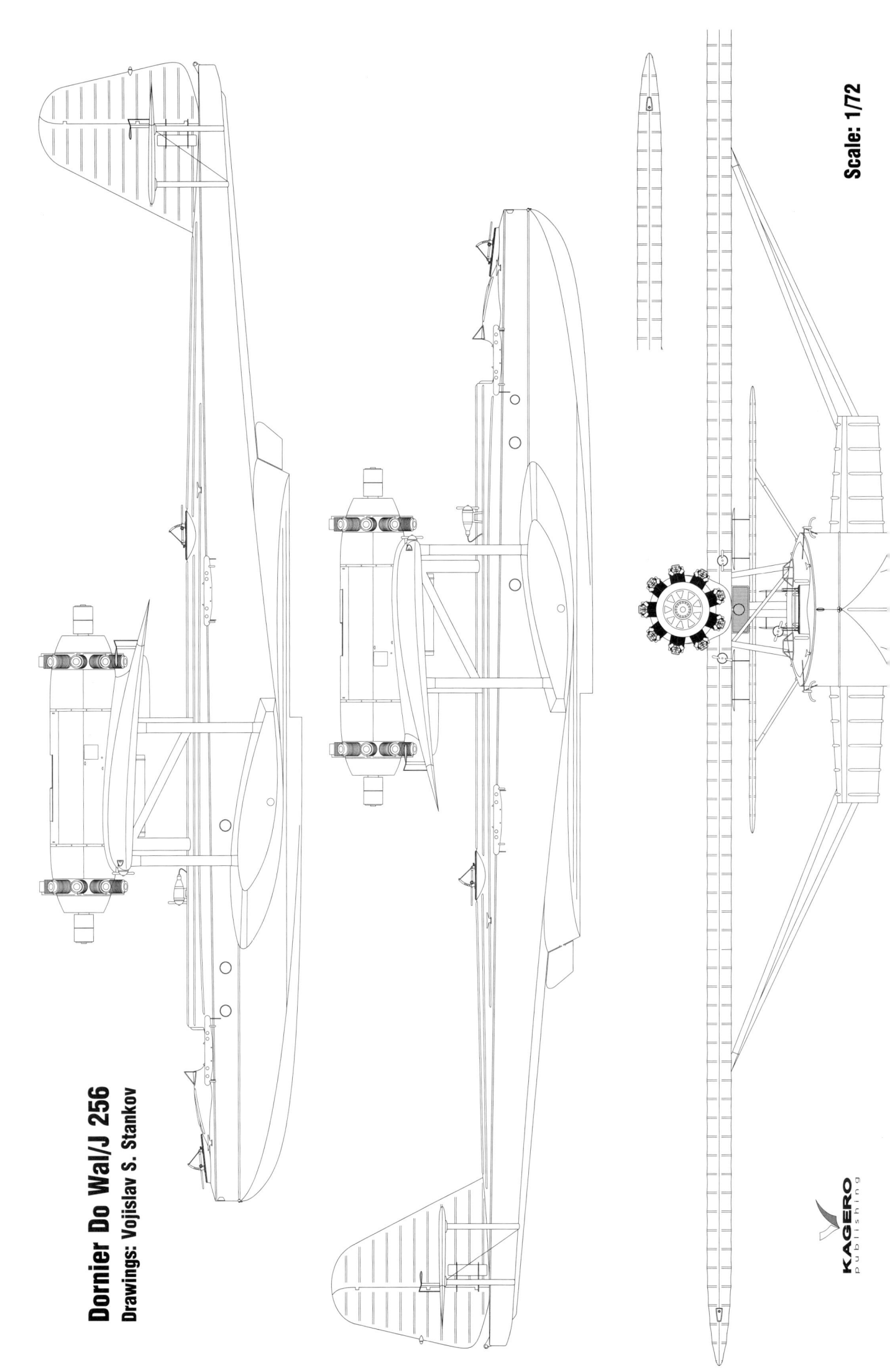

Dornier Do Wal/J 256
Drawings: Vojislav S. Stankov

Scale: 1/72

KAGERO publishing

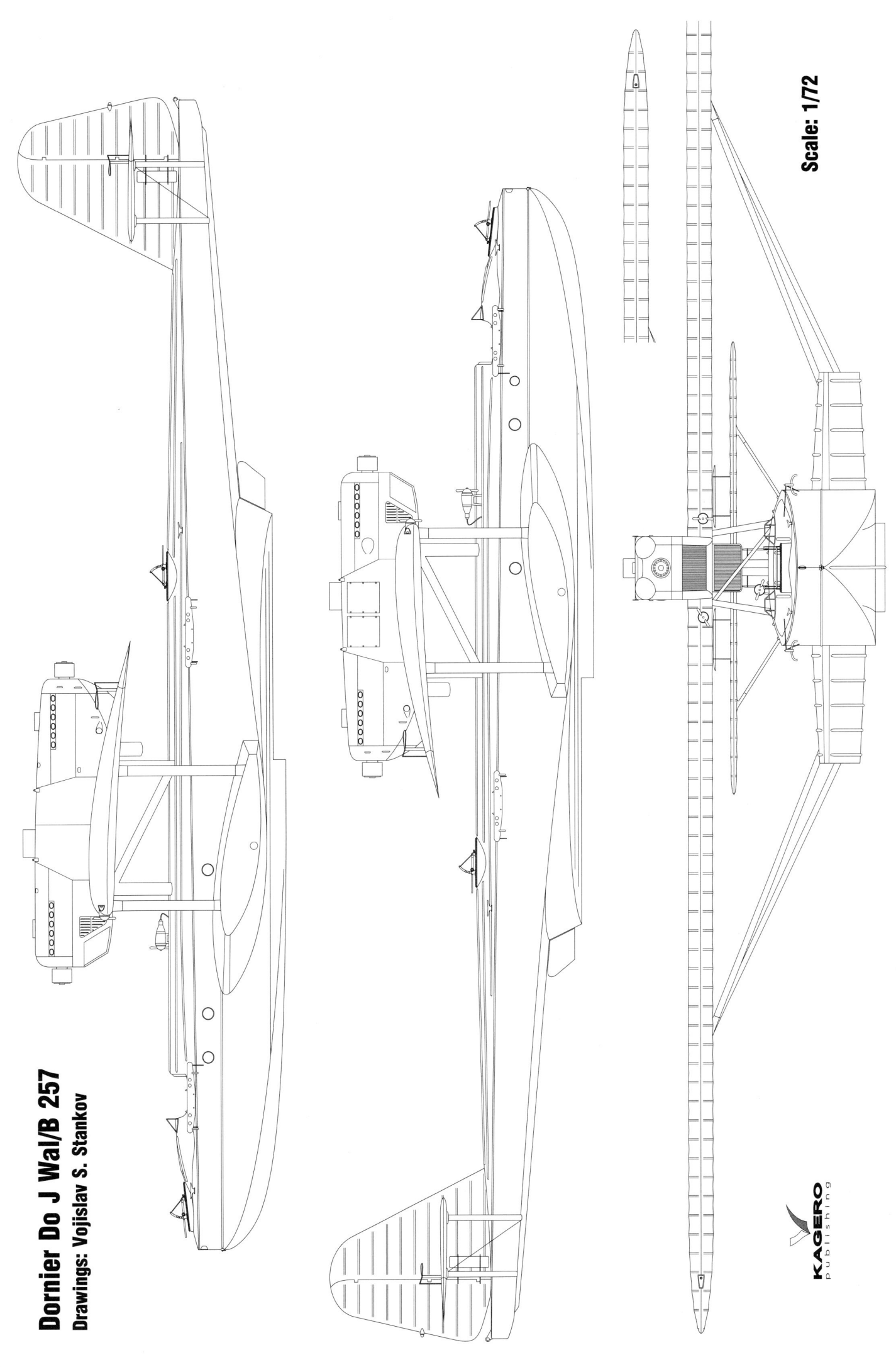

Dornier Do J Wal/B 257
Drawings: Vojislav S. Stankov

Scale: 1/72

KAGERO
publishing

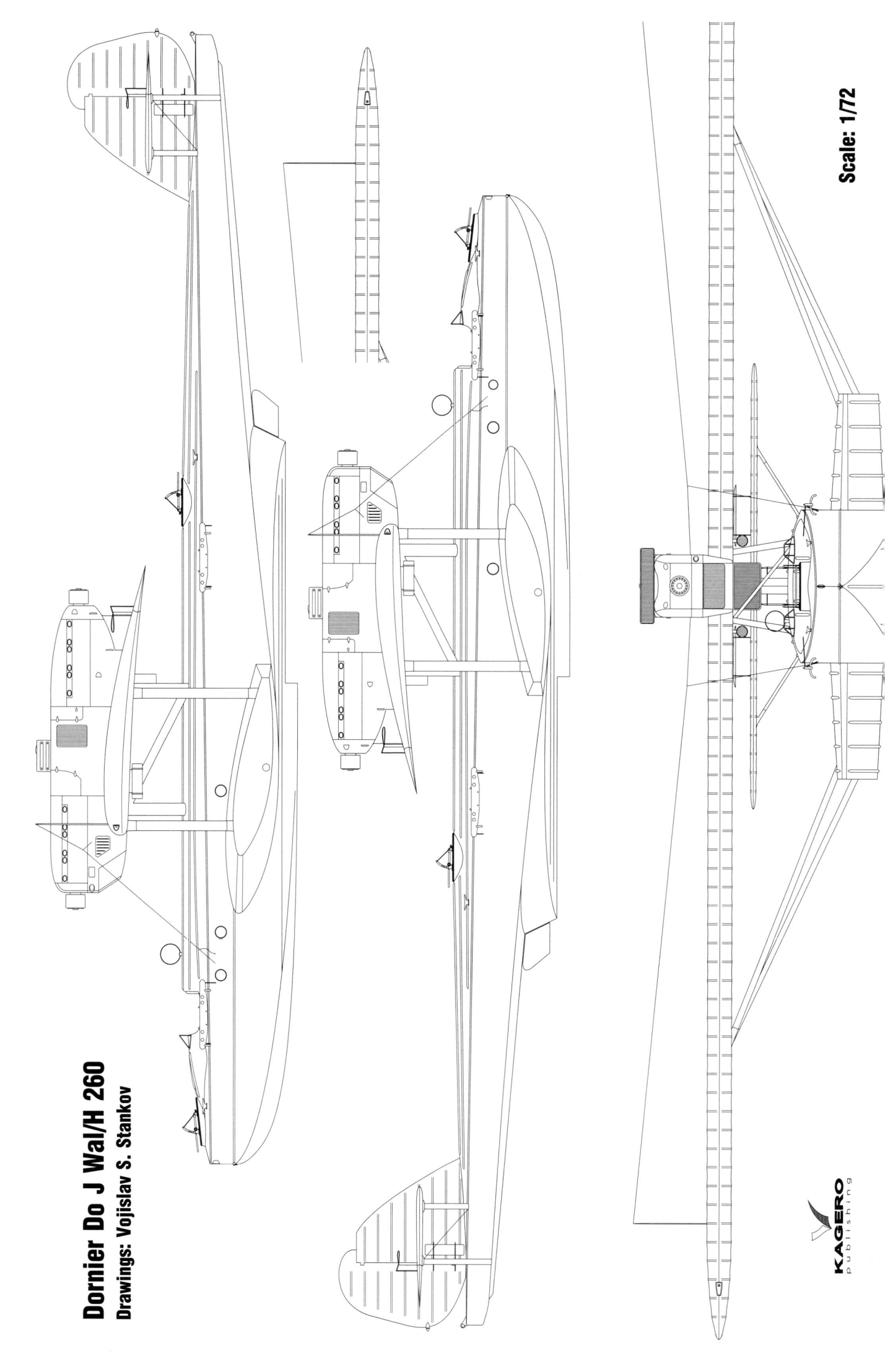

Dornier Do J Wal/H 260
Drawings: Vojislav S. Stankov

Scale: 1/72

KAGERO
publishing

Dornier Do Wal/B 200

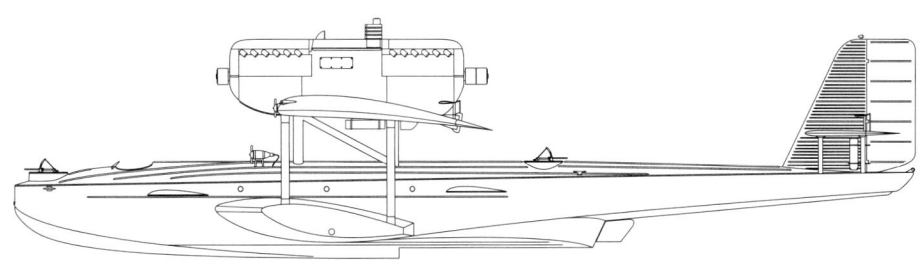

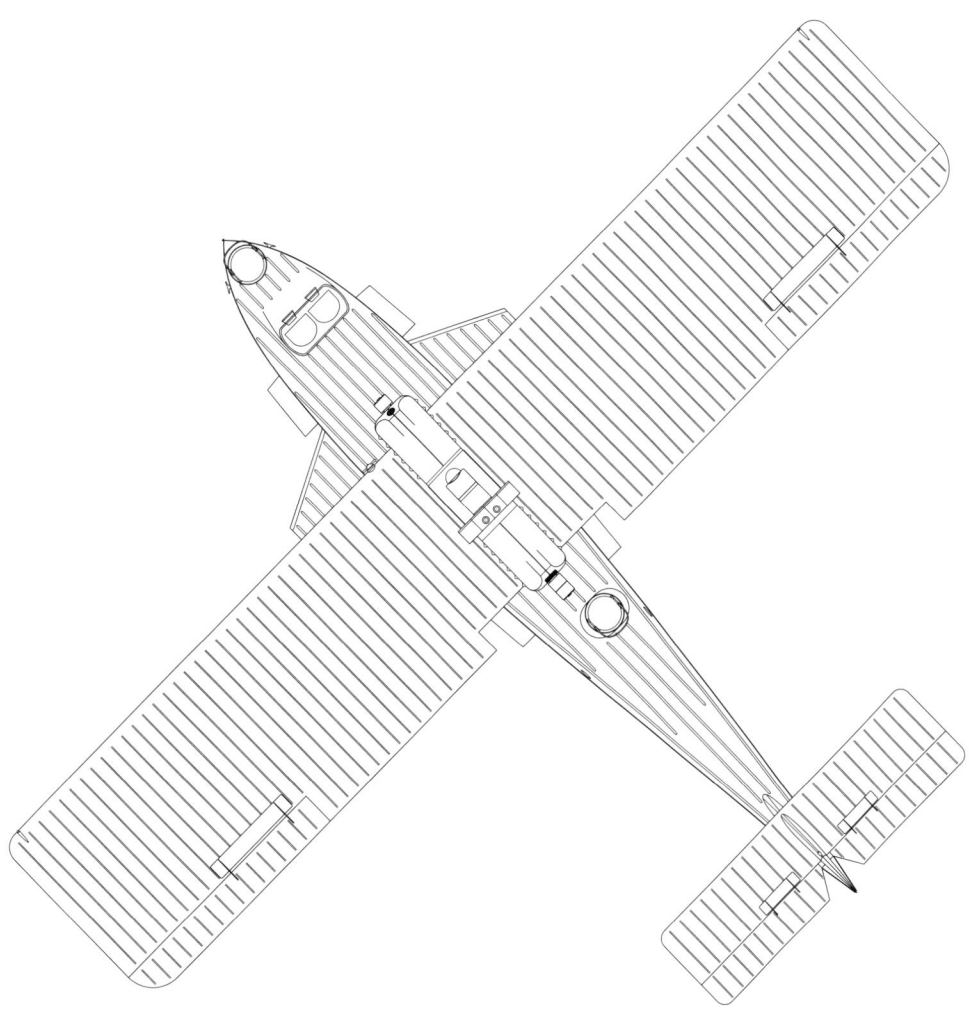

Drawings: © Vojislav S. Stankov

Scale 1/144

Dornier Do Wal/B 200

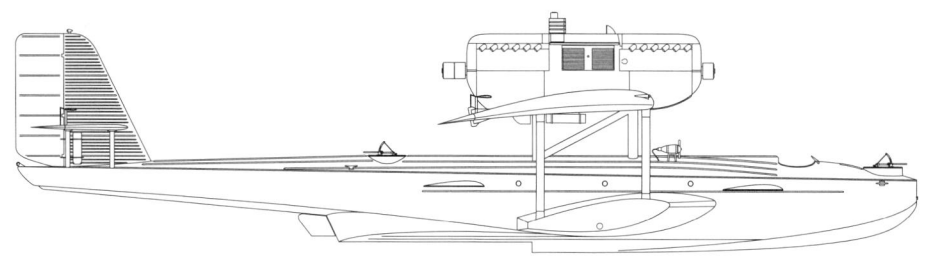

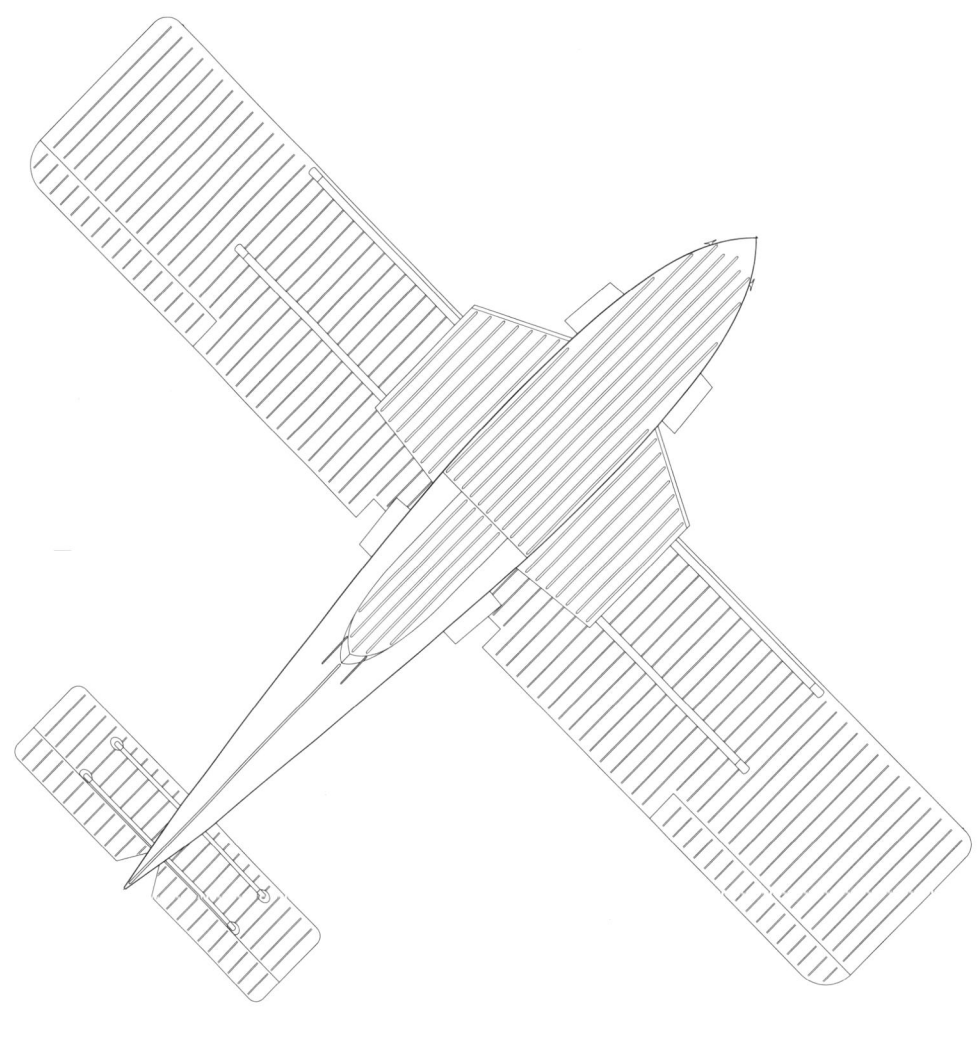

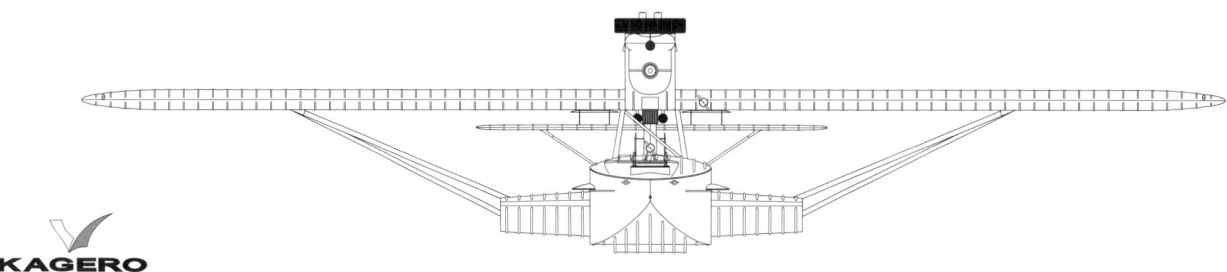

Drawings: © Vojislav S. Stankov

Scale 1/144

Dornier Do Wal/J 256

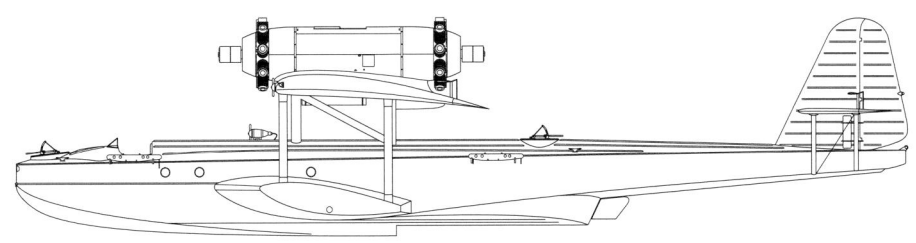

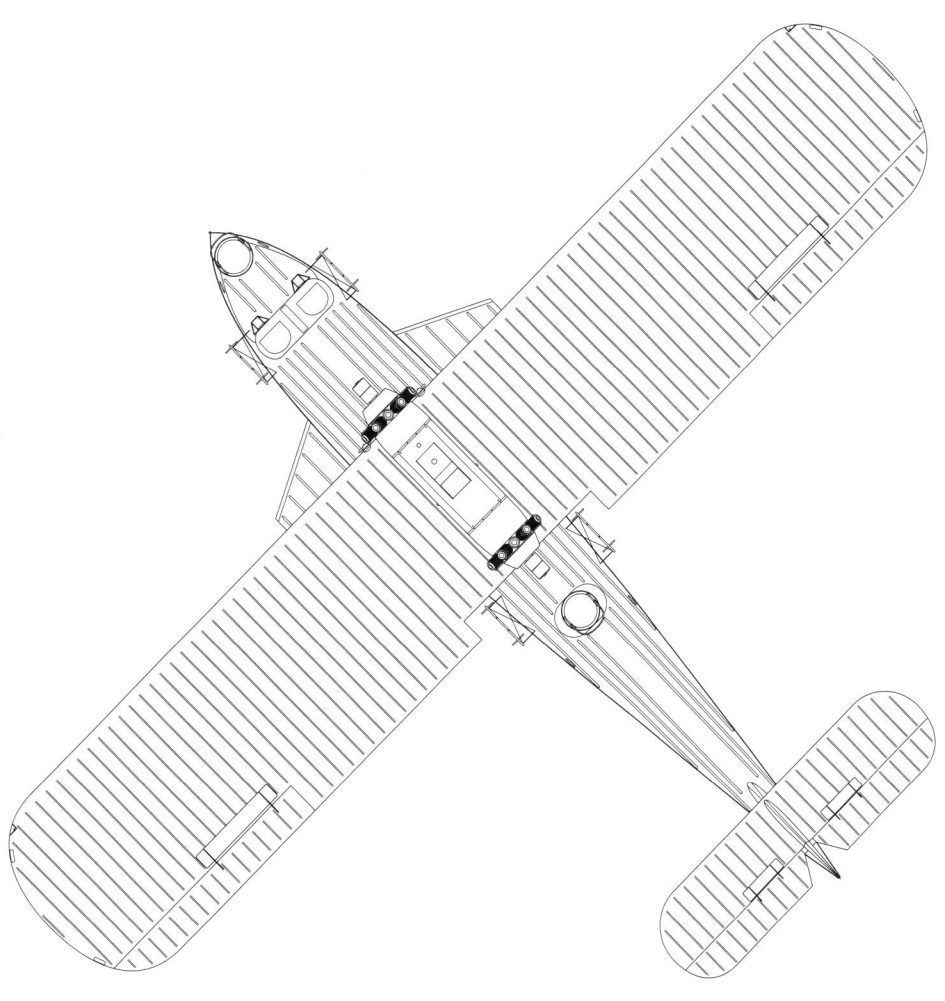

Drawings: © Vojislav S. Stankov

Scale 1/144

Dornier Do Wal/J 256

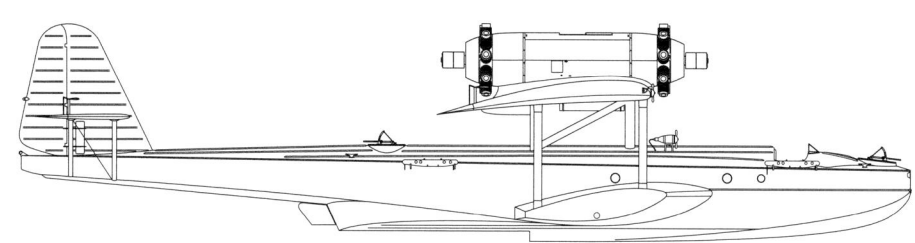

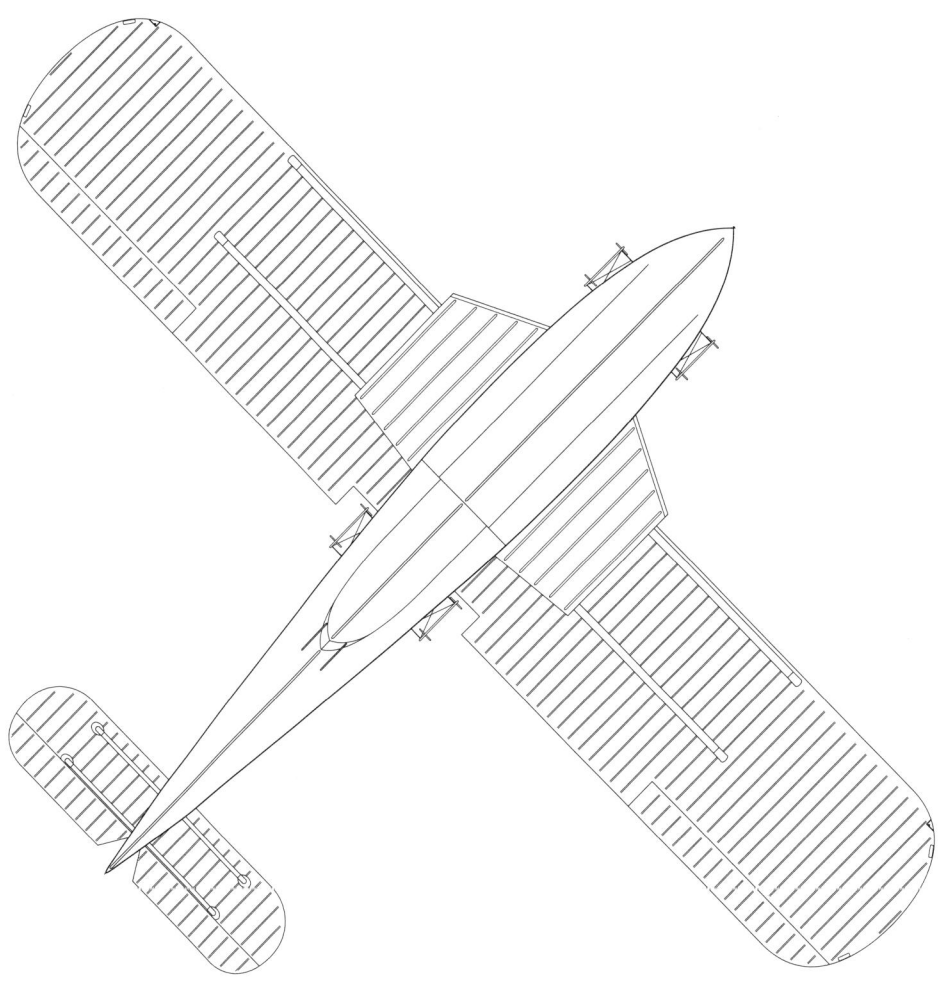

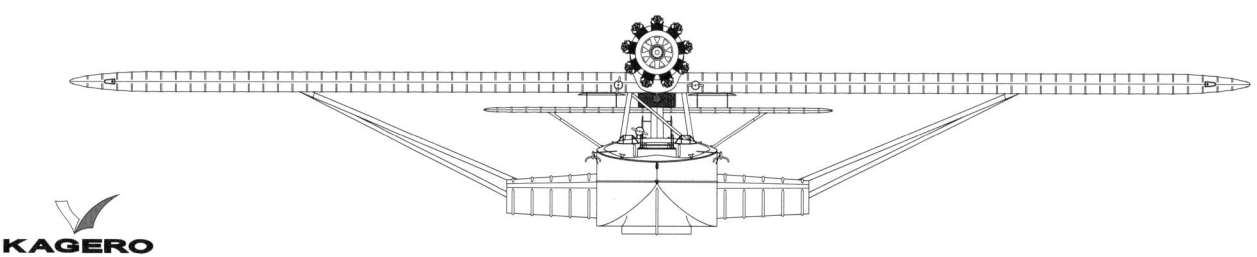

Drawings: © Vojislav S. Stankov

Scale 1/144

Dornier Do Wal/B 257

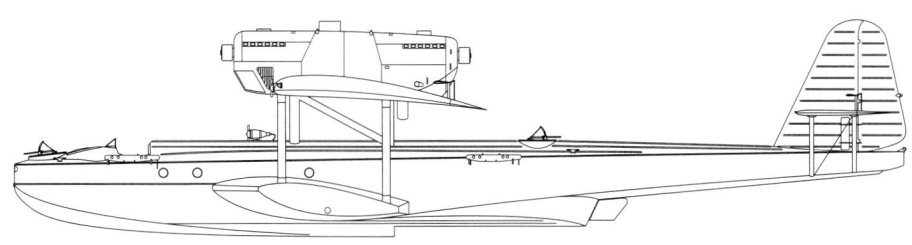

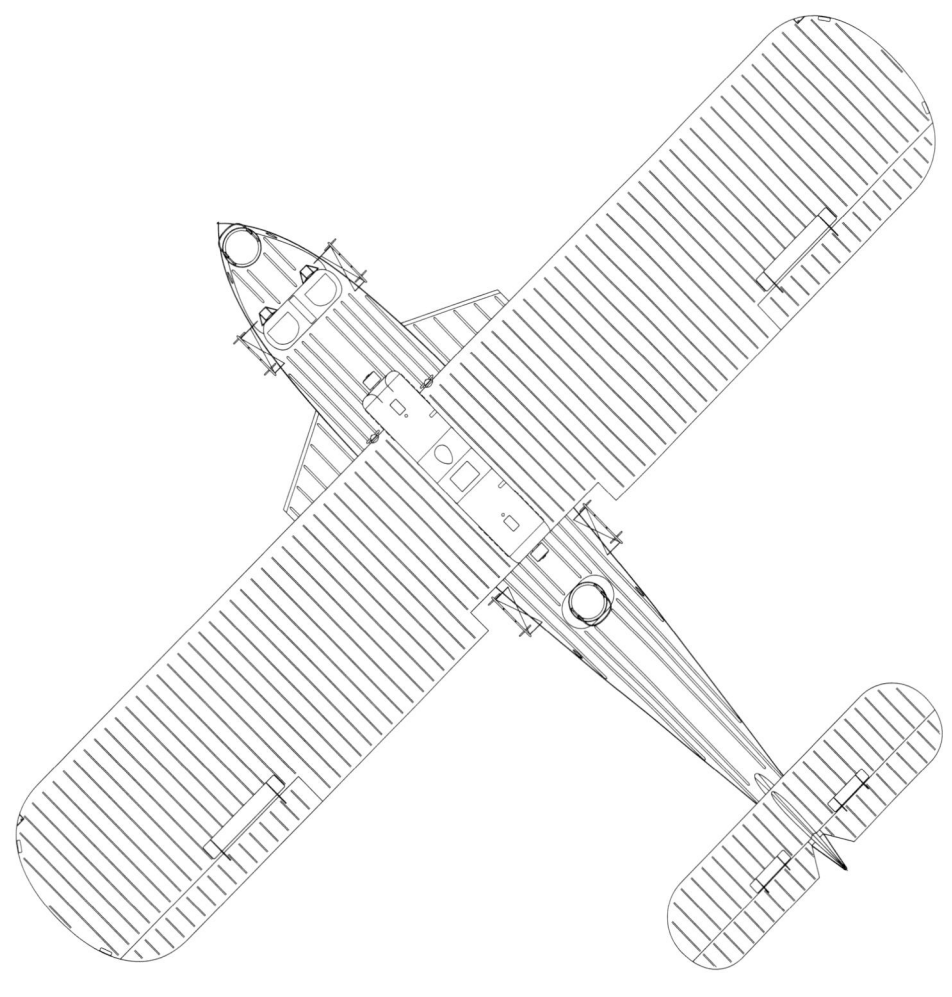

Drawings: © Vojislav S. Stankov

Scale 1/144

Dornier Do Wal/B 257

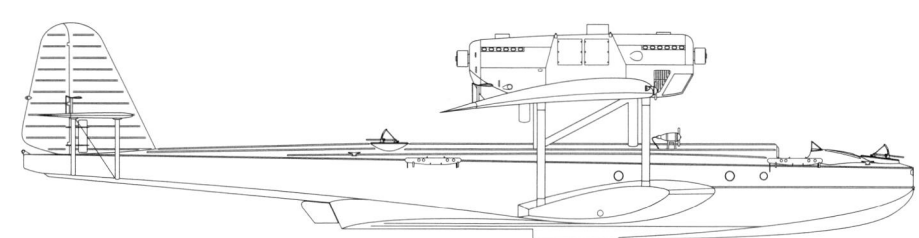

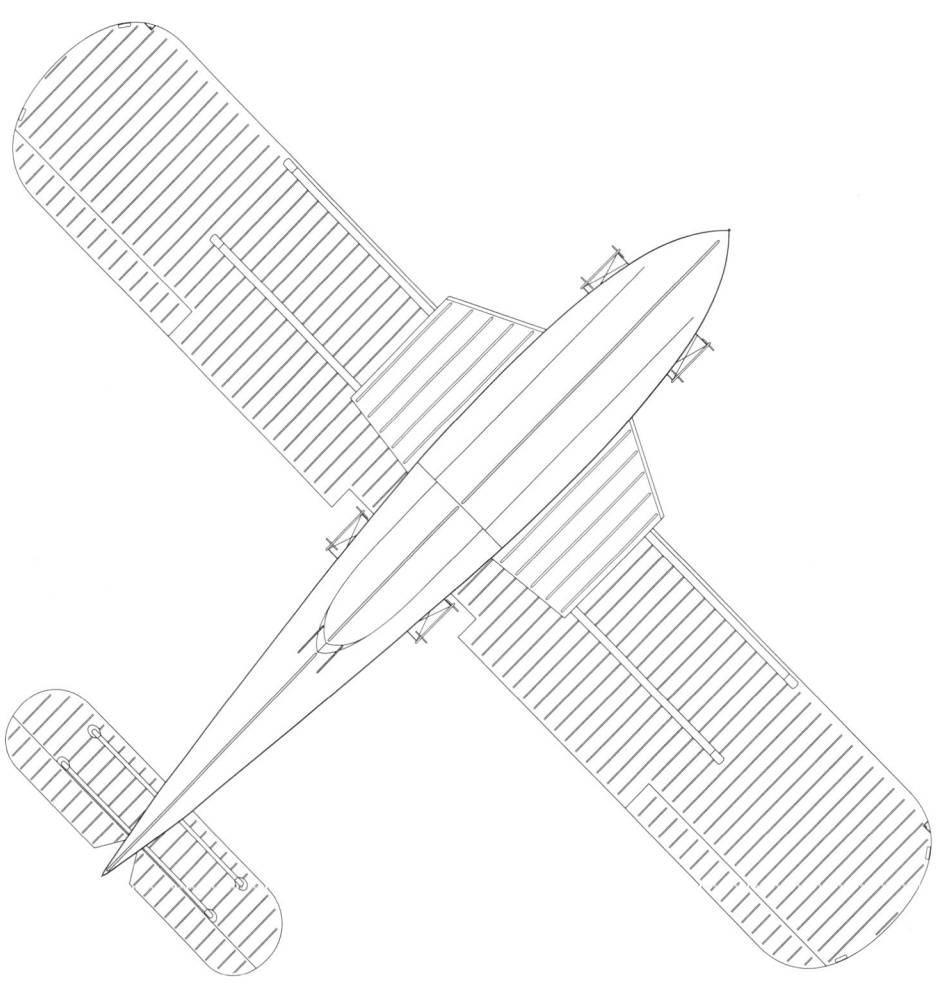

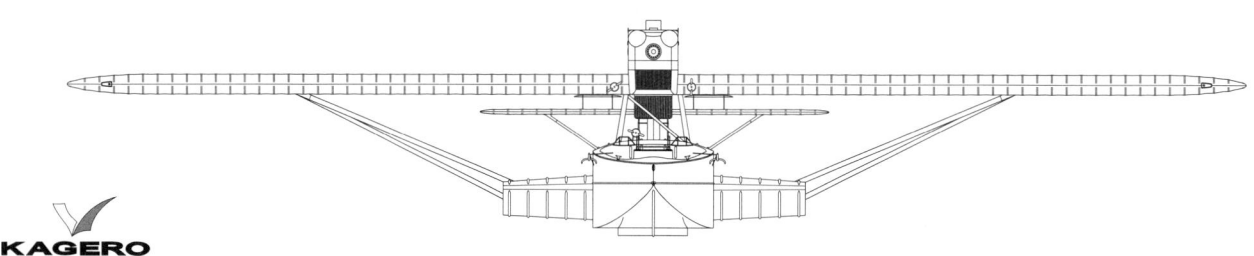

Drawings: © Vojislav S. Stankov

Scale 1/144

Dornier Do Wal/H 260

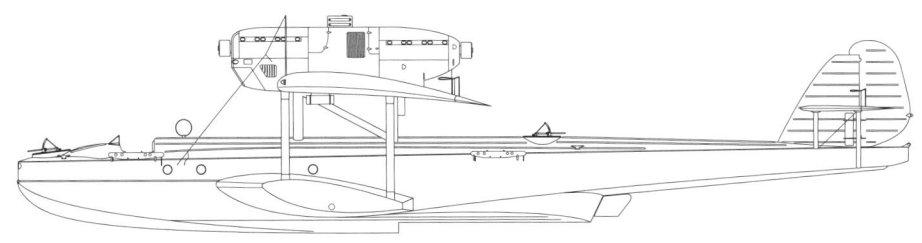

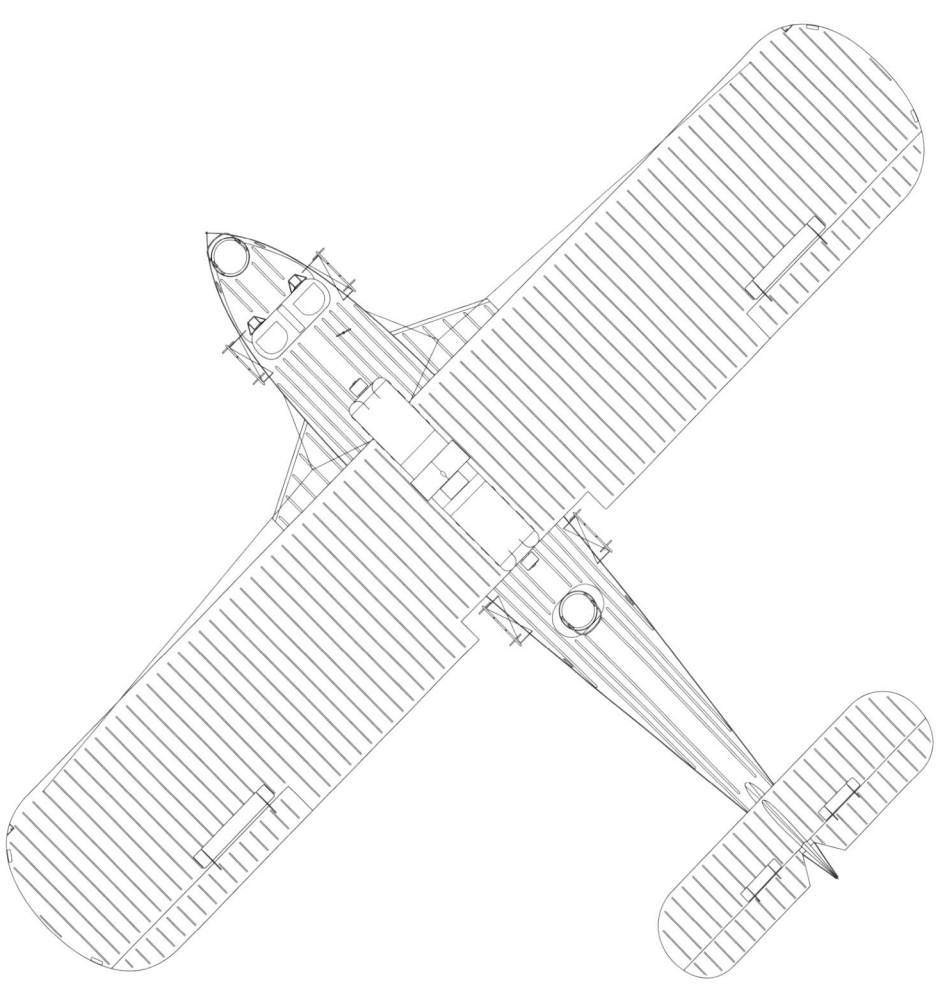

Scale 1/144

Dornier Do Wal/H 260

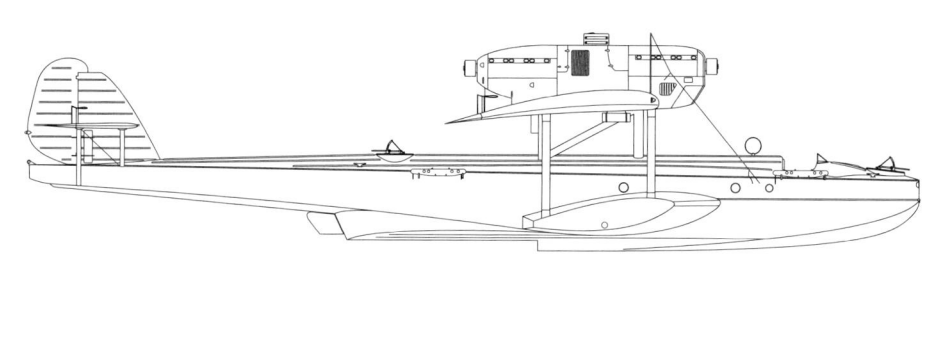

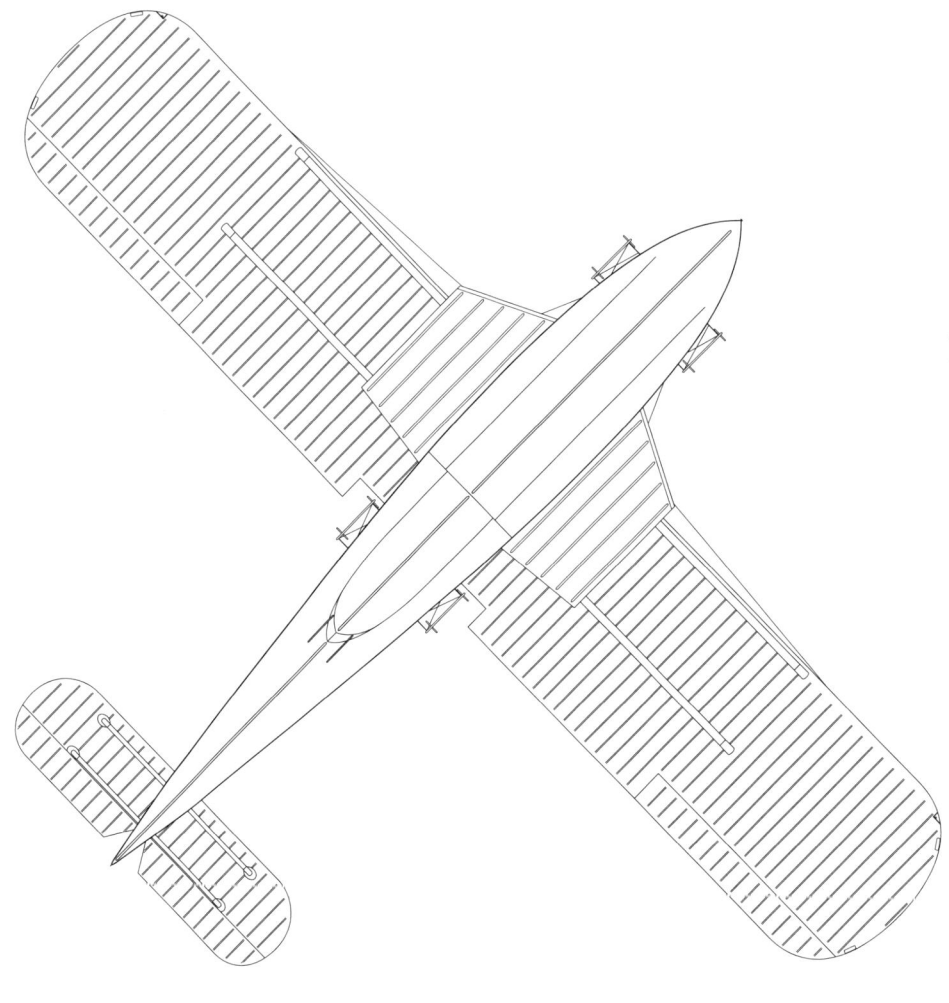

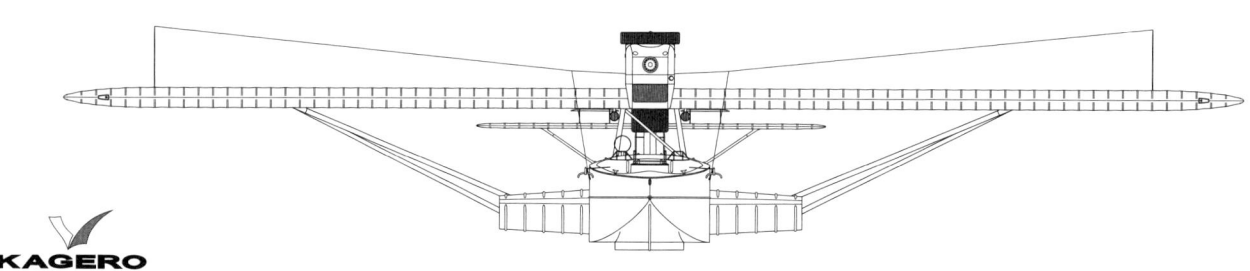

Drawings: © Vojislav S. Stankov

Scale 1/144

Dornier Do Y

Development

In the early 1930 KJ was interested in creating a fleet of multi-engine bombers, and due to the political reasons and the desire to pay for the new airplanes thorugh the war reparation funds, German companies *Dornier Metallbauten GmbH* and *Junkers* were contacted. In parallel with the two German companies, *Avia* from Czechoslovakia was also approached. The intent was to conduct comparative testing of airplane characteristic before making the final decision which future "heavy bomber" to purchase in numbers.

One of the requirements was that the airplane uses Yugoslav license built *Gnome-Rhône* Jupiter VI engine. The required performance characteristics as well as the short delivery dates caused problems for the manufaturers, since neither had available airframes in stock. An entierly new construction was not under consideration due to the high development costs and no guarantees that a follow up order would be awarded. *Junkers* resolved this problem by modifying the three engine passenger airplane G.24 to the required engines. Since the airplane was to be delivered from Germany as part of the war reparation payments, the modified G.24 was not delivered under its military designation K30, but as G.24 nao, of which two were delivered in 1932. *Avia* offered a license built version of the passanger *Fokker* F.9 in a bomber version, which was named *Avia* F.39, of which two were also delivered in 1932.

In spite of the hight costs of the new design development, at the end of 1930 *Dornier* began work on a three engine bomber airplane designated as Do Y. To control the costs, the design work was conducted in parallel with that on the twin engine Do 11 bomber developed for the *Reichswehr* (Imperial Army). Despite the inability to meet the short delivery deadline of 30 August 1931 for the first airplane, *Dornier* was still awarded the job.

The first produced Do Y, W.Nr. 232 seen here in front of the factory buildings at Manzell. Note the wooden propellers. [Airbus Corporate Heritage]

Top view of W.Nr. 232 from the factory building shows its unique engine arrangement. [Airbus Corporate Heritage]

To conceal its true purpose as a bomber, Dornier covered the nose windows and designated it as a "transport airplane" (Frachtflugzeug). [Djordje Nikolić]

Do Y W.Nr. 232 took off on its first flight on 17 October 1931. [Djordje Nikolić]

Do Y emerged as a cantilevered high wing airplane like the Do 11 with an unusual engine configuration where two engines were mounted on the wing leading edge while the third one was mounted on the pylon above the fuselage. This was a compromise solution as in this short time period a better design could not be implemented. This design was able to carry a bomb load of a maximum 1,200 kg in fuselage bomb bays.

At the time when the official Yugoslav contract was awarded on 21 May 1931, two airplanes, W.Nr. 232 and 233, were already under construction at Manzell. The airplanes were officially designated as "transport airplanes" (*Frachtflugzeuge*) because production of military airplanes in Germany was still prohibited. On 17 October 1931, W.Nr. 232 had its first flight from Friedrichshafen-Löwental airfield, while W.Nr. 233 followed in December. Test flights discovered design deficiencies partially due to the engine layout and overheating and also due to the wing shape. The newly designed wing shape used in Do Y and Do 11 was based on the Do K3 which was developed only several months earlier, and it tended to oscillate under certain flight parameters. The engine cooling was also problematic as the engines overheated and different cowlings were tested until a satisfactory design was discovered. Two blade metal propellers used during factory test flights from Friedrichshafen airfield were replaced with two blade wooden propellers prior to delivery to the KJ. At the conclusion of the test flight program from Friedrichshafen, airplanes were validated to have met the design requirements.

An international political event threatened to jeopardize the entire contract since the American president Hoover proposed that German war reparation payments be suspended for an entire year. The so-called Hoover plan came into effect on 1 July 1931 and instantly the Reich stopped all deliveries which were funded from the war reparation accounts.

Due to the lack of funds, Dornier fell into a serious liquidity crisis. Apart from two Do Ys worth 570,000 RM (*Reichsmark*), there were also six Do J II Wal worth 2 million RM under construction. Under these circumstances, *Dornier* had to rely on the willingness of KJ to pay for the airplanes on order. In the meantime, the airplanes were showcased in Germany, but *Reichswehr* chose more economic twin engine bombers instead.

With the agreement finally reached between *Dornier* and KJ, the required sum of money was received, and the delivery preparations progressed. Kap Ik Hinko Hubl

The engine cooling proved problematic during testing as the engines overheated, hence different cowlings were tested until a satisfactory design was discovered. Note the metal propellers which replaced the wooden ones. [Djordje Nikolić]

Do Y W.Nr. 233 in between test flights and prior to delivery. [Airbus Corporate Heritage]

Prior to delivery to KJ, both W.Nr. 232 and W.Nr. 233 received ferry markings D-3 and D-6. Shown here prior to a promotional flight. [Airbus Corporate Heritage]

The first Do Y, W.Nr. 232 not long after the delivery had this number applied in Black color on the fuselage and top of the wings. Note the absence of any other markings or insignia. [Šime Oštrić]

Curious onlookers around a Do Y following its arrival to KJ. German envoy in Belgrade, Viktor von Heeren, is standing in front of the airplane wearing a suit. [Aleksandar Ognjević]

arrived to Friedrichshafen-Löwental to attend trials on behalf of the Yugoslav government. In the first days of January 1932 both Do Y were assigned with the ferry markings D-3 and D-6 and were shown to Romania, Hungary and Austria but no orders were secured. During these flights the airplanes were flown unarmed and

W.Nr. 232 shown here at Novi Sad with its W.Nr. on the fuselage removed and with newly applied "Kosovo cross" insignia below the wings and the Yugoslav tri color flag on the rudder. [Šime Oštrić]

Both Do Y at Novi Sad Jugićevo airfield. [Šime Oštrić]

Preparing for take off. [Šime Oštrić]

Romanian prince Nicholas visiting Belgrade airfield and inspecting two Do Y on 20 September 1933. [Aleksandar Ognjević]

Romanian delegation inspecting Do-Y up close. [Djordje Nikolić]

After arrival to Kraljevo on 22 September, prince Nicholas in flying outfit greets Yugoslav officer. [Aleksandar Milošević]

Taking off the warm flying outfits following the flight to Kraljevo. [Djordje Nikolić]

Airmen from 261.VG posing in front of a Do Y in Zemun, 1936. [Šime Oštrić]

General Milan Nedić following a flight in a Do Y. [Aleksandar Ognjević]

with covered glass windows in the nose. With the glass covers removed, both Do Y took off from Friedrichshafen airfield on 10 January and landed in Vienna for refueling following which they continued towards Belgrade.

It is interesting to note that Flight magazine in its 23 November 1933 issue indicated, with significant delay, that "*Dornier works Altenrhein, Switzerland, have just completed the Do Y, a three-engine cantilever high wing monoplane. The machine is designed to carry freight and has a fuselage with plenty of space.*" The German intent to hide the true purpose of this airplane clearly succeeded.

A fine shot of Do Y in flight somewhere over the Kingdom. [Šime Oštrić]

Do Y and two Avia F.39 during take off. [Djordje Nikolić]

All six bombers from 261.VG seen here in low pass escorted by no fewer then nine Avia BH-33 fighters. [Djordje Nikolić]

Introduction into Service

All three bomber types were ferried over to Zemun military airfield where they joined 6.VP (*Vazduhoplovni Puk* – Aviation Regiment) where full testing by the VVKJ began. Kap Hubl, who was also an instructor for Do Y, commanded the unit. During the first year he trained other pilots on Do Y, who later became instructors themselves. It is here that the airplanes eventually received their military equipment and weapons. Based on the evaluation of the three bomber types, *Junkers* was declared as the winner, however no follow-on orders for either bomber or the license production of *Avia* bomber took place.

In September 1933 VVKJ formed a special three engine bomber unit at Zemun airfield, 261.VG (*Vazduhop-lovna Grupa* - Aviation Group) (CO puk Zdenko Gorjup) consisting of 426.E (CO kap Ik Gavro Scrivanić) and 427.E (CO kap Ik Milivoje Mišović).

Apart from the official roles, Do Y were showcased at various airshows throughout the country and during foreign delegation visits. During one such visit by the Romanian prince Nicholas, two Do Y took part on 20 September 1933 at a large military parade from Zemun airfield. During this occasion Do Ys were escorted by Avia BH-33 biplane fighters. Two days later they ferried the prince and his entourage to Kraljevo and Novi Sad.

In 1933 por Dimitrije Kneselac flew a training mission to the bomber school practice grounds at Bela Crkva.

On 15 October 1934 during the funeral preparations for king Aleksandar I, Do Y escorted the train

Do Y and two Junkers G.24 during a flight above Oplenac, where king Aleksandar I was laid to rest. [Djordje Nikolić]

A trio of Do Y in a flight above the Parliament building in Belgrade. [Dornier Post 06/07 1937]

W.Nr. 233 at Pančevo in the summer of 1940. Note the overall green camouflage. [Šime Oštrić]

Do Y in a flight above Sava river. [Šime Oštrić]

Do Y at an airshow at Petrovgrad. [Aleksandar Ognjević]

Flying over Banjica, Belgrade. [Šime Oštrić]

W.Nr. 233 in a flight somewhere over the Adriatic coast. [Šime Oštrić]

with the late king's casket from Zemun to Belgrade and then on 18 October conducted a flyover above Oplenac monastery where king was buried.

In 1935 the airplanes finally received their defensive armament which consisted of 7.7 mm *Darne* machine guns in the nose and in the fuselage turrets, and in autumn took part in military maneuvers. The same year 261.VG was renamed *Nezavizna bombarderska grupa tromotoraca* (Independent trimotor bomber group).

Frequent flights across all parts of the country including the coastal areas continued in 1936. The commanding structure of 261.VG changed with kap Ik Milutin Dostanić taking the leadership of 426.E whereas kap Ik Kosta Simić took the leadership of 427.E. At the end of the summer 1936, a large aviation exercise was ordered by gen Milan Nedić and 261. VG with its Do Ys was to take part. Do Ys together with *Avia-Fokker* F.39 102 were to form "South" force while the other *Avia* and two *Junkers* G.24 were to form the "North" force. The "South" force was to be based at Skopje and Niš airfields. Due to the hesitance

of gen Nedić to hold the exercise, he was replaced by gen Dušan Simović, and Do Y lost out on the opportunity to take place in a serious military exercise. The last important event of the year for the Do Y was on 6 September when they took lead part in the parade which celebrated the birthday of prince Peter.

Do 15

In addition to the bomber version, Do Y, *Dornier* developed a civilian version designated as Do U. In the third quarter of 1932, *Dornier* offered this airplane to *Lufthansa* in a reworked version, Do 15, which according to the official documents was *"derived from Do K and should serve the same purpose as the Ju 52"*. In parallel, *Dornier* sent on 14 January 1933 an offer for "two high speed Do 15 airplanes" to the Army Weapons Department (*Heereswaffenamt*) Wa.Pr8. Meanwhile, the construction of two Do 15 already began at Manzell with W.Nr. 243 and 244 assigned. The first airplane

W.Nr. 555 immediately following completion of assembly at Altenrhein. Note the four blade wooden propeller. [Staatsarchiv St. Gallen]

W.Nr. 555 in front of the assembly building. The airplane is still unpainted. [Staatsarchiv St. Gallen]

Head on view of the factory fresh Do Y. [Staatsarchiv St. Gallen]

Prior to delivery, airplanes were painted and received Swiss ferry markings. Shown here is W.Nr. 556 as HB-GOF. [Airbus Corporate Heritage]

HB-GOF in front of the assembly building at Altenrhein. Note the Swiss flag on the rudder and opened access panels to the engines. [Staatsarchiv St. Gallen]

W.Nr. 555 as HB-GOE at the same location. The airplane has aerodynamically faired wheel covers. [Staatsarchiv St. Gallen]

One of the Swiss factory workers prior to the delivery flight. [Airbus Corporate Heritage]

was to be a delivered as a civilian airplane for *Lufthansa*, while the second was intended for the *Reichswehr*. Nevertheless, the project was still being worked on until March 1933, when Herman Göring, the acting *Reichsluftfahrtminister* (Reich Minister of Aviation), now in charge, suspended the work on the Do 15 project. The number 15 is of interest as it belongs to the number series beginning with 10, since until then, *Dornier* had designated the various types of airplanes only with letters.

Less than two months later, the work on Do 15 stopped in favor of the more advanced and modern Do 17. This left *Dornier* with two started airplanes and an investment which needed to be paid off. The discontinuation of the Do 15 project understandably caused the *Dornier* company financial difficulties. In November 1933, an article appeared in German magazine *Flugsport* with "Dornier transport airplane Do Y". The airplane was shown in pictures of configuration built in Manzell for KJ. The first sentence of the article read "*From A.G. for Dornier airplanes*

HB-GOE prior to take off. [Airbus Corporate Heritage]

Last preparations prior to the ferry flight. [Airbus Corporate Heritage]

Both Do Y are rolling past the Altenrhein assembly buildings. [Airbus Corporate Heritage]

Ready for take off towards the first stop, Vienna. [Airbus Corporate Heritage]

Do Y W.Nr. 555 HB-GOE after landing in Vienna. [Airbus Corporate Heritage]

Altenrhein (Switzerland) built 3 engine high wing airplane with high load capacity". In reality, in 1933 no Do Y production took place in Altenrhein, since at the time all of the assemblies and parts for the two Do 15 were being moved from Manzell to Altenrhein in order to free space for other orders. If the potential buyer was found, the completion of Do 15 would take place there.

Finally, the quest for a customer ended in November 1935 when VVKJ placed an order during the Yugoslav delegation's visit to the factory for acceptance of the spare parts for the Do Y. Following signing of the purchase order, the work at Altenrhein continued and the airplanes were assigned new W.Nr. 555 and 556. Based on the Yugoslav request, the airplanes were equipped with more powerful engines, *Gnome-Rhône*

129

Following a brief stop at Vienna, both airplanes continued towards Belgrade. [Airbus Corporate Heritage]

Prime minister Milan Stojadinović following a flight in a Do Y at Zemun, 1937. [Šime Oštrić]

W.Nr. 555 with escadrille number 170 on the fuselage following the delivery to KJ. [Josip Novak]

Do Y during an exercise most likely at Mostar along with an Avia F.39 in the foreground. [Djordje Nikolić]

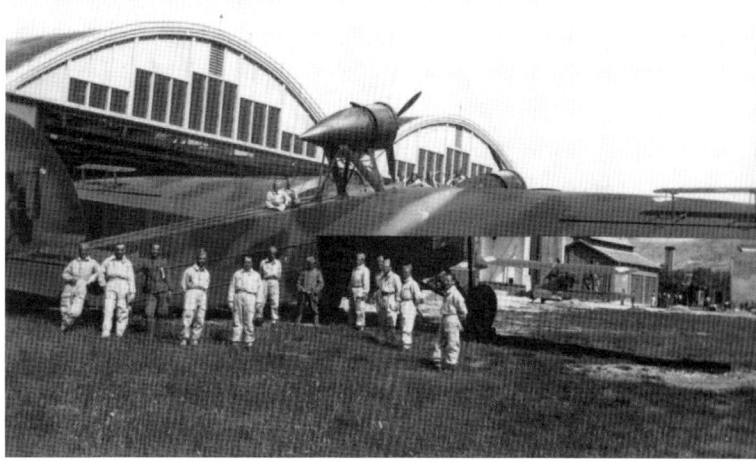

Unknown Do Y with new camouflage and insignia at Pančevo. [Milan Micevski]

Mistral 9 Kers. This time NACA (National Advisory Committee for Aeronautics) cowlings were used to shroud the engines as these were more efficient than the original nacelles used on the first two airplanes.

First flight of W.Nr. 555 with *Dornier* pilot Egon Fath at the controls took place on 1 July 1936, and the request to issue certificate of airworthiness was filed on 6 July. The four blade wooden propellers which were used for factory test flights at Altenrhein were exchanged for three blade metal propellers prior to the delivery to the KJ. Due to the tendency to vibrate under certain flight parameters, these two airplanes had shortened wingspan as well as metal skin below the outer sections. Additionally, new windows were added to the nose as well as on the fuselage.

Por Dimitrije Kneselac was sent on behalf of VVKJ to Switzerland to test these airplanes prior to delivery and on 22 October 1936 he flew in W.Nr. 555 at 100 m altitude measuring the airspeed. The next day he took off on another flight which ended prematurely since the engines overheated at 6,300 m, thus the airplane was unable to attain the required 8,000 m altitude.

Following in process modifications, these airplanes performed well during further factory testing, but as the assembly work and administrative paperwork progressed slowly, the airplanes were ready for delivery over a year after the order was placed. The permission for the ferry flight was issued on 25 February 1937 with the validity of two months. In March 1937 the airplanes received Swiss ferry registrations HB-GOE and HB-GOF and flew on 8 March from Altenrhein to Vienna and the next day to their final destination, Zemun, with one of the pilots being Egon Fath. HB-GOE flew back to Vienna on 22 March to return German technicians.

Two More Enter Service

After the arrival of W.Nr. 555 and 556 to Zemun, on 13 March Prime Minister Milan Stojadinović

Do Y at Pančevo next to a Bücker Bü-131 training airplane. [Vladeta Vojinović]

Do Y with crews and mechanics posing in front. [Aviation Museum – Belgrade]

Unknown Do Y during maintenance at Pančevo, 1940. [Aleksandar Ognjević]

Soldiers posing next to the canvas covered fuselage mounted engine on either W.Nr. 232 or W.Nr. 233. [Mario Raguž]

Same soldiers now at the tail. Note the covered fuselage turret behind the engine. [Mario Raguž]

expressed a desire to fly in one of the newly arrived airplanes and took off at 11:55 in W.Nr. 555.

Mid August 1937 one Do Y was in overhaul while the other three were operational. In September of the same year, five heavy bombers took part in maneuvers in Slovenia, including two Do Y. The first two Do Y, 232 and 233, were transferred on 8 October to VTZ (*Vazduhoplovno-tehnički Zavod* - Aviation Technical Depot) Kraljevo for a general overhaul, where they remained for a while due to lack of spare parts. At the end of 1937, three Do Ys were operational, and the last airplane returned to service by August 1938.

At the beginning of 1938 due to the reorganization 261.VG was renamed 81.VG and its escadrilles 261.E and 262.E accordingly.

W.Nr. 555 and 556 were overhauled in May 1939 at VTZ Kraljevo where they had domestic IAM (*Industrija aeroplanskih motora* - Aero Engines Industry) *Gnome-Rhône* 9Kers engines installed and W.Nr. 555 also received a semi-autopilot manufactured by *Askania*. W.Nr. 232 received instrument flying equipment, while W.Nr. 233 had its engine replaced. Eventually all four airplanes received instrument flying equipment.

All three tri-motor types in service were intended to be struck off charge from the bomber role in 1939 but due to lack of airplanes in service their use was prolonged until the Italian *Savoia-Marchetti* Sm.79 arrived in September 1939. In the summer of 1940 W.Nr. 555 was used in Pančevo at *Vazduhoplovna Akademija* (Air Academy) to familiarize new recruits, while W.Nr. 556 was used by 81.VG from Mostar until May 1940. During the military drills which took place in the summer at Mostar, Do Y was still used

in navigation, bombing, machine gunning as well as transport and liaison roles.

As *Vazduhoplovna Akademija* moved in the beginning of the year from Pančevo to Sarajevo, Do Y were transferred to Kraljevo. As a result, three engine airplanes were transferred to transport and paratrooper drop assignments, which is supported by the fact that on the port side of the fuselage below the wing a new large door was installed on both W.Nr. 555 and 556.

April War

By April War, four Do Ys accumulated around 1,200 flight hours. Due to their obsolescence and secondary role, Do Y did not take part in the April War and were not attached to any active units. W.Nr. 555 and 556 were discovered by the advancing German troops on the grass strips of Kraljevo airfield and were later moved closer to the airfield buildings. Due to the importance of DFA (*Državna Fabrika Aviona* – State Airplane Factory) and VTZ Kraljevo to the German war effort after the war, this airfield was spared from bombing and strafing attacks which most likely saved two Dorniers as well. Two of the initial airplanes, W.Nr. 232 and 233 were discovered in the partially sabotaged DFA factory at the same airfield. These two airplanes had their wings and wing mounted engines disassembled from the fuselage and all paint stripped off. W.Nr. 233 with letter "Ћ" on the vertical stabilizer had its third engine above the fuselage also removed. The large Yugoslav tri-color flag on the rudder indicates that the

As German troops rolled into Kraljevo, they discovered two Do Ys, W.Nr. 555 and 556 at the grass strips. [Djordje Nikolić]

Another shot of the two Do Y as the German forces approached them. [Harald Schiess]

One of the two Do Y shown here in Kraljevo. Do Y was not used during the April War due to its obsolescence. [Jan van den Heuvel via Aleksandar Ognjević]

Two curious German soldiers inspecting on of the captured Do Y. [Harald Schiess]

Do Y with barrels of fuel next to it, likely this airplane was repaired and sold to the NDH where it was briefly used. [Djordje Nikolić]

airplane was sent for overhaul prior to the application of the new marking system in April 1940.

Dornier documents indicate that as of 11 September 1941, there were 48 airplanes at DFA which were intended for ZNDH (*Zrakoplovstvo Nezavisne Države Hrvatske* – Air Force of the Independent State of Croatia), of 56 which were selected from various airfields in KJ. Amongst them were Do Y, labelled in the documents as Do Y/J 3221 and 3222 and Do Y/kg 3223 and 3224 respectively. The document indicates that all of the airplanes were to be provided to NDH (*Nezavisna Država Hrvatska* Independent State of Croatia) free of charge but that the Croats must pay for their repairs which included labor and parts. This document therefore indicates that

Both Do Y were eventually towed to the partially destroyed DFA factory buildings. [Djordje Nikolić]

despite their condition at the time of capture, all four Do Y were delivered to ZNDH.

It is known that one of the captured Do Y which was repaired was handed over to ZNDH was marked by its new owners with number 1601. In January 1942 it was assigned to 9. Jato BZ. It flew only several times and in the summer of 1944, it was placed at Rajlovac airfield as a decoy for USAF (United States Air Force) bombers and fighters.

German soldier inspecting one of the captured Do Y. [Djordje Nikolić]

Do Y in front of one of DFA factory buildings. [Airbus Corporate Heritage]

Inside the partially sabotaged DFA buildings, W.Nr. 232 and 233 were found without wings and engines. Note the incomplete Do 17K fuselage with belly machine gun position. [Jan van den Heuvel via Aleksandar Ognjević]

W.Nr. 232 still retains one of its engines. Note the wing next to the airplane. [Jan van den Heuvel via Aleksandar Ognjević]

Construction features

Do Y was a cantilever high-wing monoplane. Its three spar wing consisted of three sections: a center section extending laterally beyond the engine nacelles and two outer wing halves. The wing spars were made from stamped duraluminium profiles while ribs were made from drawn ones. The spaces in between the spars and transverse beams were stiffened with bracing cables. The wing shape was semi-elliptical, with a straight trailing edge which was slightly inclined forward to the middle axis of the airplane. The wing tips on W.Nr. 555 and 556 were cut back in the course of flight testing to improve performance. The wings were covered entirely with fabric with exception of the areas behind the propeller wash, which were covered with duraluminium and which could be walked on for servicing.

Monocoque fuselage was built from duraluminium and was subdivided. The fuselage framework made from longerons and the outer skin formed the load bearing structure. Subdivisions included the observer, radioman and machine gun turrets in the front section, cockpit with a boarding hatch at the bottom, bomb bay, storage area, fuselage top turret and belly machine gun as well as the fuselage space. In the cockpit there

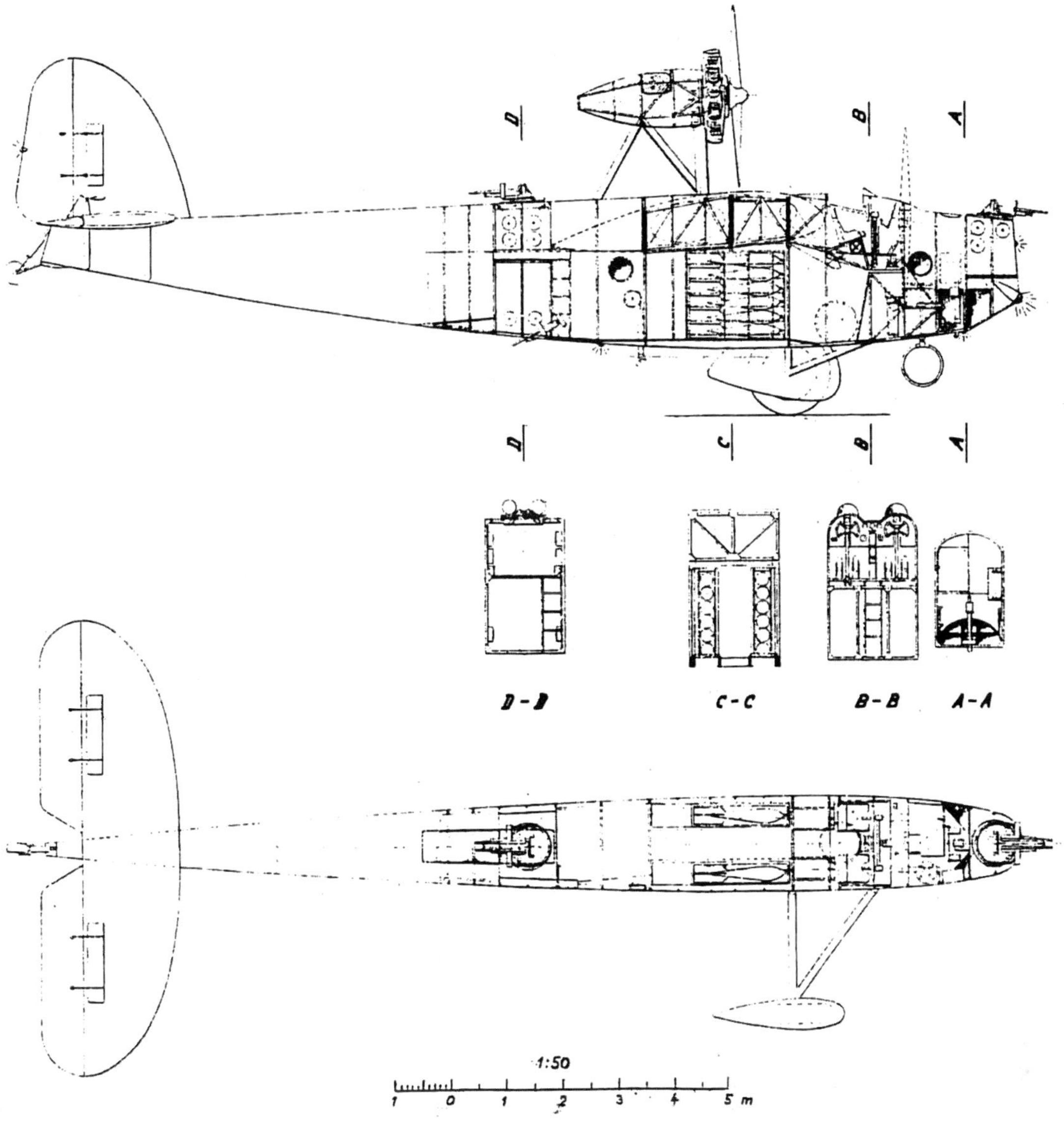
D
B
A
D
C
B
A
D-D
C-C
B-B
A-A
1:50
1 0 1 2 3 4 5 m

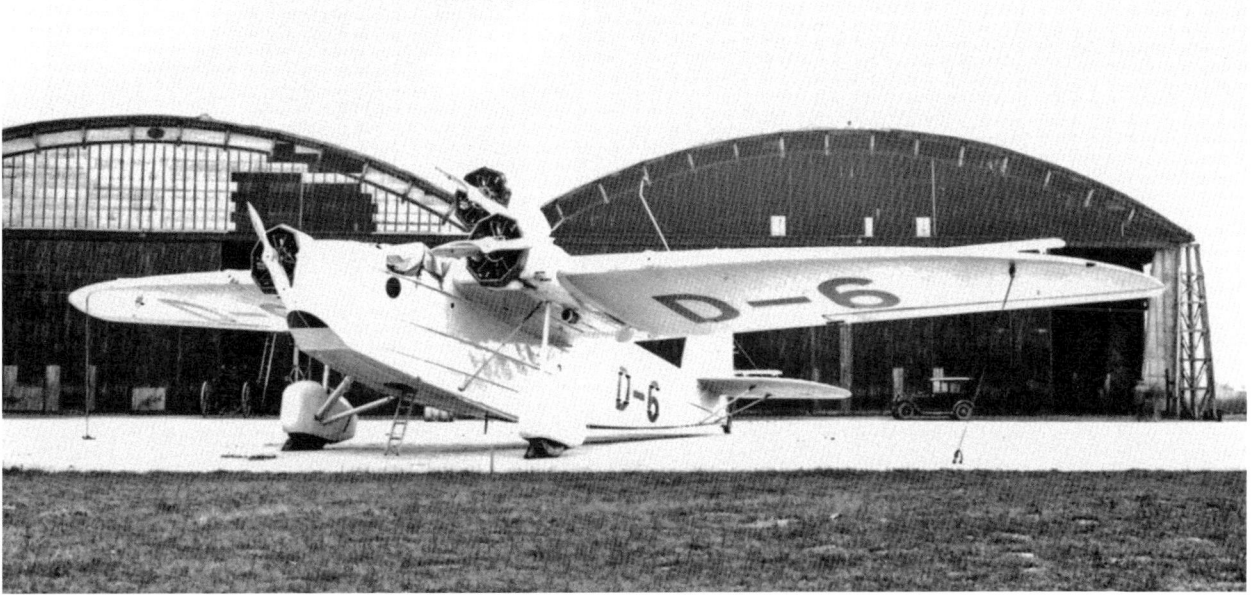

Factory drawing of a Do Y with engine without nacelles and a radio antenna below the nose instead of on top. The cross sectional view offers insight into the nose, cockpit and bomb bay arrangements. [Airbus Corporate Heritage]

D-6
D-6

Factory fresh D-6 showing the unique three engine arrangement. [Airbus Corporate Heritage]

W.Nr. 555 at Altenrhein without the paint shows the monocoque fuselage was built from duraluminium. [Staatsarchiv St. Gallen]

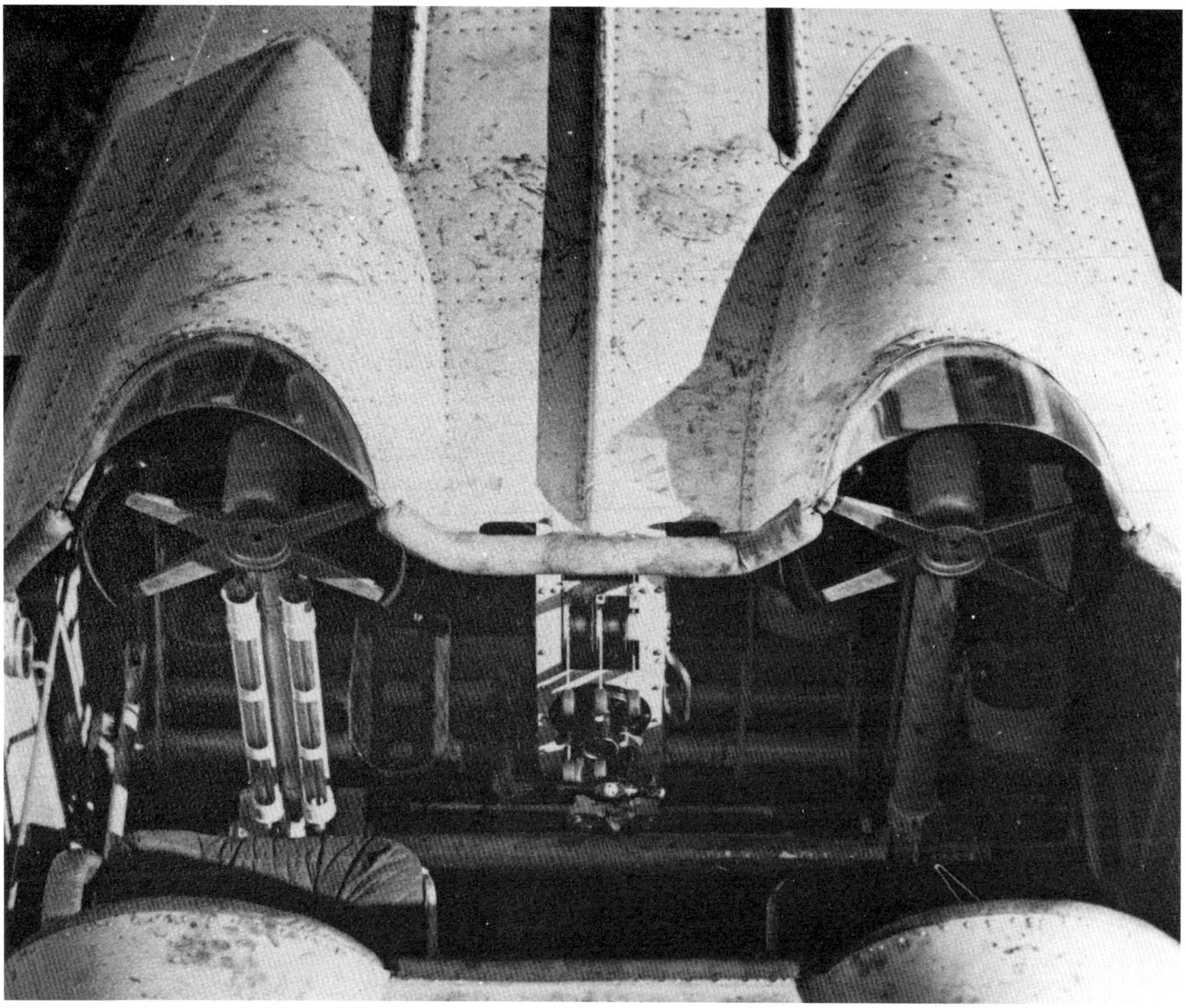

View of the cockpit with pilot and co-pilot yoke. [Djordje Nikolić]

A detail of the pilot instrument panel. All inscriptions were in Serbian language. [Šime Oštrić]

were two seats and dual controls for the pilot and the co-pilot which were protected with rudimentary windshields. Both seats were adjustable in height and position. The co-pilot acted as the observer and bomb aimer. As such with help of a lever he could control the airplane rudder from the nose position to correct the approach. To enable easier bomb aiming the nose had large windows built in the front and the sides. The access to the nose area was located below the instrument panel in the cockpit. The access to the cockpit was through a hatch in the bottom of the fuselage which housed a ladder. Towards the end of service life, according to the available photographs of W.Nr. 555 and 556 from Kraljevo, additional windows were installed in both sides of the fuselage whereas a door was installed on the port side only. The purpose remains unknown, however it is likely that these were

added for the intended transport and paratrooper dropping role.

The tail assembly was constructed from duraluminium profiles and was covered with fabric. Trimmers were installed on the rudder as well as the elevator to balance the forces on the control surfaces. Cables and bar linkages were used to transfer the commands from the control columns to the control surfaces.

The airplane was equipped with non-retractable landing gear. The main landing gear legs were attached to the front wing spar and were braced with V struts and a set of twin cables to the fuselage. The telescopic main landing gear legs housed a shock absorber. Main wheels measured 1,500 x 350 mm and had hydraulic breaks installed. Airplanes were supplied from the factory with aerodynamically faired wheel covers, however these were removed soon after entering VVKJ service.

Radio equipment in the nose. [Airbus Corporate Heritage]

Tail wheel assembly was mounted at the very end of the fuselage and it could swivel 90° in either direction.

Do Y was powered by three air cooled engines of which two were installed in the wing's leading edge, while the third one was installed on struts above the fuselage section which was strengthened to carry the load. W.Nr. 232 and 233 were originally equipped with 450 hp *Bristol-Siemens* Jupiter VI engines but before delivery VVKJ requested that these be replaced with the domestically license produced 480 hp *Gnome-Rhône* Jupiter 9Ak. These two airplanes used two blade wooden propellers during factory testing which were prior to delivery replaced with two blade metal

ones. W.Nr. 555 and 556 were factory installed with *Gnome-Rhône* 9Kers rated 625 hp each and were later retrofitted with the domestically produced equivalent. During the production of W.Nr. 555 and 556 four blade wooden propellers were used and were prior to delivery replaced with three blade metal ones instead. W.Nr. 232 and 233 had polygonal cowlings whereas W.Nr. 555 and 556 were equipped with streamline NACA cowlings.

Two 1,045 liter fuel tanks, which were made from brass, were housed in the wing center section. Three oil tanks with 70 liter individual volume supplied each engine. The oil tanks feeding the wing mounted en-

Detail of the engine and landing gear configuration. [Djordje Nikolić]

mestically produced 100 kg *Stanković* bombs were only used. Even though factory provisions accommodate for 1,200 kg of bombs to be carried, it remains unknown if such loads were ever carried by Do Y in VVKJ service. Bombs were stored against both fuselage walls, one on top of the other and were released by unilaterally unlocking the bars on top of which the bombs were stored, by pulling the lever located at the observer's seat in the fuselage nose. Simple bomb bay doors were located at the bottom of the fuselage.

Defensive armament was located in three machine gun positions, one twinned machine gun turret in the nose, one twinned machine gun turret on top of the fuselage and one single machine gun in the belly behind the bomb bay. The machine gunners in the two turrets were exposed to the elements. *Darne* 7.7 mm machine guns, common weapon within VVKJ, could be installed at all three positions, although no photographs of Do Y show that these were installed at any point in service.

Standard equipment consisted also of a radio, goniometer and high altitude oxygen equipment for the crew.

gines were located in the airplane nose while the oil tank for the fuselage engine was housed on the upper side of the nacelle.

Bomb bay was located between the main front and rear spar. It could house a total of 12 100 kg or alternatively six 200 kg bombs, but in practice do-

Aft view showing the trimmers installed on the rudder and the ailerons. [Šime Oštrić]

141

Technical Specifications

Technical specifications Dornier Do Y W.Nr. 232 and 233	
Quantity used:	2
Crew:	4
Years of Service:	1932-1941
Span:	28.0 m
Length:	18.2 m
Height:	6.80 m
Wing area:	113.4 m²
Engine:	Three 450 hp Gnome-Rhône Jupiter VI
Empty weight:	5,200 kg
Maximum weight:	8,500 kg
Maximum speed:	245 km/h
Cruise speed:	210 km/h
Service ceiling:	4,820 m
Maximum range:	1,500 km
Armament:	100 kg Stanković bombs (max. payload 1,200 kg) and four 7.7 mm Darne machine guns

Technical specifications Dornier Do Y W.Nr. 555 and 556	
Quantity used:	2
Crew:	4
Years of Service:	1937-1941
Span:	26.62 m
Length:	18.2 m
Height:	7.3 m
Wing area:	108.8 m²
Engine:	Three 625 hp Gnome-Rhône 9Kers
Empty weight:	6,360 kg
Maximum weight:	8,500 kg
Maximum speed:	300 km/h
Cruise speed:	250 km/h
Service ceiling:	8,300 m
Maximum range:	1,500 km
Armament:	100 kg Stanković bombs (max. payload 1,200 kg) and four 7.7 mm Darne machine guns

Camouflage and Markings

During initial testing in Germany airplanes were painted in overall DKH L40/52 Light Grey color and carried no markings. This was done to hide their true bomber role and also to market them in civilian role.

Prior to delivery, ferry markings D-3 and D-6 were applied on the fuselage, below and on top of both wings and airplane was unsuccessfully marketed to several countries with this livery.

Immediately after the delivery of the first two airplanes, W.Nr. 232 and 233, their ferry markings D-3 and D-6 were painted over and numbers analogous to the W.Nr. were applied on the fuselage and on top of

both wings in Black color. At this time the airplanes still lacked any national insignia on the wings or the rudder. Soon thereafter large "Kosovo cross" insignia were applied on top and below the wing and Yugoslav tri-color flag was applied across the entire rudder.

Prior to delivery of the second two airplanes, W.Nr. 555 and 556, Swiss ferry registrations HB-GOE and HB-GOF were applied in Black color on the fuselage, on top and below the wings along with a Swiss flag on the rudder. Soon after the arrival these registrations were painted over and in their place escadrille numbers were applied. Like the first two airplanes, large "Kosovo cross" insignia were applied on top and below the wing and Yugoslav tri-color flag was applied across the entire rudder.

Factory fresh W.Nr. 232 at Manzell with covered nose windows. The airplane was painted overall in DKH L40/52 Light Grey color. [Djordje Nikolić]

In VVKJ service, the rudders were painted in the color of Yugoslav tri color and the airplane type and W.Nr. were written on the vertical stabilizer. [Aleksandar Ognjević]

W.Nr. 232 with this number applied on both wings and the fuselage in rather large font. [Airbus Corporate Heritage]

Large "Kosovo cross" insignia was applied below the wings. [Vladeta Vojinović]

143

W.Nr. 232 was painted in overall *Sivo-maslinasta boja* (SMB – Gray-Olive color) and had White escadrille number 173 applied on the fuselage. Note the Cyrillic letter "Ђ" in White color on the vertical stabilizer. [Mario Raguž]

W.N.r. 555 had Black color fields applied over the topside of the engine nacelle as well as the entire wing behind the wing mounted engines. [Šime Oštrić]

W.Nr. 233 painted in overall SMB color. [Šime Oštrić]

Do Y at Kraljevo were captured wearing the three color camouflage scheme which consisted of Ochre Yellow, Dark Green and Dark Brown patches. [Jan van den Heuvel via Aleksandar Ognjević]

In VVKJ service, the first two Do Y carried service numbers 232 and 233 in accordance with their W.Nr. After the second two Do Y arrived in 1937, airplanes received service numbers 170 (555), 171 (556), 172 (232) and 173 (233) while at the beginning of 1940 they were renumbered one more time to EvBr. 3221 to 3224.

W.Nr.	Ferry marking	Escadrille No.	EvBr.
232	D-3	172	3221
233	D-6	173	3222
555	HB-GOE	170	3223
556	HB-GOF	171	3224

Factory numbers for the first two airplanes, 232 and 233, analogous to the W.Nr. were written on the vertical stabilizer below the airplane type Do Y in Black letters. On the second two airplanes, 555 and 556 were written in line with the airplane type also on the vertical stabilizer. Large escadrille numbers on the fuselage were applied from 1937 in White color.

Do Y W.Nr. 232 was during its service painted overall *Sivo-maslinasta boja* (SMB – Gray-Olive color) as it was officially known, which was in essence a shade of a Green color. It had no factory numbers but instead had a large Cyrillic letter "Ђ" applied in White color on the vertical stabilizer.

W.Nr. 555 interestingly had Black color fields applied over the topside of the engine nacelle as well as the entire wing behind the wing mounted engines, presumably to hide the exhaust stains.

New three tone camouflage consisting of Ochre Yellow, Dark Green and Dark Brown patches was applied on the wing topsides and fuselages of W.Nr. 555 and 556 in April 1940, similar to the modern Dornier Do 17K bombers then in service. The undersides remained the original DKH L40/52 Light Grey. Large "Kosovo cross" insignia below the wings were retained, fuselage escadrille numbers were removed and in place of the large rudder flag, a small Yugoslav war flag was applied. Only top starboard wing had a reduced size "Kosovo cross" insignia applied.

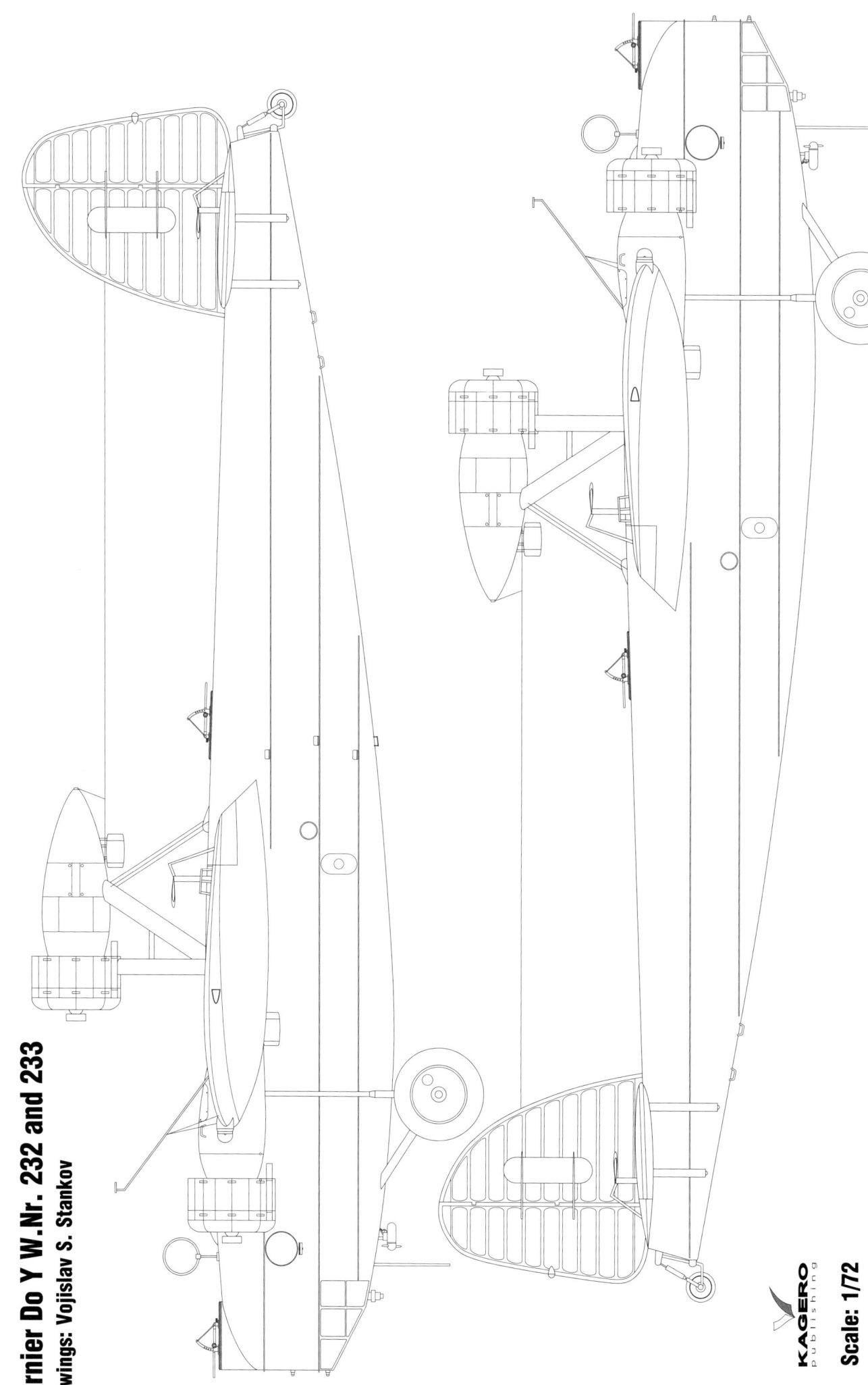

Dornier Do Y W.Nr. 232 and 233
Drawings: Vojislav S. Stankov

Scale: 1/72

Dornier Do Y W.Nr. 232 and 233

Drawings: Vojislav S. Stankov

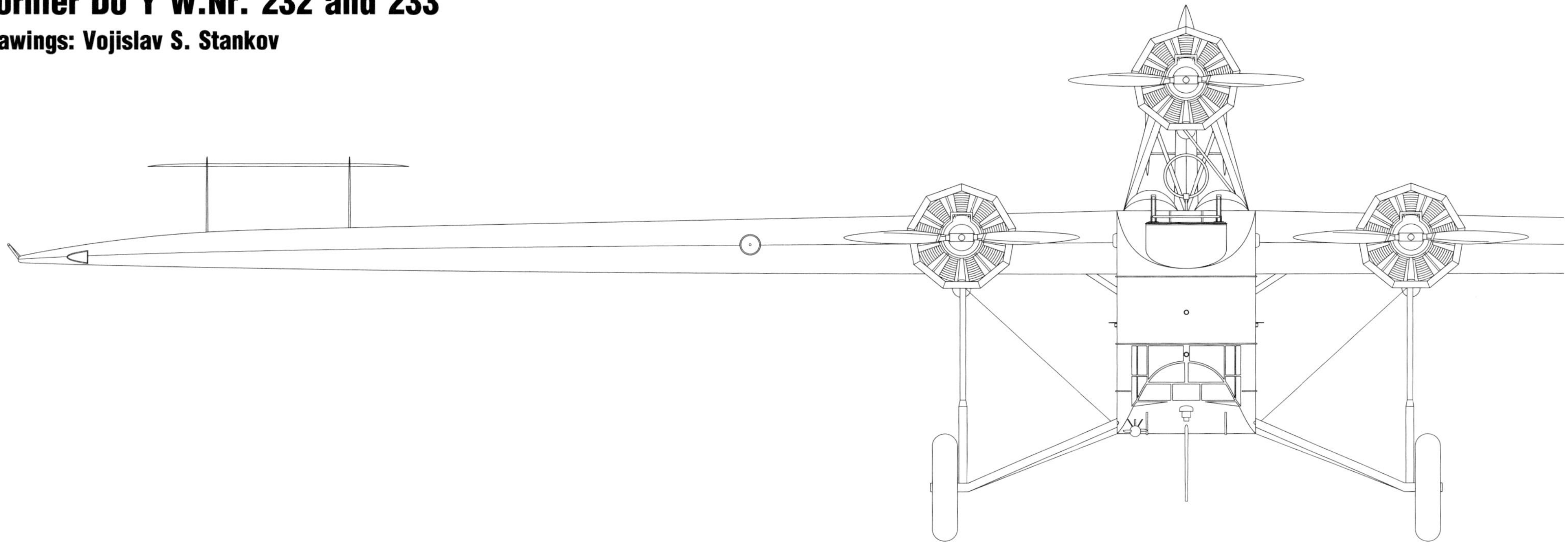

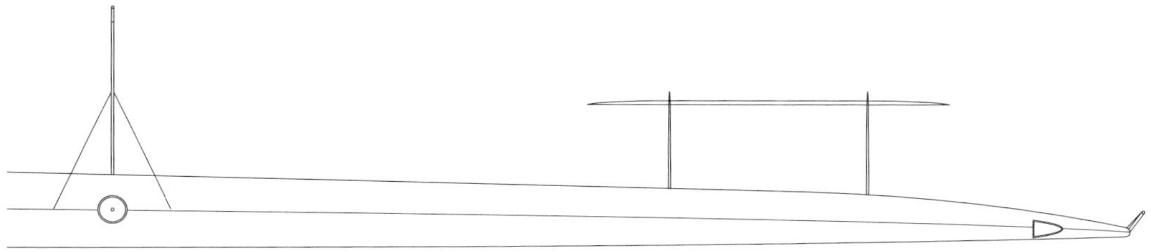

Scale: 1/72

KAGERO
publishing

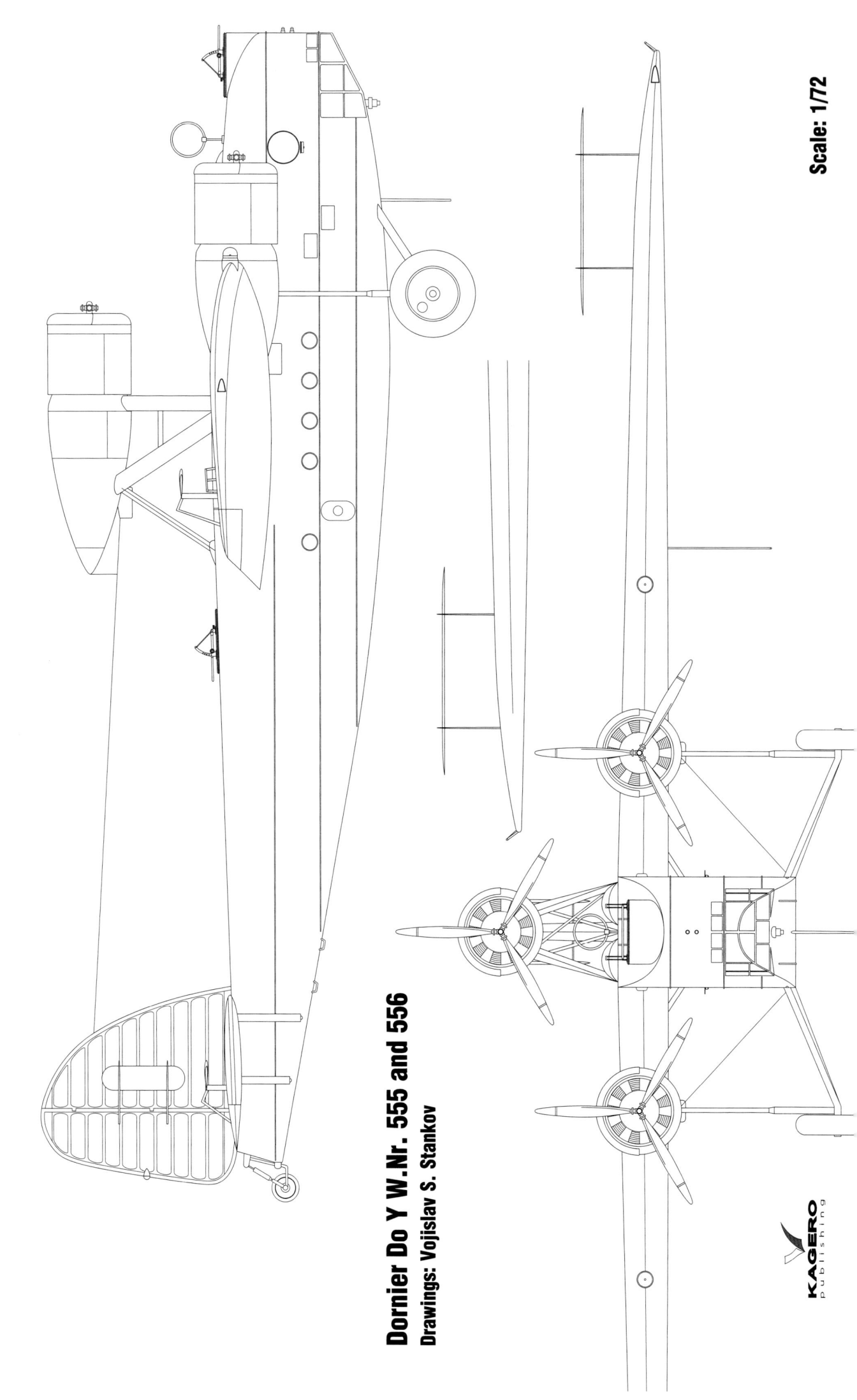

Dornier Do Y W.Nr. 555 and 556

Drawings: Vojislav S. Stankov

Scale: 1/72

KAGERO
publishing

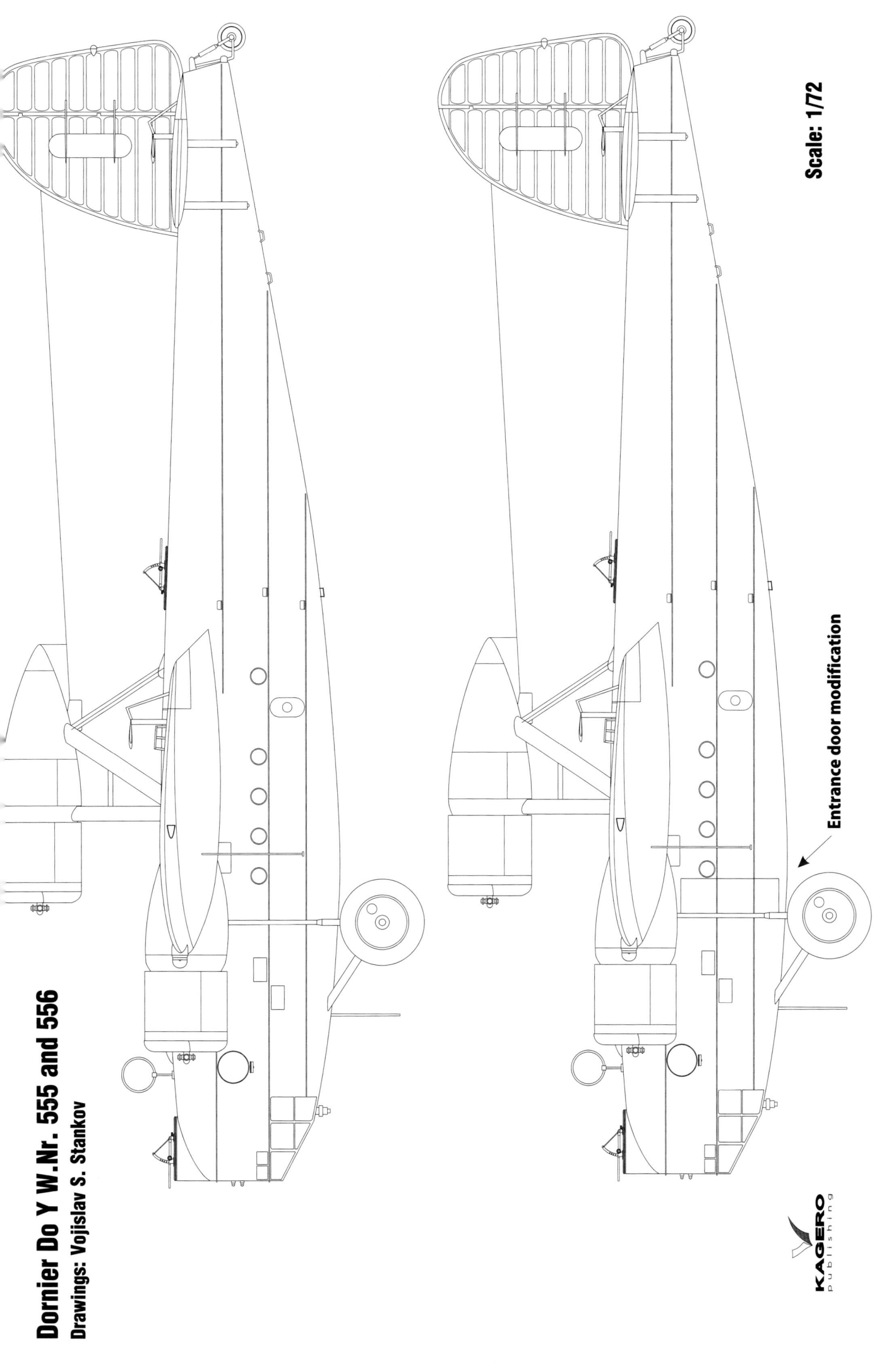

Dornier Do Y W.Nr. 555 and 556

Drawings: Vojislav S. Stankov

Entrance door modification

Scale: 1/72

KAGERO publishing

Dornier Do Y W.Nr. 232 and 233

KAGERO publishing

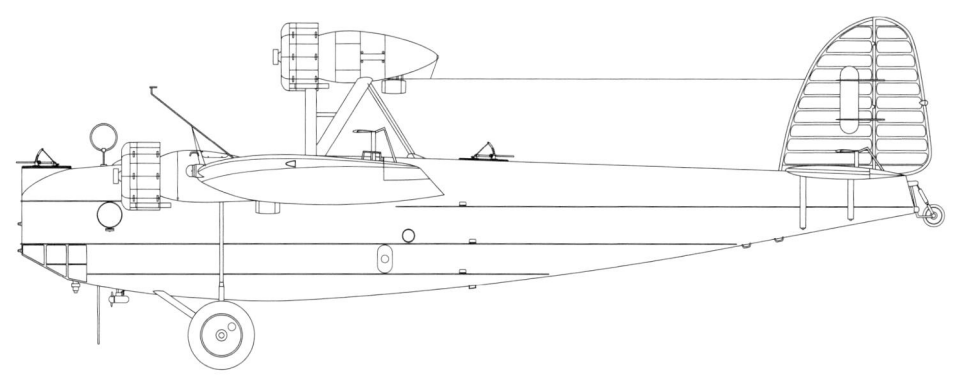

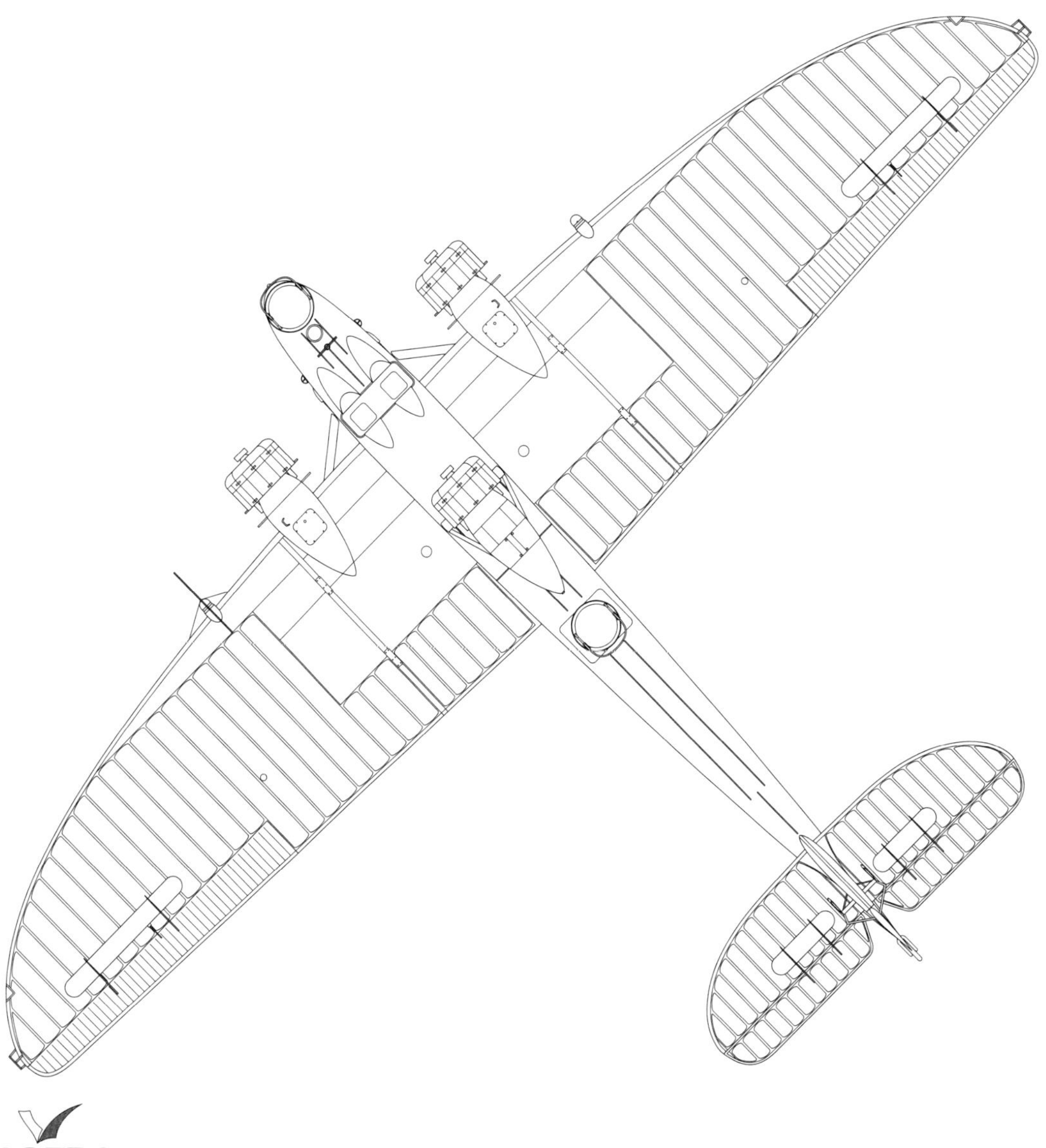

Drawings: Vojislav S. Stankov

Scale 1/144

Dornier Do Y W.Nr. 232 and 233

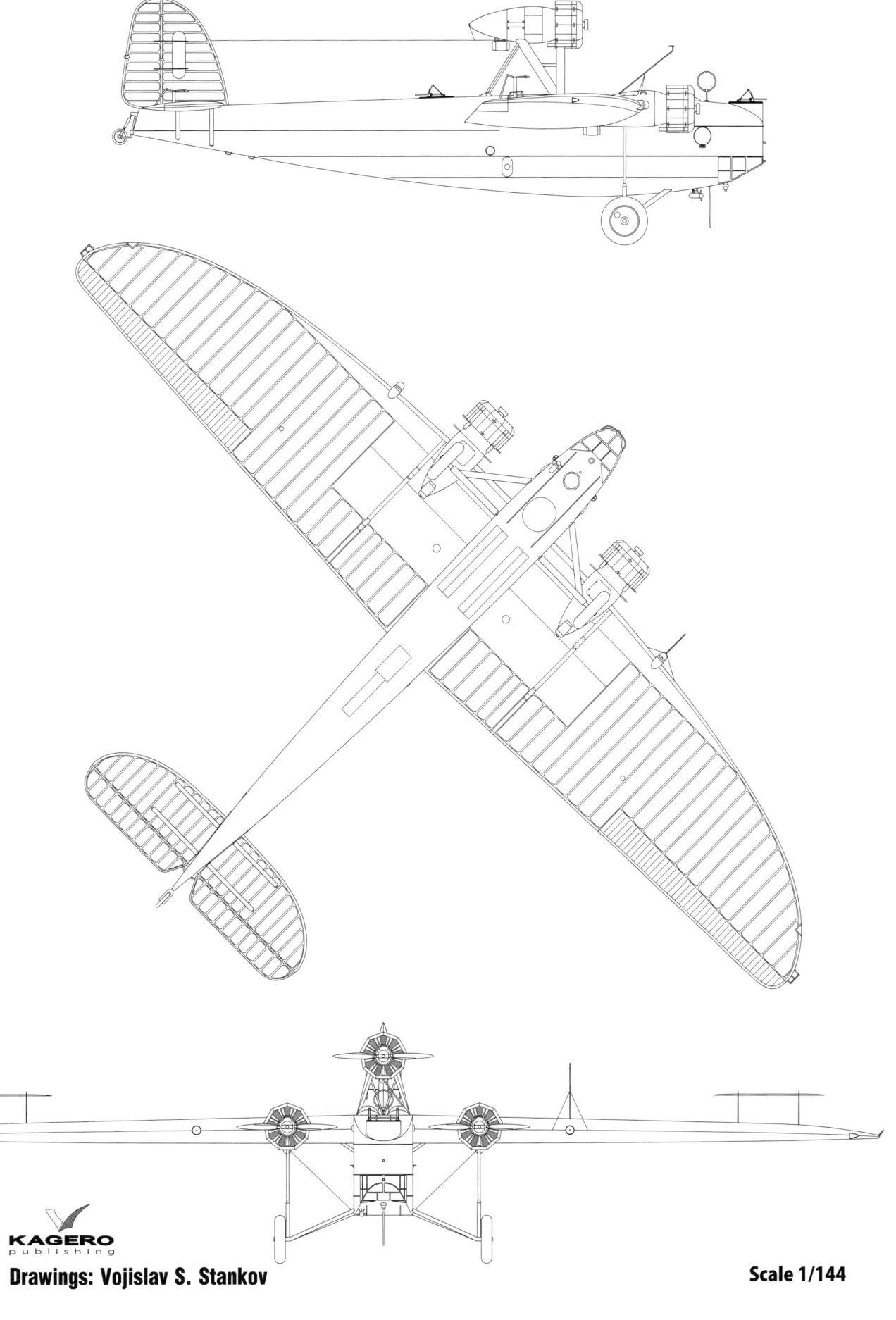

Drawings: Vojislav S. Stankov

Scale 1/144

Dornier Do Y W.Nr. 555 and 556

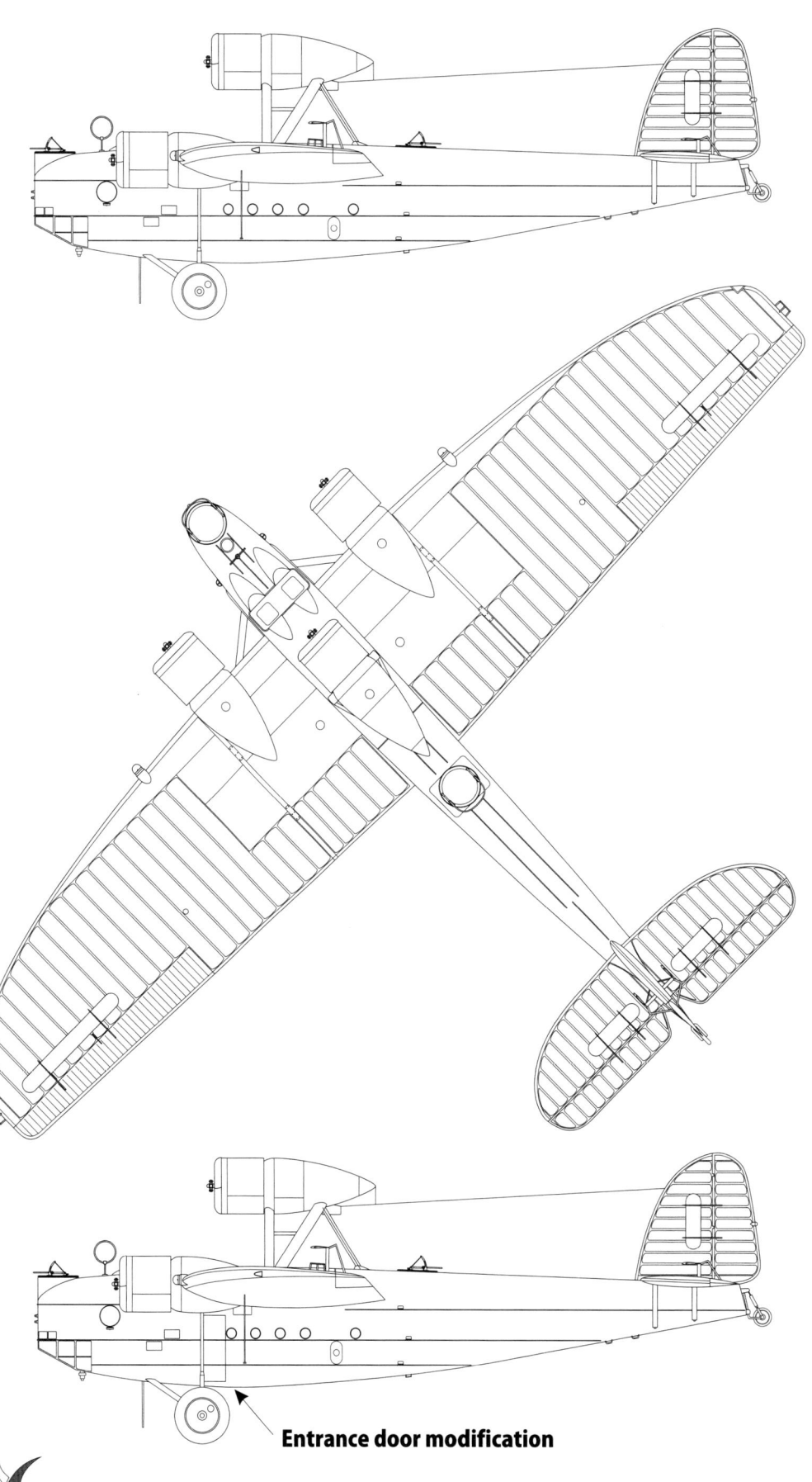

Entrance door modification

Drawings: Vojislav S. Stankov

Scale 1/144

Dornier Do Y W.Nr. 555 and 556

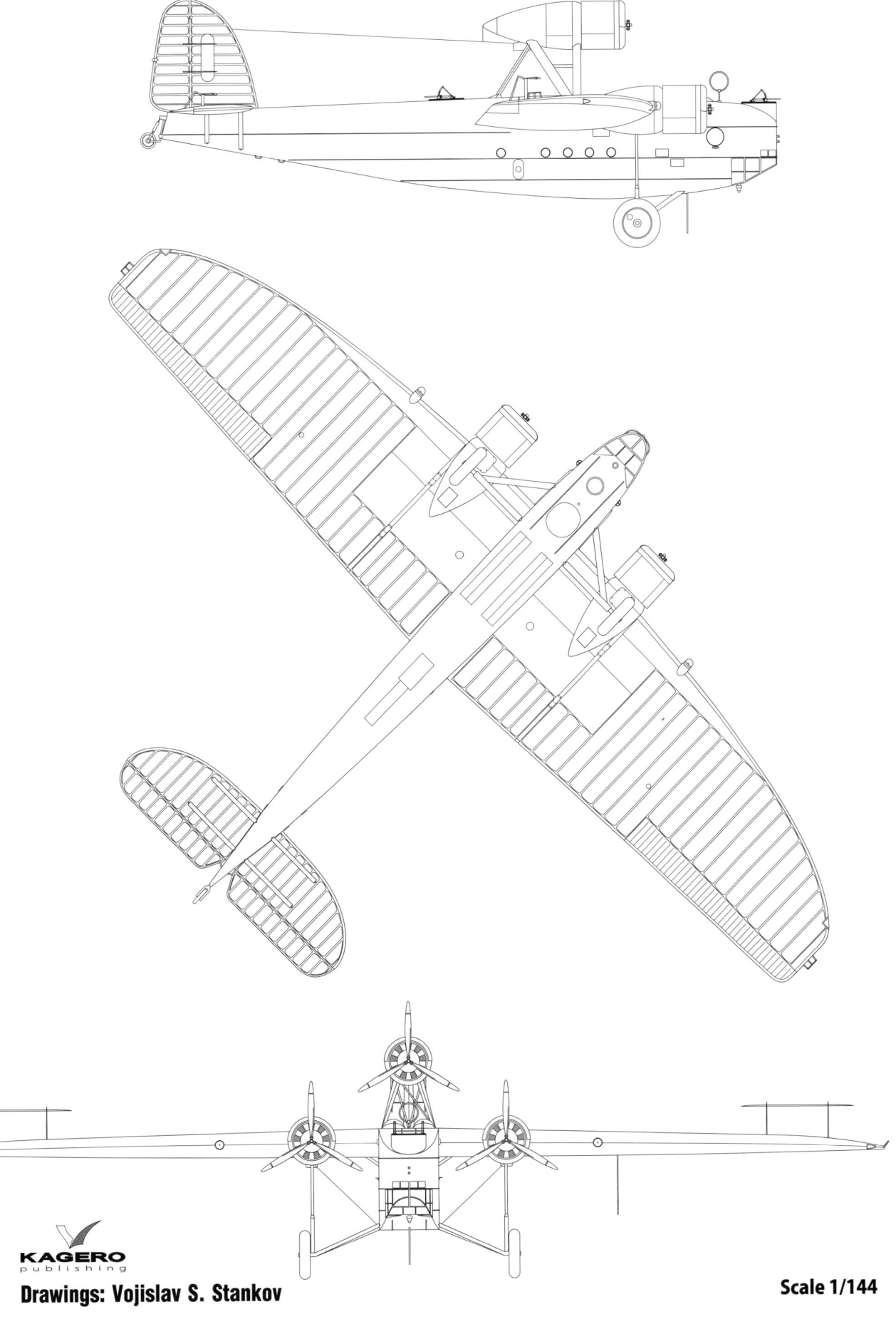

KAGERO
publishing

Drawings: Vojislav S. Stankov

Scale 1/144

Dornier Do 22

Do 10, Do C3 and Do C2A

The story of Do 22 starts with Do 10, a two-seat braced high-wing monoplane. Do 10 development was initiated at the beginning of 1930 by the RVM and the work was assigned to *Dornier* factory in Manzell. At the very beginning of the development process the airplane was given a designation Do C 1 and its real use was disguised by proclaiming it was developed for postal service.

Do 10 first flew on 25 July 1931 with pilot Egon Fath at the controls. Only two airplanes were com-pleted, the first one being W.Nr. 226 (fuselage code D-1592) and the second one W.Nr. 227 (fuselage code D-1898). Both were handed over to *RDL* (*Reichver-band der Deutschen Luftfahrtindustrie* - Reich Society of Airplane Manufacturers) airport at Staaken where numerous tests were conducted, including evaluation of different engines such as the *Rolls-Royce* Kestrel IIIs, *Hispano Suiza* Xbrc and BMW VI. A four-blade wooden propeller manufactured by *Dornier* was used with all of the tested engine configurations.

Despite the significant time and resource invest-ment, Do 10 did not meet climbing speed expectations.

The first prototype Do 10 W.Nr. 226 (D-1592) first flew on 25 July 1931. [Djordje Nikolić]

Do 10 with engine axis pointing upwards. The intent of this modifi-cation was to improve climbing performance. Testing concluded that no significant performance improvements were at-tained. [Dornier Museum Friedrichshafen (Airbus Group)]

Do C2A during take off. In 1932 Dornier offered the Kingdom of Yugoslavia to purchase the type, however the offer was rejected due to poor performance. [Djordje Nikolić]

Do C3 W.Nr. 229 hoisted onto the calm Bodensee surface early in the testing phase. Note the observer's position which is covered with canvas fairing. [Djordje Nikolić]

To improve performance, testing was conducted with engine axis rotated upwards and with an additional surface added to the wing leading edge to improve lift. This testing resulted in no significant performance improvements. Due to strong vibrations experienced during flights, on one test flight specifically over the Altenrhein airfield in Switzerland, an aileron broke but luckily pilot Egon Fath, owing mostly to his skill, managed to land the airplane intact. As a result of the poor performance, no further airplanes were constructed and project was officially cancelled.

In parallel with the Do 10 development, in 1931 *Dornier* factory in Manzell began to develop a next generation military floatplane based on the Do 10. The wing and fuselage were taken from the Do 10 however the engine was changed and instead of the 710 hp BMW VI engine, the new two seat reconnaissance floatplane designated as the *Dornier* Do C3, was powered by the French made 725 hp *Hispano Suiza* 12Y engine coupled with a four-blade wooden propeller of *Dornier* design.

The first Do C3 W.Nr. 229 flight took place on 18 September 1931. With maximum speed of only 235 km/h, service ceiling of 5,500 m and a range of 700 km it did not meet design expectations. As it became clear that the lower wing did not provide the expected lift gain, it was removed and with this modification the airplane also changed designation to Do C2A while retaining its original W.Nr.

Do C2A was a braced high-wing monoplane. The wing, fuselage, tail assembly and the floats remained unchanged from the Do C3, but a new, more powerful *Hispano Suiza* 12 engine with 740 hp was installed. The airplane retained the four-blade wooden propeller of *Dornier* design.

Contrary to the Do C3, Do C2A had an additional fuel tank behind the engine bulkhead in addition to the main fuel tank in the fuselage floor. The observer's position could be equipped with dual controls if required, a feature which would find its way in Do 22 in the future.

As seen from the table, owing to the somewhat stronger engine and removal of the lower wing, the Do C2A was able to increase its maximum take off weight and improve overall speed, however it was not able to achieve maximum flight altitude and climbing performance of the Do C3.

	Do C2A	Do C3
Crew	2	2 to 3
Powerplant	Hispano Suiza 12 with 740 hp	Hispano Suiza 12 with 725 hp
Wingspan	15 m	15 m
Overall length	12.8 m	12.7 m
Overall height	5.1 m	4.7 m
Wing surface area	32.4 m²	44.6 m² with lower wing
Empty weight	2,550 kg	2,700 kg
Maximum load	650 kg	600 kg
Max. take off weight	3,200 kg	3,300 kg
Maximum speed	250 km/h	235 km/h
Maximum altitude	4,400 m	5,500 m
Rate of climb	26 min to 4 km	13 min to 3 km

In 1932 *Dornier* made an offer to the KJ to replace or supplement the Do D and Do Wal in service with Do C2A, which was presented at Divulje naval base. Do C2A W.Nr. 229 did not meet Yugoslav expectations as the Yugoslav experts found many deficiencies, hence this floatplane was not purchased.

Despite this rejection, with all the effort put into design and testing of these floatplanes, *Dornier* persisted in its efforts to make Do C2A a commercial success. Germany was not interested in the type as it

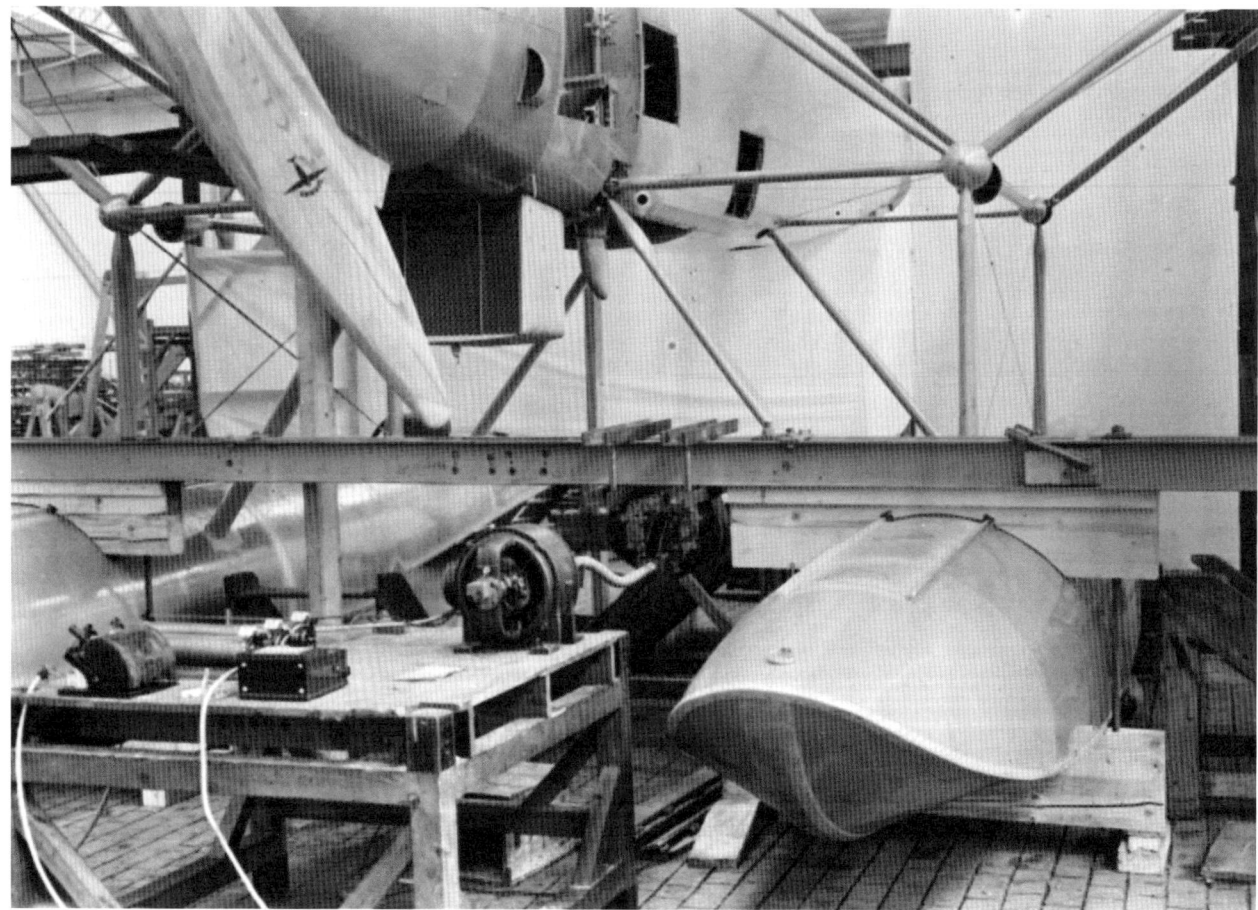

Do 22 prototype W.Nr. 259 under construction at Altenrhein. Note the four blade propeller with Dornier logo. The tail assembly is still missing. [Staatsarchiv St. Gallen]

W.Nr. 259 with its tail assembly installed. Note the sand bag weights on top of the horizontal stabilizers as well as hanging from the elevator hinges. [Staatsarchiv St. Gallen]

Following completion at Altenrhein, factory photographer took the opportunity to capture the completed floatplane in sunny autumn weather. [Dornier Museum Friedrichshafen (Airbus Group)]

did not meet RLM (*Reichluftfahrtministerium* - Reich Air Ministry) expectations. Eventually, this floatplane and the only other completed Do C2A, W.Nr. 242, were sold to the *Fuerza Aérea Colombiana* (Colombian Air Force). Interestingly, the *Dornier* archives W.Nr. list indicates that both airplanes were intended for SCADTA (*Sociedad Colombo Alemana de Transportes Aéreos* - Colombian-German Air Transport Company).

Another official factory photograph following completion. Do 22 prototype is on a sliding ramp which transported completed floatplanes from the manufacturing plant all the way to the lake. [Airbus Corporate Heritage]

Do 22 Prototype

Despite the latest *Dornier* floatplane rejection by the KJ, Yugoslav Navy representatives visited *Dornier Metallbauten GmbH* factory. The funds for the purchase of a single *Dornier* C series floatplane were already approved in 1932 through the Government Order No. 9516/32. Following the conclusion of negotiations and presentation of requirements and necessary modifica-

tions between the Navy representatives and *Dornier*, an order followed on 30 November 1933 for one example of the reconnaissance/bomber version prototype Do 22. In Yugoslav documents this floatplane was referred to as the Do C3 and is not to be confused with the original Do C3 floatplane from 1931. It is highly likely that the confusion with the Do C3 designation originates from the fact that the KJ approved payment in 1932 for the original biplane Do C3, however as this float-

Top view showing the three man cockpit. Rear facing type Do 1a turret for the machine gunner is clearly visible. [Luftfahrt-Archiv Hafner]

Another view showing the mock 50 kg bomb as well as the Darne 7.7 mm machine gun. Do 22 prototype was unarmed until immediately before the 1941 April War. [Airbus Corporate Heritage]

During the factory trials, numerous weapons options were evaluated. Shown here is a wooden mockup of a 50 kg bomb. [Luftfahrt-Archiv Hafner]

Do 22 lowered onto the Bodensee lake surface prior to its first flight. [Djordje Nikolić]

Immediately following completion, Do 22 prototype taxies in water. [Djordje Nikolić]

Do 22 prototype in low level flight above Bodensee. [Michel Ledet]

An idyllic photograph of the prototype in the evening sky above the clear and calm lake water. [The Museum of Flight]

Do 22 prototype performing one last flyover prior to landing following an evening flight. [Airbus Corporate Heritage]

This photograph was taken on 30 January 1935, not long after Do 22 prototype arrived to KJ. This Do 22 was assigned to 25.HE at Divulje. [Jože Meden – Tomaž Perme collection]

plane even in the modified version, Do C2A, did not meet the expectations and requirements, the sale never materialized. Therefore, when the new floatplane was developed in accordance with the Yugoslav requirements, documents on the Yugoslav side were never

corrected. All German documents reference only the Do 22 prototype.

During the design phase, Do 22 prototype was tested in the wind tunnel at the *Dornier* plant, with various iterations of the parts such as the wings, fuselage,

Do 22 prototype 302 at Divulje naval airbase in 1937. [Mario Raguž]

Do 22 prototype tied to a buoy in the background of the 25.HE airmen. On the left side is a Rogožarski PVT-H 52 floatplane. [Aleksandar Kolo]

Overhead view of Do 22 prototype following the return from a flight in 1936/1937. [Aleksandar Ognjević]

A number of Slovenian nationality airmen served in the Royal Yugoslav Naval Aviation. Next to the Do 22 prototype standing from left to right are: Milan Šegula, unknown, Ivan Grbec, unknown, Alojz Ručigaj and Jelenc. [Mario Hrelja]

floats and engines to obtain the best possible aerodynamic profile. With every iteration, calculations were updated until a desired outcome was reached which satisfied the design requirements.

While undergoing construction, Do 22 prototype, assigned with W.Nr. 259, was inspected by members of Swiss *Eidgenössische Luftamt* (Federal Air Office) on 26 April, 30 May and 25 June 1934. The construction was completed in August 1934, following which another inspection took place on 25 September. The next day engineers from the Federal Air Office conducted the first flight reaching an altitude of 8,000 m. The floatplane was subjected to extensive factory testing, including flights on 9 October and 12 October to determine its maximum altitude and range respectively conducted with the participation of Yugoslav commission members, to ensure all the performance requirements set by the KJ were met in full so as not to repeat the commercial failure of Dornier's Do C3 and Do C2A floatplanes.

On 20 November 1934 a flight took place in Friedrichshafen with chief *Dornier* factory test pilot Richard Wagner at the controls. The main purpose of this flight was to test the flying characteristics and usefulness of the floatplane in its indented role. Three flights were completed that day with a total duration of 40 minutes.

Two days later, after a series of factory test flights, a crew of three led by Dipl-Ing. Hans Simon from RDL Naval Test Center at Travenmunde conducted another test flight with the Do 22 prototype. Test flight was most likely conducted without prior or any knowledge of the Yugoslav side which ordered this floatplane. The purpose of this flight was not to investigate if Yugoslav requirements were met, but to evaluate if Do 22 is suitable for *Kriegsmarine* (German Navy) aviation use. For that reason, the other two crew members were Dipl-Ing. Festner and Mettig from RLM Berlin. Dipl-Ing. Festener wrote a post flight report which was used in the *Geheime Kommandofache* (Secret Command Document) dated 22 November 1934 describing in detail the observations made during their flight that day. The report concluded that further development of the Do 22 to meet the RLM requirement is advised against as the floatplane does not meet the requirements of the RLM's *Projekt Seemehrzwecke* (Muliti-purpose floatplane project).

During testing, it was determined that at the maximum allowable take off weight was 3,350 kg in reconnaissance configuration and that the highest attained air speed was 287 km/h at sea level and 317 km/h at 4,000 m altitude. The tested performance was somewhat

The same Do 22 prototype on another occasion. Notice that the propeller blades are covered. Also, float support trolley is clearly visible. [HPMS]

After the purchase of the 12 production Do 22s, the Do 22 prototype was renumbered from 302 to 301. [VPM 272-5801 via Milan Micevski]

lower than the declared one, which was calculated under the ideal conditions.

Take off time was noted during factory testing as 14 seconds on average. Climbing time was measured in 1,000 m intervals as shown in the table below.

Altitude gain	Time to complete	Altitude gain	Time to complete
0 - 1,000 m	2 min 22 s	0 - 1,000 m	2 min 22 s
1,000 – 2,000 m	2 min 08 s	0 – 2,000 m	4 min 30 s
2,000 – 3,000 m	2 min 04 s	0 – 3,000 m	6 min 34 s
3,000 – 4,000 m	2 min 20 s	0 – 4,000 m	8 min 54 s
4,000 – 5,000 m	2 min 48 s	0 – 5,000 m	11 min 42 s
5,000 – 6,000 m	3 min 29 s	0 – 6,000 m	15 min 11 s

All of the above tabulated performance values were validated during service in the KJ during the acceptance trials. Combat radius for the Do 22 prototype was declared by the factory as 1,000 km whereas the Yugoslav side claimed it to be somewhat lower, 930 km.

Due to the number of observations noted both by the factory pilot as well as RLM commissioners during the end of November flights, Do 22 prototype was not delivered to KJ in 1934 as was previously believed. Furthermore, an official report by Director of Swiss Federal Air Office, Herr Isler, was submitted on 7 December 1934. Therefore, it is most likely that after improvements were implemented, Do 22 prototype was delivered to KJ in January 1935 where it was subjected to an extensive test program through 1936. Along with the Do 22 prototype, *Dornier* engineer Josef Goetz as well as several mechanics arrived. During testing and evaluation several minor modifications were made on the spot. First flights resulted in some equipment breakdowns such as the water rudders.

Do 22 prototype taxiing in water prior to take off. [Aleksandar Ognjević]

Do 22 prototype heading on a flight sometime in 1936/1937. [Aleksandar Ognjević]

As the testing neared completion at the end of 1936 and beginning of 1937, the final modifications were made and the Do 22 prototype officially entered the PVKJ registry.

Initially, following the arrival to the KJ, Do 22 prototype was numbered 302 since sequentially this was the next number available after the *Šmolik* Š-16J (300) and the *Heinkel* He-8 (301), which were already in service. As soon as the production floatplanes started arriving in 1938, the Do 22 prototype was renumbered to 301 and *Heinkel* He-8 to 192.

The *Bauberschreibung Nr. 1200* (Construction Specification Document No. 1200) for the Do 22 prototype was completed on 28 July 1935, after the delivery to KJ. As there are no records in existence of an official Operations and Maintenance manual, it is possible that this document was written for the same purpose instead. Whether it was handed over to KJ is not known, however the document has survived the war in the Dornier archives in Friedrichshafen currently under Airbus Group patronage.

During its use, the most notable achievement in PVKJ was attained with Do 22 prototype, an almost unthinkable flight altitude of 13,000 m. Nar Ik Franjo Lolić, who described this feat, was a test engineer during this flight and he recorded all of the data which allowed verification of the results. Such an achievement did not go without consequences as crew bled from the ears and had to seek medical attention. It is interesting to note that not even factory pilots attempted to attain such altitude during any of the test flights nor the factory calculated it was possible. In the official factory documents the absolute maximum altitude was listed as only 9,500 m! It is certain that airplane had all the

Italian soldier in front of the hangars at Kumbor. Clearly visible behind him, a Do 22 prototype and another Do 22, most likely 304. Do 22 prototype metal panels at the front of the fuselage indicate some deliberate damage was possible. [Roberto Gentilli via Boris Ciglić]

unnecessary equipment removed and that in addition to nar Ik Lolić there was only one more crew member on board, the pilot.

On 5 April 1937, KM submitted a request to the Trade and Industry Ministry to allow the spare parts for the repair of Do 22 302, which have already arrived, to be excluded from the import duties. In this document, the only spare part listed is one example of the fuselage. It is not known why the fuselage needed to be replaced, when it was conducted and how long the airplane was out of service. As it appears initially no action was taken and the letter was resent again on 22 July 1937 to the Trade and Industry Ministry with the note of utmost urgency.

During its entire service, this floatplane was used for training however immediately prior to war it was armed with offensive and defensive machine guns. The *Dornier* factory did make provisions according to the information from Construction Specification Document Nr. 1200 for the installation of single French *Darne* 7.7 mm machine gun for the pilot as well as twin *Darne* 7.7 mm machine guns for the observer. In bombing role, Do 22 prototype was to be equipped with British-made *Wimperis* Mk. II bomb sight and could carry either 1 kg, 15 kg, 100 kg or 250 kg incendiary,

illuminating or explosive bombs up to a total weight of 542 kg. *Dornier* factory also calculated that a single torpedo could be carried up to a maximum weight of 750 kg. For communication, it was equipped with *Telefunken* 262 F radio. A 70 m long wire antenna was released from the fuselage during flight to enable crew to send and receive telegraph messages. Additionally, the observer used *Carl Zeiss* 13 x 18 cm camera with 70 cm focal length to take photos during reconnaissance missions.

It is interesting to note that *Dornier* considered using radial engines for Do 22 prototype such as the 870 hp *Gnome-Rhône* K-14 and 850 hp BMW 132 engines. These of course were never specified by the customers and they progressed no further than calculations and potentially some preliminary drawings.

As of 1 May 1938, Do 22 prototype was listed in active service in a document which tracked all PVKJ compass inventory by airplane type to which they were issued to. From this document, some light is shed on the equipment installed, being that it lists the pilot compass as *Ludolph* F.K. 5 s/n 2635 and navigator compass *Campbell* s/n 1152.

By 31 January 1941, Do 22 prototype completed over 345 flights and flew a total of 222 hours and 45 minutes. During the month of January, it only made one flight lasting 20 minutes. The *Hispano Suiza* 12 Ydrs engine s/n 485322 logged 193 hours and 20 minutes total working time, which indicates that the floatplane had its engine replaced some time during its service. Comparing the number of flight hours on production Do 22s, which were received starting in 1938, it can be deduced that the engine was replaced sometime around this time or potentially as early as the end of 1937 during the replacement of the fuselage discussed earlier.

On 5 April 1941 just prior to the start of war, the Do 22 301 along with three other Do 22s from 25.HE flew to the wartime dispersal location at the island of Zlarin which was prepared ahead of time for their arrival. After this dispersal location was discovered by *Regia Aeronautica* (Italian Air Force), 301 was flown to Kumbor. During the short course of the April War, this floatplane did not take action and was captured in Kumbor along with two other Do 22s. It is unknown if it was sabotaged or not, Italian reconnaissance photos show two Do 22s, one of them 301 in front of the hangars at the naval base. At the time, the Italian soldiers entered the base several days later, Do 22 prototype was still at its original location and no attempt was made to fly out, damage or destroy it altogether to avoid capture.

From Prototype to Production

Since the exodus of German airplane manufacturers to the Baltic, *Dornier* factory in Altenrhein, with its enormous size covering 17 hectares laid dormant. An intense propaganda attempt was undertaken to obtain foreign orders in addition to those from the KJ at the last Stockholm Air Show, ILIS (*Internationella Luftfartsutställningen i Stockholm* – International Aviation Exhibit in Stockholm) 1936, where various products from *Dornier Metallbauten GmbH* and in particular a model of the Do 22 prototype were presented to visiting Military delegations.

Even before the successful conclusion of the Do 22 prototype testing in the KJ, negotiations took place between the KM and *Aerometal A.G.* from Zürich for the purchase of 12 additional floatplanes. Do 22 was selected in preference to the local airplane manufacturing companies which applied to the Government tender as well. Contract negotiations were not finalized until 1937.

Finally selected for PVKJ service, *Dornier* floatplane beat competition from the British made *Fairey* Swordfish and *Blackburn* Shark, owing most likely to the excellent relations between the *Dornier* and the KJ due to long term and successful use of other *Dornier*

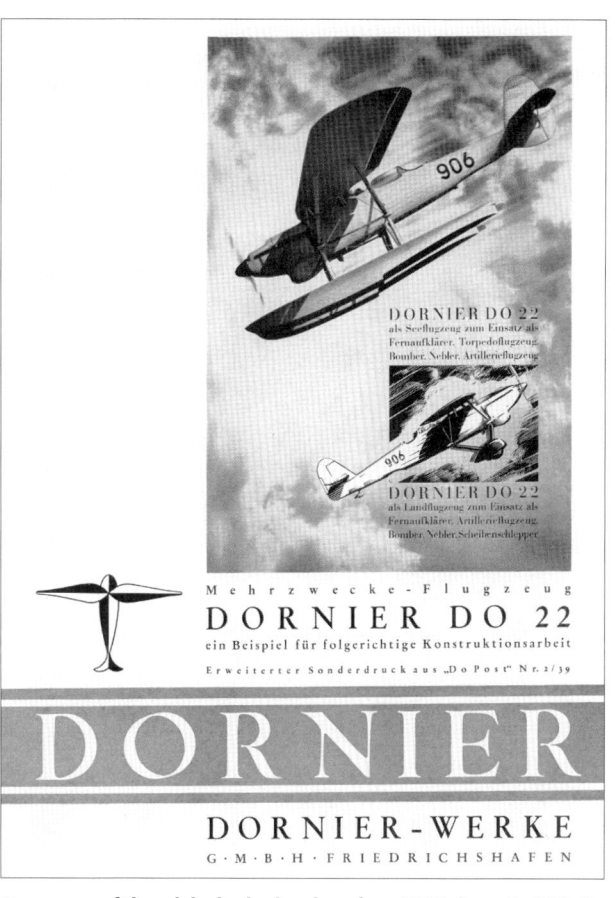

Cover page of the original sales brochure from 1939 shows Do 22 in its floatplane and landplane configurations. [Djordje Nikolić]

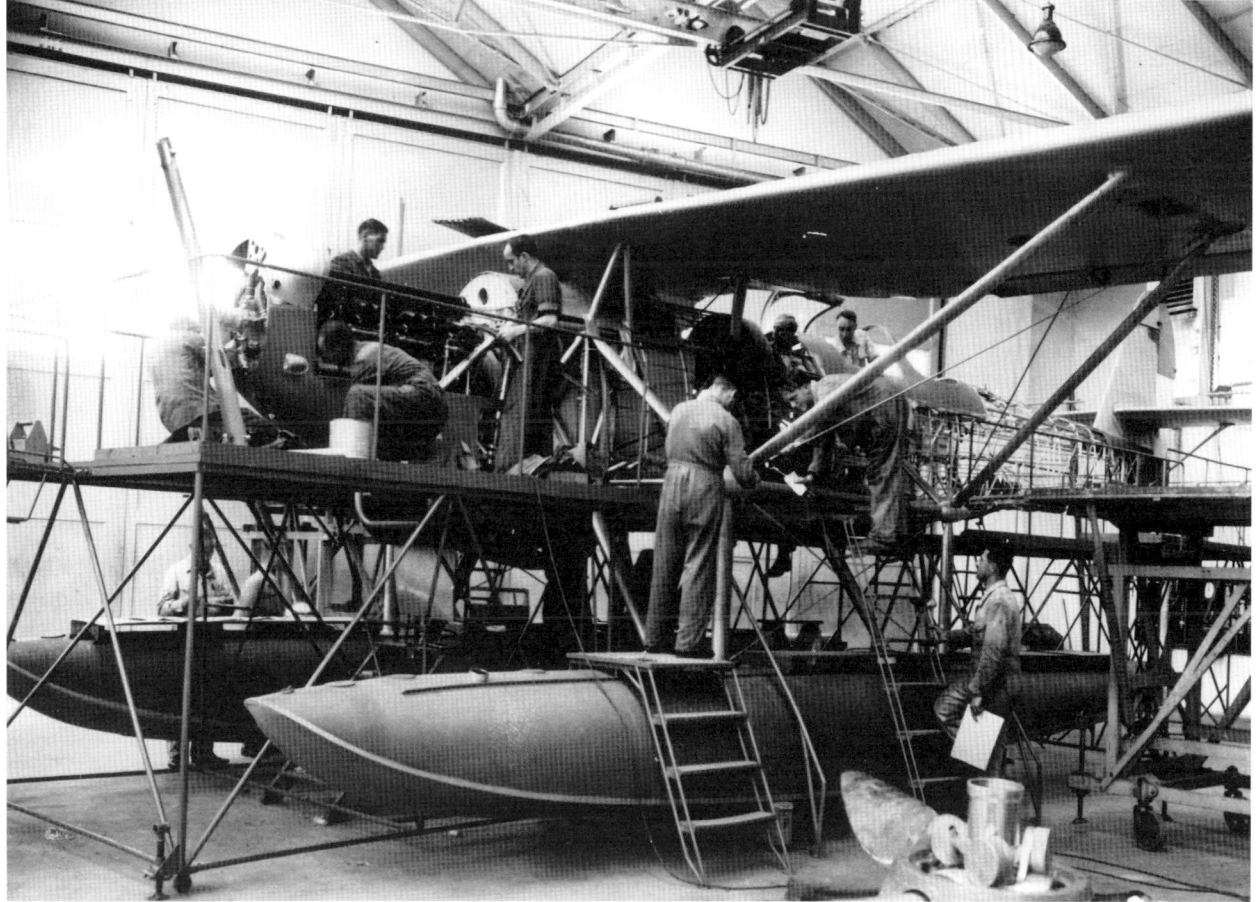

Do 22Kj with no fewer than 11 Dornier employees working hard prior to a lunch break. Time clock reads 12:00. [Airbus Corporate Heritage]

Do 22Kj during assembly at Friedrichshafen. The Hispano Suiza engine and the internal structure are clearly visible. In the background on the wall, a time clock for Dornier employees reads 12:50. Lunch time! [Airbus Corporate Heritage]

First production Do 22 302 at the production hangar at Friedrichshafen lifted by a crane. At this time 302 had an early type exhaust manifold installed which followed the fuselage and terminated below the machine gunners cockpit. [Robert Čopec]

Do 22 302 lowered on the surface of Bodensee lake by a crane directly from the factory building. Note that the exhaust manifold has been removed. [Airbus Corporate Heritage]

An interesting photograph from a production building at Friedrichshafen showing Do 18G flying boat in the foreground with Yugoslav Do 22 307 on the left and an unknown Greek Do 22 on the right. [Airbus Corporate Heritage]

302 heading out on a factory test flight with only the pilot at the controls. [Djordje Nikolić]

Take off! 302 about to lift its floats off the Bodensee surface. [Djordje Nikolić]

Do 22 302 taking off on another flight. Note that the observer Plexiglas canopy is not installed. [Airbus Corporate Heritage]

The first production Do 22 302 resting on calm waters of Bodensee. [Michel Ledet]

302 during take off on a factory test flight. [Airbus Corporate Heritage]

Do 22Kj under tow following a completed test flight. [Airbus Corporate Heritage]

types. It is notable to say that the series production airplanes were based on the Yugoslav requirements which included 900 hp *Hispano Suiza* 12Y-21 engine, an engine which was already in use in Yugoslavia in Do Wal. The same engine type, although somewhat less powerful, was used on the Do 22 prototype and it met all performance and reliability expectations. In April 1937, a contract for six engines was signed, however due to additional requests commonly seen with Yugoslav orders, the delivery was prolonged until 1938.

The series production line was established at the *Dornier* factory in Friedrichshafen, and not in Alten-

A propaganda photograph of Do 22 with fuselage number 906 in a dive. [Airbus Corporate Heritage]

rhein, which originally developed the type. Numerous improvements were implemented over the Do 22 prototype. The summary of major differences between the prototype and the production Do 22 is as follows:

1. The most significant difference was the engine. The production Do 22 adopted the most powerful *Hispano Suiza* 12Y-21 with three blade wooden propeller by *JU Hamilton* for PVKJ ones whereas the Do 22 prototype had a four-blade wooden propeller by *Dornier*.

2. The very silhouette of the floatplane changed, as the radiator was now firmly positioned within a cowling under the engine and was not retractable.

3. The hood over the observer's position was larger and more streamlined on production Do 22.

4. The section of the vertical stabilizer below the fuselage was also significantly larger on production Do 22.

5. Rear machine gun installation on the prototype had a fully rotating ring turret but had a fixed mount with pivot on the production Do 22.

6. The location of the wing position lights was moved further aft on the wings of the production Do 22.

7. Aileron counterweights were installed on production Do 22.

8. Pitot tube was moved from the starboard wing on the prototype to the port wing on production Do 22.

9. Engine cooling louvers were removed from the nose.

10. Venturi generator was removed from the starboard side of the cockpit on production Do 22.

11. Forward facing machine gun was installed on the production Do 22 at the factory whereas the prototype was delivered without offensive armament.

171

A fine photo of two Yugoslav Do 22 undergoing testing at Bodensee. [Djordje Nikolić]

It was a standard German practice at the time to add a suffix "K" to the airplane designations manufactured for export. As a result, Do 22s exported to the KJ were designated by *Dornier* as Do-22Kj (j for Jugoslawien) and in Yugoslavia they were better known as *Dornier* Do H due to the use of *Hispano Suiza* engines.

The first series of 24 Do 22s was officially released to production under the Delivery Program No. 5 on 1 April 1937. The first four floatplanes were completed in August 1938 and the last of the 24 was completed by March 1939. 12 were exported to the KJ and 12 to the Kingdom of Greece, with 11 delivered in 1938 and 13 in 1939 to these two countries. Four additional floatplanes were intended for Latvia, however they never

reached their destination due to the start of World War II, and eventually ended in Finland.

Introduction into Service

KM concluded the contract with *Aerometal AG*, in accordance with the KJ Government decree M.S No. 19/37, for the delivery of 12 Do 22s to the KJ in February 1937 with deliveries intended to start in October 1937 and to conclude in July 1938. The total cost of purchase was 1,242,000 RM with a 50% advance payment.

Immediately after the contract for the Do 22 was signed, it was necessary to secure engine supplies to ful-

Men of 3.HK following the formation of 1. and 2. HG mid-October 1939. Note the white Maltese, a group mascot, between the two FN Browning machine guns. [Mario Raguž]

Five Do 22Kj flying overhead in a formation. [Mario Raguž]

fill the requested delivery dates. As a result, six *Hispano Suiza* 12Y 21 engines, gearboxes, compressors, RBD-12 magnetos, *Avia* spark plugs, AM fuel pumps, *Viet* hand cranks, 1200 W generators and tooling were ordered on 13 May 1937. The list price on the contract for each engine was 281,000 Swiss Francs. The delivery date for the above listed equipment was set for 9 January 1938 and the engines were to be delivered to *Dornier* factory at Rorschach, Switzerland. Prior to delivery, each engine was checked on a factory test stand, with witnesses from the PVKJ present. Upon delivery to Rorschach, the same representatives checked the engine crates for contents. The contract required that the *Hispano Suiza* factory guaranteed 200 hours of engine run time within 18 months of delivery. As a side clause of the contract, *Hispano Suiza* was to cooperate with *Dornier* and investigate the prospect of adapting new *Fuseau-Motoeur* Type 36D radiators to the engines included in the signed contract.

To ensure the required quality standards were upheld, to learn about the Do 22 and to report on any production delays, numerous Yugoslav representatives were dispatched to *Dornier* factory. Senior military

technical clerk Anton Arko was sent from 22 February to 7 June 1937 to Friedrichshafen for in process quality control of the first Do 22Kj batch assembly. He was replaced on site by engineer maj Stevan Stojanović who arrived on 6 June 1937 and remained through 6 October 1937. Anton Arko returned to Friedrichshafen on 28 October, where he remained until January 1938. He was replaced by pbb Ik Dušan Djukić on 22 January 1938 who remained until 17 June 1938. pbb Oton Čermak and kf Dražen Delia and pbb Ik Augustin Grošelj spent a month in Friedrichshafen from 6 August 1938 until 5 September 1938 conducting the factory acceptance of the first batch prior to delivery. Engineer maj Nikola Baskijević and kbb Nikola Nardeli arrived at Friedrichshafen on 20 October 1938 and remained until 22 November 1938 performing factory acceptance of the second batch of completed Do 22Kjs. They were followed by pbb Ik Vladeta Petrović from 23 October through 20 February 1938. Kf Većeslav Dujšin was the VVKJ HQ representative at the *Dornier* factory from 10 March 1937 until 3 March 1938.

Of all the representatives, perhaps pbb Danilo M. Hubmajer provided the best insight into what was taking place at *Dornier-Werke*. He was sent to *Dornier-Werke* in Friedrichshafen between 15 July 1938 and 17 February 1939 to supervise and inspect manufacturing and factory acceptance of Do 22 floatplanes ordered under contract Pov. M.V. br. 3096/37. It is important to note that 15 July 1939, coincided with the date of the first flight and the start of the factory acceptance trials of Do 22 according to the preserved *Dornier* records. Periodically, he reported the progress updates to the KM HQ and the PVKJ CO. He noted that two facts affected the manufacturing quality and the fabrication speed:

Do 22s were ordered in Germany and the engines, due to the technical reasons at a time, in France. In order to make a cohesive assembly, Dornier placed an order in France for all other components necessary for the engine function such as the radiator, vacuum pumps, etc. This proved to be unsatisfactory for several reasons. The sub-suppliers could not honor the lead times for the delivery of these components since the French airplane industry was fully committed to satisfying domestic orders. This negatively affected the assembly speed of the Dornier floatplanes. Additionally, once the equipment was delivered, it was prone to defects, which were only discovered as the floatplane was undergoing its factory acceptance testing. Especially, radiators proved to be very troublesome.

The ordered floatplanes were expected to attain better performance. Dornier believed that the performance issues were caused by the engines, while *Hispano Suiza* had the opposite opinion. For this reason, it should be concluded that it is extremely important to purchase the airplane along with the standard engine from the country of origin if technical and political factors allow it.

Quality control during individual component fabrication at *Dornier* was very difficult because factory was split into separate departments, which were located around Bodensee in a radius of 30 km. Dornier did provide transportation to facilitate easier quality control.

Additionally, pbb Danilo M. Hubmajer made the following interesting observation:

Anti-aircraft artillery was located around Bodensee and it was under orders to open fire at any airplane or floatplane with foreign markings overflying the area. I found this by chance since at the time of my arrival to the factory on 28 September, I noticed that the Yugoslav flag on the tail and the wing markings on two Do 22s were painted over and on top of them a Swastika was added. When I inquired about the reasons for this action, one of the Dornier clerks informed me about the reasons which I mentioned above.

PV airman posing next to a Do 22 float. Photo was most likely taken in the early 1939 or 1940 since the Do 22 is still equipped with an exhaust manifold. [Aleksandar Ognjević]

Even though most of the railway transpiration was reserved for military needs during this time period, to appease the Yugoslav side due to the delays in the delivery of the floatplanes, Dornier-Werke was able to arrange for transport of four Do 22s by train which passed factory acceptance testing to Divulje.

To ensure seamless delivery of engines, which were desperately needed to prevent the delayed delivery of Do 22 from slipping even further, several personnel were sent to the *Hispano Suiza* factory in France. Engineer maj Milan Milutinović was the production supervisor at the *Hispano Suiza* factory from 1 January until 31 January 1938. Pbb Ik Anton Brumen and kf Danilo Hubmajer conducted factory acceptance of the first batch of *Hispano Suiza* engines at the factory in France between 6 January and 7 February 1938. Engineer maj Dragutin Matić was the production supervisor at *Hispano Suiza* factory from 10 June through 31 July 1938. Pbb IIk Ivo Buljan and pbb Ik Alfred Malbohan conducted factory acceptance of the second batch of engines at the factory in France from 6 June through 3 August 1938 intended for the first Do 22 batch.

Vladeta Petrović, the 2.HK CO, being one of the most capable PVKJ pilots, was sent in 1939 to *Dornier* factory at Friedrichshafen to supervise the manufacturing of the remaining eight Dorniers on order. After his return, he completed training of the entire 20.HE to the newly arrived floatplanes. In this time period, he developed a detailed strategy for using the Do 22 in combat and successful trained crews how to dive bomb. Since Do 22 was the most modern floatplane in PVKJ service, it was intended for use exclusively in missions involving long distance reconnaissance and bombing.

The first completed production Do 22 flew on 15 July 1938. In September 1938 the first two were delivered, which were followed at the beginning of 1939 with the remaining ten. Prior to April 1941, there were two HK. 2.HK 25.HE at Divulje was equipped with six Do 22s. 3.HK 20.HE at Kumbor had five Do 22s on strength. Floatplanes with even escadrille numbers were assigned to 25.HE based at Divulje, whereas those with odd numbers were assigned to 20.HE based at Kumbor.

Regular wear and tear replacement parts were purchased as required from *Dornier*. Minor repairs were conducted frequently at both Kumbor and Divulje with Yugoslav mechanics becoming very proficient and skilled in maintaining their Dorniers in perfect working order.

Decree 6656/ad/38 from 18 July 1938 approved the purchase of 36 FN 7.92 mm machine guns from FN (*Fabrique Nationale Herstal* – National Factory Herstal) for the newly delivered Do 22s. The purchase price of each machine gun was 37,015 Dinars with the delivery date of four months following a 25% down payment. Enough machine guns were delivered to arm all of the purchased floatplanes. The following year, cutaway models were purchased on 24 August 1939 to better train the maintenance personnel on machine gun function and servicing.

On 26 July 1940 as part of the PVKJ reorganization initiative, which intended to finally purge the obsolete equipment from use, a request was made to the KM for the purchase of additional 12 Do 22s. These plans never materialized.

Due to frequent use, accidents were bound to happen. The first one took place on 1 August 1940 when Do 22 311 was performing an engine test run following a minor repair, which led to a damaged engine compressor rotor. Until this incident, the engine had only 90 hours and 35 minutes of working time and was due for it's 100-hour checkup hence this opportunity was used to replace the damaged rotor.

Only several days later, on 6 August 1940 Do 22 305 caught fire in mid-flight. Floatplane was perform-

Spectacular underside view of a Do 22 in flight. [Aleksandar Kolo]

ing test runs at sea level and then climbed slowly to 3,600 m, checking the engine function each 1,000 m altitude gain. After 40 minutes of flight time, a fuel pipe cracked, and a small detonation soon followed. The engine immediately caught fire and the pilot closed the fuel isolation valve and activated the fire extinguishing system, however the flame could not be extinguished. Pilot nv IIIk Andrija Arapović ordered his flight mechanic nar Ik Nikola Batalo and radio operator pnar Dragutin Bašić to bail out. The pilot dove sharply hoping to extinguish the fire however at 150 m altitude, after sustaining heavy burns, he bailed out of the floatplane, which eventually crashed into the sea. Do 22 was recovered the same day, it was 100% damaged whereas engine sustained 60% damage according to the preserved records. The commission formed to investigate this accident determined that a pipe joint in the left carburetor fuel intake broke at the location where it connects to the fuel manometer. The fuel ignition likely took place when fuel came in contact with hot exhaust pipes. This connection broke most likely due to vibration or poorly welded joint. The decision to remove 305 from the register was finally approved on 7 January 1941. Before scrapping, *Goertz-Boykow* bomb sight s/n 804, FN machine guns s/n 15 and 18 respectively, spent ammo casing bags, ammo boxes, gun sights and other useful parts were stripped from this Do 22.

To enhance cooperation between VVKJ and PVKJ between 13 and 14 October 1940 an exercise took place in which Do 22 was used for laying smoke curtains. Three Do 22 participated in this action successfully. Smoke generators were mounted beneath the fuselage. Such exercises were rare and did not significantly contribute to the overall cooperation effort, which would already be evident the following year during wartime.

With Europe ravaged by war, KJ understood the dangers which were drawing nearer to its boarders each day, and started to prepare for war, albeit too late. On 31 January 1941 2.HK reviewed the average estimated monthly wartime usage of fuel per floatplane type. For 12 Do 22s in service, based on an average hourly consumption of 250 liters per hour, 3,000 liters would be necessary daily. This meant that in total of 120 tons per month of 93 octane fuel were required. 93 Octane fuel reserves for these airplanes allowed for 100 flying hours per plane in wartime conditions.

The last pre-war accident occurred in the morning of 1 March 1941, when Do 22 310 was conducting mock dive bombing against a torpedo boat near Kaštelan Bay. The pilot began the dive recovery too late,

misjudging the distance to the water due to the sun's reflection off the water surface and as a result Do 22 struck the water. Luckily the crew managed to escape without injury. 310 sank to a depth of 15 meters. It was recovered in the afternoon of the same day but was never repaired and put back into service prior to the outbreak of the war. According to the preserved records this Do 22 was brought back to 2.HK, but it is unclear if it was used as a source of spare parts or for another purpose. No photos of this accident are known to be in existence.

Following this accident and immediately prior to the German invasion, there were ten operational Do 22s, all delivered floatplanes but 310 and 305.

Just three days prior to signing the accord to join the Tripartite Pact, on 22 March 1941, in a last effort to secure spare parts for its most prized naval floatplane, KM submitted a list of *Hispano Suiza* spares to *Avia A.D.* in Czech Republic because France was already occupied by Germany and no engine or spare parts purchase was possible. Additionally, *Avia A.D.* was approached, as this factory manufactured the *Hispano Suiza* 12Y 21 engine under license. Spares were supposed to be delivered in two shipments, one within three months of order and the other shipment four and half months later. *Avia A.D.* additionally offered a complete engine from stock, which could have been delivered within 15 days of order. The beginning of the war prevented any further deliveries from taking place.

The last several days prior to the war, apart from reconnaissance, were spent patrolling the skies above the minelayers, which were operating at various locations in the Adriatic Sea. Supporters of the KPJ (*Komunistička Partija Jugoslavije* – Yugoslav Communist Party) at 25.HE at Divulje considered flying over a Do 22 308 to the Soviet Union in protest of the KJ joining the Tripartite Pact. Following the 27 March coup d'état, this plan was abandoned because the country needed defending. In the last flight prior to war, on 5 April 25.HE, sent three Do 22s and a sole Do 22 prototype to Zlarin for wartime dislocation. It is interesting to note that Do 22 prototype had machine guns installed immediately prior to war, as it served until then purely for unarmed reconnaissance.

PVKJ airmen were well aware of the immediate dangers of war with Germany and Italy. However, they did not anticipate the strength of their opponents and the storm that was approaching, since across the Adriatic Sea *Regia Aeronautica* was much better equipped; it had 262 bombers, 295 fighters and 109 tactical reconnaissance airplanes on strength.

25.HE and 20.HE. flights per Do 22 as of January 1941										
2.HK 25.HE										
Airplane number	Current month		Total		Flights per month	Total flights in service	Engine S/N	Current month		Total	
	Hours	Minutes	Hours	Minutes				Hours	Minutes	Hours	Minutes
301	–	20	222	25	1	345	485322		20	193	20
302	1	5	152	55	1	259	485722	1	25	157	30
303	–	–	112	35	–	180	485713	–	–	116	35
304	–	45	125	55	1	179	485715	1	5	130	15
306	–	–	158	20	–	227	485716		20	162	40
308	7	–	126	10	7	228	485721	7		132	25
310	8	25	135	50	9	204	485719	8	45	140	25
312	3	50	109	35	4	177	485712	3	50	113	45

Note:
1. 301 originally had engine S/N 485292. Low hours on engine S/N 485322 further confirm airplane had its engine replaed during service.
2. 303 and its engine were unserviceable at the time.
3. There were three spare Hispano Suiza engines 12 Y 21 S/N 485724, 485728 and 485729

3.HK 20.HE										
Airplane number	Current month		Total		Flights per month	Total flights in service	Engine S/N	Current month		Total	
	Hours	Minutes	Hours	Minutes				Hours	Minutes	Hours	Minutes
305	–	–	98	–	–	126	485714	--	–	102	–
307	4	5	118	10	9	172	485717	4	5	122	50
309	–	45	103	10	1	148	485723	–	45	107	25
311	4	5	113	40	7	160	485718	4	5	116	55
313	–	–	122	5	–	216	485720	–	–	126	20

Note:
1. 305 was involved in an accident on 6 August 1940 and the decision to scrap it was approved on 7 January 1941.
2. 311 was unserviceable at the time.
3. There were three spare Hispano Suiza 12 Y 21 engines S/N 485725, 485726 and 485737 with 4 h 15 min run time each.

The only known photograph of 312 at a hangar in Divulje during a ceremony. [Mario Raguž]

April War

On the morning of 6 April 1941 the following Do 22s were on strength within the two Do 22 equipped HE:

1.HK 25.HE Zlarin, Visovac Lake	2.HK Divulje base workshop	3.HK 20.HE Kumbor
301, 302, 306, 308, 312	303*, 304, 310**	307, 309, 311, 313

* undergoing repairs
**damaged beyond repair

At 00:30 a dispatch received at Kumbor indicated that Germany would most likely attack KJ on the morning of 6 April 1941. At 2:00 and 4:00 the 3. HK CO raised the alert status accordingly. At 6:40 dispatch from forward observation station at Mljet island received an alert that a number of unknown airplanes were flying towards Mostar airbase. Immediately thereafter, dislocation of 20.HE was ordered. The first airplanes attacked Boka Kotorska at 07:05 from 2,000 – 3,000 m altitude. Three of the fighter escorts, Fiat G.50s, broke their formation and swept in for a strafe run against the 20.HE floatplanes, which ended unsuccessfully. Nv Jože Meden was in one of the Do 22s during the attack:

While flying low, the Italian Fiat G.50s (Macchis – a.c.) attacked railway station and strafed all buildings in their path. Above my wing tracer rounds started flying, landing in water 20 meters from my airplane. I squeezed myself into the narrow Dornier cockpit while the mechanic found cover between the floats with head just above the water. I remained still, waiting for another burst.

Following this attack, the 2.HG. CO ordered all remaining floatplanes to dislocate. Once the floatplanes were moored at their dislocation points, their machine guns were removed and brought on shore for improvised anti-aircraft defense.

On 6 April, 25.HE was based Zlarin and Visovac islands. Floatplanes at Zlarin were well camouflaged, being taken on shore and covered with fishing nets and branches. On the same day, having escaped Italian bombing of Divulje unscratched, three Do 22s flew over to Zablaće where crews hurried to mask their floatplanes and avoid discovery. In the afternoon, 25.HE sent three Do 22 to auxiliary airfield at lake Visovac.

Poor weather and lack of orders from KM resulted in low flying activity on 7 April. Despite this, at 17:00, 307 (nv Obradović, pf Filipović, nar Barl) performed reconnaissance over Boka Kotorska – Mljet route and 309 (nv Meden, pf Konjević, nar Bačić) flew over Boka Kotorska – Ulcinj, both flights were uneventful with Do 22s returning to base at 19:00.

First flights on 8 April were conducted by Dorniers from 20.HE, when 311 (nv Meden, pbb Kocijančić and nar Bašić) took off at 09:00 to conduct reconnaissance of Korčula – Lastovo – Mljet route due to reported Italian presence in the area. Having found nothing, they returned to base two hours later. Commander of the Naval base at Boka Kotorska ordered Do 22 313 (nar Besarabić, pbb Zobundžija and nar Čuhraj) to take off and verify the alleged reports that German airplanes have landed at Župa and that German crews were defending themselves against the local gendarmes with their machine guns. Do 22 took off at 10:00 and, while flying over that location, crew determined that the airplane in question was a VVKJ *Savoia Marchetti* SM.79 which was shot down by friendly anti-aircraft fire. 313 returned to base within the hour. In the afternoon of the same day, first combat patrols took place finding no enemy activity. In the late afternoon of the same day, between 17:00 and 20:00, 307 (nv Obradović, pbb Zobundžija and nar Čuhraj) and 309 (nv Meden, pf Konjević and nar Barl) conducted reconnaissance of

Bari and Bridisi harbors identifying three destroyers at the former, with no activity noted at Brindisi due to poor weather. The only 25.HE mission of the day took place at dawn of 8 April, a single Do 22 was sent over from Zlarin to conduct reconnaissance over the Italian port of Ankona. Do 22 was to fly within 20 km of target at 7,000 m altitude. At that instance, the pilot was to reverse the throttle and descend to 4,000 m, take photos of the port and return to base. Two attempts were made to accomplish this mission in the afternoon, both times Do 22 flew halfway across the Adriatic only to experience irregular operation of flying instruments and the loss of oil system pressure. It is not excluded that there was a deliberate act of sabotage involved due to the nature of the problem and timing of the occurrence.

The next day, Do 22 313 (nv Koroša, pbb Kocijančić and nar Bačić) flew to Brindisi, this time loaded with two 50 kg bombs and attacked ships in the harbor. However due to poor weather the crew was unable to observe the effects of their attack. Italian anti-aircraft defense failed to act altogether, and floatplane returned safely to base. During the return flight, crew discovered an Italian convoy heading towards the Croatian coast. A decision was made that this convoy was too good of a target to miss, hence it was attacked in the evening of 9 April. Due to the very strong anti-aircraft defense, crews failed to observe the effect of their attack. In the late afternoon, that is at 17:00, the same Dornier (nv Valić, pbb Kocijančić, nar Čuhraj) was sent to conduct reconnaissance of Durrës harbor in Albania. Following the successful mission, the crew brought back photographs and gave a full report concerning the location of enemy ships. Using this information, a decision was made to conduct a nighttime bombing raid with three Do 22s,. 307 (nv Koroša and pbb Zobundžija), 309 (nv Obradović and pf Filipović) and 311 (nv Meden and pf Konjević). These floatplanes took off at 19:30 approximately ten minutes apart. They reached their target at dusk and strong although inaccurate anti-aircraft defenses opened fire. All three Do 22s dropped their 100 kg bombs from an altitude 600 – 1,200 m but could not observe the effect of their attack.

On 10 April, 307 (nv Besarabić, pf Zobundžija and nar Čuhraj) took off at 10:30 to conduct photo reconnaissance of Brindisi harbor. On their way, crew discovered an Italian convoy consisting of four merchant ships and two destroyers. After taking photos over Brindisi from 1,000 m altitude, the crew continued towards Otranto harbor where they took more photographs. On the same morning at 10:30, 309 (nv

Valić, pbb Kocijančič and nar Bačić) took off for photo reconnaissance over Bari. Along the way, they spotted a single Italian tanker, which was attacked with two bombs resulting in a near miss however the tanker stopped its engines. Following photo reconnaissance over Bari, Do 22 returned back to base at 13:30.

The same day, due to the worsening situation, two Do 22s from 25.HE, 302 and 304 flew over from Visovac to Kumbor. 303, which had repairs completed at Divulje workshop, also attempted to fly over to Kumbor at 8:00, however on its way in passing it was shot at by an *Regia Aeronautica* bomber. Having sustained damage to the engine it was forced to make an emergency landing. Crew remained in their Do 22 until the next day when KB *Jaki* came to tow them back to base while *Rogožarski* Sim XIV flew overhead to ensure there were no enemy submarines in the vicinity.

On 11 April, 20.HE Dorniers were dispersed near the base and were protected by strengthened anti-aircraft defenses. The photo reconnaissance of Bari and Brindizi harbors did not take place due to the worsening weather conditions. Weather conditions improved the next day, 12 April, which enabled two Do 22s, 311 (nv Klešnik, pf Filipović and nar Kikanović) and 313 (nv Sinkovec, pf Konjević and nar Čuhraj) to set out on the photo reconnaissance mission however, both returned due to engine problems one of the Dorniers experienced in flight.

On 13 April, the three remaining Do 22s from 25.HE, 306, 308 and 312 flew to Orahovac to join the 20.HE. The same day all Dorniers were dispersed again and camouflaged to prevent detection and potential damage from strafing.

On 14 April, Do 22s from 20.HE conducted their last combat missions. Under the orders from 3.HK CO, three Do 22s, 307 (pbb Petrović, pf Filipović and nar Čuhraj), 309 (nv Koroša, pbb Kocijančič and nar Bačić) and 311 (nv Obradović, pf Konjević and nar Bašić) took off at 10:15 on a mission to determine which airfield the Italian fighters were originating from. While at 5,000 m the crew of 307 observed five *Fiat* G.50 fighters at the altitude of 4,000 m flying towards the Italian coast in south-west direction. The Italians did not notice 307, 309 and 311 did not observe any Italian airplanes at the Albanian airfields. The same evening, at 17:00, 313 (nar Besarabić, pf Jovanović and nar Barl) took off on a photo reconnaissance mission of Durrës which was conducted successfully.

During the ten days of hostilities, Do 22s successfully completed assigned missions while avoiding interception by the Italian fighter planes such as *Macchi* C.200, *Breda* 65 and *Fiat* G.50, which were superior to Do 22, due to the higher speed and better maneuverability. These *Regia Aeronautica* fighters easily crossed the Adriatic Sea and prowled the

The fuselage of 310, which crashed on 1 March 1941, seen here at a damaged hangar at Divulje naval airbase. It would seem that as the fuselage is on the wooden supports, it was in the early stages of repair. [Stato Maggiore Aeronautica]

Yugoslav coast, which put Yugoslav Do 22s crews at a significant risk. Only Do 22 303 was damaged by an enemy airplane, owing to the strong protection by the shore and ship based anti-aircraft artillery from Boka Kotorska bay in Montenegro within the area of responsibility of the 3.HK.

At the end of the hostilities, Italy captured three Do 22s: prototype 301 and production 303 and 304. At the time of war, 303 was in one of the hangar workshops at Divulje as it crash landed in June 1940 damaging its struts and floats and was in need of a major repair in addition to the damage sustained from a *Regia Aeronauica* bomber. 301 and 304 were captured in front of hangars at Kumbor and were photographed during aerial reconnaissance by the *Regia Aeronautica* prior to the arrival of ground troops. 304, which was kept during the war at Divulje for exclusive use by the PVKJ HQ, was flown to Kumbor on 10 April by the 25.HE CO pbb Ivo Rebula, who made no effort to continue southward towards Greece after the first failed attempt. Similarly, no effort was made to move these Do 22s and when the Italian troops finally arrived on 17 April, they were found in identical location. The fuselage of 310, which crashed on 1 March 1941 was captured at the damaged hangar at Divulje naval airbase. As Germans captured several *Caproni* Ca 310 airplanes along with 440 hp *Piaggio Stella* P.VII C16/35 engines, Italy offered to exchange captured German made Do Wals and Do 22s. The deal was rejected by the German side. Since the three Do 22s did not end up in Italian service, it is highly likely that they were scrapped for being unserviceable or damaged beyond repair prior to the Yugoslav surrender. Had they been serviceable, or had it been possible to bring them back to service, they would have almost certainly been transferred to the Italian mainland like the *Rogožarski* Sim XIV, which were captured in Montenegro.

In total, 20.HE completed five bombing and 16 long distance reconnaissance missions lasting a total of 37 hours and 45 minutes. Do 22s from 25.HE were not as active as their floatplanes spent only around 10 hours and 40 minutes on missions which involved changing their dispersal locations, single failed mission to the Italian port of Ankona and finally retreating south to join the 20.HE. Towards the end of the conflict, a plan was formulated to bomb Italian convoys in groups, however due to a deteriorating war situation, it never materialized. Instead on 15 April no flights took place and maintenance crews worked hard to prepare their floatplanes for the upcoming adventure.

20.HE combat mission flight time				
Date	Take off	Landing	Flight time	Airplane number
7 April 1941	17:00	19:00	2:00	307
7 April 1941	17:00	19:00	2:00	309
8 April 1941	9:00	11:00	2:00	311
8 April 1941	10:00	11:00	1:00	313
8 April 1941	17:00	20:00	3:00	307
8 April 1941	17:00	20:00	3:00	309
9 April 1941	10:00	13:00	3:00	313
9 April 1941	17:00	18:00	1:00	313
9 April 1941	19:30	21:00	1:30	307
9 April 1941	19:30	21:00	1:30	309
9 April 1941	19:30	21:00	1:30	313
10 April 1941	10:30	13:30	3:00	307
10 April 1941	10:30	13:30	3:00	309
11 April 1941	No flights due to weather			
12 April 1941	10:30	12:00	1:30	311
12 April 1941	10:30	12:00	1:30	313
13 April 1941	Dispersal and camouflage			
14 April 1941	10:15	12:10	1:55	307
14 April 1941	10:15	12:10	1:55	309
14 April 1941	10:15	12:10	1:55	311
14 April 1941	17:00	18:30	1:30	313
15 April 1941	Maintenance			

Egypt Exile – 1941 to 1942

The situation facing KJ Army changed from bad to worse. The capital city, Belgrade, was captured on 12 April which prompted the all but defeated Army to start negotiations for a general cease fire. Germans did not accept this, as a complete capitulation was requested. As a result, King Peter II and factions of the Government escaped the country for Greece between 14 and 16 April.

On the evening of 15 April, a decision was made for all serviceable Do 22 to fly to Corfu. An alternate escape route to the Soviet Union was also considered, however it was not possible due to the limited range of Do 22. The crews from 20.HE did not want to accept the armistice and capitulation, instead they opted to continue fighting for the Allies. To prevent any acts of sabotage, armed guards protected their floatplanes the entire night.

At 05:00 in the morning of 16 April, the entire group of Do 22s took off. Immediately thereafter it flew over Boka Kotorska at low level, making an example for other Yugoslav servicemen to follow, encourage them to keep fighting and not accept the capitulation. As Dorniers flew south towards Greece at 200 m altitude, they experienced extremely harsh weather with heavy wind and rain. They flew in close formation, which was intended to make it easier to protect the group from any potential Italian airplane attacks.

Escadrille No.	Pilot	Navigator	Gunner
302	pbb Ilk D. Stanić	pf R. Jovanović	pf M. Karić
306	pbb Ilk M. Protić	Army Officer Živković*	pf V. Šijaković
307	nv Ik I. Koroša	pf Ž. Filipović	nar M. Bačić
308	pbb Ilk I. Beran	pf B. Ivković	por F. Lolić
309	pbb Ik V. Petrović	pbb Ilk L. Zobundžija	por N. Batalo
311	nv IIIk R. Besarabić	pbb Ilk M. Kocijančič	nar J. Kikanović
312	pbb Ik A. Grošelj	ppuk U. Popović	maj V. Simić
313	ppor L. Ankon	pf M. Vales	nar E. Ceraj

* Joined the crew during the second attempt to fly over to Corfu

The 25.HE CO, pbb Ivo Rebula flew his Do 22 304 to Kumbor and allegedly attempted to fly over to Corfu but was forced to turn back due to poor weather. He never reattempted to escape to Greece with the rest of the crews.

While landing at Corfu harbor at 07:05 after their two hour flight, Do 22s 302, 307, 308, 309 and 311 were mistaken for enemy airplanes and anti-aircraft fire opened up. Fortunately, they were identified shortly as friendly and anti-aircraft fire subsided without any of them being hit. They were able to land approximately five minutes later. Crews were greeted by the Greek Commander of Corfu and Yugoslav consul. They were allowed to rest in the best hotel on the island before their next leg of the journey.

306 returned to Boka Kotorska as it stalled en route and crew lost orientation. It took off again at 14:00 together with 312 and arrived at 15:40 to Corfu with 312 following at 16:00. 313 made an emergency landing off the coast of Albania due to overheating engine. The Albanian forces opened fire, which forced the crew to distance themselves from the coastline while cooling down the engine. The crew was finally able to arrive to Corfu at 15:45.

After an overnight stay at Corfu, on 17 April Dorniers continued their journey at 09:20 towards Patras where they landed to refuel at 10:35. Crews were again greeted by the locals with music, a speech by the Municipality President and were told they could eat anywhere they liked free of charge, including buying

309, still wearing the original camouflage during the early 1941 at Aboukir, taxing for an anti-submarine mission. Note the RAF Sunderland in the background. [Aleksandar Ognjević]

This poor quality photograph shows 312 not long after its arrival to Egypt. [Mario Raguž]

King Peter II paid a visit to Yugoslav airmen on 7 June 1941, inspecting the line up of floatplanes at the beach. [Flight Magazine]

PV airmen pulling the 309 to shore. Second from the left is pbb Ilk Beran who flew the 308 to Aboukir. Pf Karić is closest to the floatplane. On the right side the first person is pf Jovanović and the third person is Por Franjo Lolić. [Mario Hrelja]

306 armed with RAF Mark VIII 250 lb anti-submarine bombs prior to a mission. [Boris Ciglić]

306 taxiing in choppy waters of Aboukir bay. Prolonged exposure to saltwater induced corrosion and made maintenance difficult. [Bojan Dimitrijević]

306 taking off on a mission. [Bojan Dimitrijević]

Two Do 22Kjs from No. 2 Yugoslav Squadron in formation flight over the Mediterranean. [Mario Hrelja]

307 resting in Aboukir bay while crews are performing checks. [Mario Raguž]

Unidentified Do 22 armed with anti-submarine bombs taxiing back to the base following a long mission over the Mediterranean. [Djordje Nikolić]

Two officers, pilot and an observer, still wearing their Yugoslav issued flying outfits pose in their dependable Do 22, Aboukir, Egypt. [Danijel Frka]

Mechanics going for a swim next to Do 22 309. Radovan Kljajić from Bajina Bašta is third from the left. [Aleksandar Ognjević]

new clothes. Despite these opportunities, crews worked tirelessly to refuel their Dorniers before continuing with their journey. That afternoon at 17:15, Do 22s left Patras and flew over to the Greek naval base at Salamina, where they arrived 45 minutes later. On 20 April the British instructed them to fly to Souda Bay, Crete. At Salamina, the Greek Do 22 base, Dorniers were refueled and the oil and water were topped off. Athens RAF Commander assigned an officer who was familiar with recognition signals to the Yugoslav group in order to ensure safe passage to Crete and to Aboukir.

On 19 April, Dorniers were still being serviced, so their crews used this time to visit Athens and to purchase some last-minute necessities prior to leaving the Greek mainland.

The next day, 20 April, at 07:25 all Dorniers took off except 313, which experienced a malfunctioning engine during take off, but were caught in strong anti-aircraft fire over Pireus aimed at German *Messerschmitt* Bf 110s which were bombing the harbor. Having dropped just four bombs, Germans retreated after seeing the Yugoslav group. As a precaution, all Do 22s returned to Salamina for inspection and took off again at 07:35. Just under two hours later, they landed at Souda Bay base in Crete. Fortunately, the RAF Officer signaled the local defenders and Dorniers avoided being "greeted" by the RAF *Gloster* Gladiator squadron based on the island. While performing general maintenance, they witnessed an evening raid by three Bf 110s from extremely low altitude.

On 21 April, bombings continued but this time by Italian airplanes. Due to the frequent air attacks the fuel, although lower octane, finally arrived that day. Crews were busy modifying the engines for proper operation with this type of a fuel. The next day, 22 April, all seven Do 22s took off at 08:30 for Aboukir where they landed at 11:50 having traveled around 790 km. Aboard pbb Ik Petrović's Dornier 309, a British Officer was boarded in order to properly signal the Allied forces prior to landing and avoid friendly fire. 302 suffered an accident due to a pilot, pbb IIk Stanić, error as he forgot to adjust the propeller pitch for landing. As a result, his Dornier suffered 35%, damage but fortunately the crew was left unharmed.

On 23 April Do 22 313 took off from Salamina for Crete with Greek observer on board. Four days later, on 27 April 313 finally arrived at Aboukir and joined the remainder of the group.

Landing at Aboukir in Egypt at the end of April and having covered a distance from Boka Kotorska to Alexandria of around 2,000 km was a great achievement. Immediately upon arrival, the RAF intended to take over the Yugoslav floatplanes and to intern the crews in Aman, Jordan. Yugoslav crews did not take lightly to this idea, they all wanted to continue fighting under the leadership of pbb Ik Vladeta Petrović. Two orders from the Government in exile were received to the effect of handing over Do 22s, but they were both ignored.

Date	Take off	Landing	Flight time	Airplane number	Take off from	Land at
			Timetable of Royal Yugoslav Do 22 retreat to Egypt			
16 April 1941	5:00	7:10	2:10	302	Orahovac	Corfu
16 April 1941	5:00	6:00	1:00	306*	Orahovac	Corfu
16 April 1941	5:00	7:10	2:10	307	Orahovac	Corfu
16 April 1941	5:00	7:10	2:10	308	Orahovac	Corfu
16 April 1941	5:00	7:10	2:10	309	Orahovac	Corfu
16 April 1941	5:00	7:10	2:10	311	Orahovac	Corfu
16 April 1941	5:00	15:45	2:10	313**	Orahovac	Corfu
16 April 1941	14:00	15:40	1:40	306	Orahovac	Corfu
16 April 1941	14:00	16:00	2:00	312	Orahovac	Corfu
16 April 1941	14:00	14:30	0:30	304***	Orahovac	Orahovac
17 April 1941	9:20	10:35	1:15	302	Corfu	Patras
17 April 1941	9:20	10:35	1:15	306	Corfu	Patras
17 April 1941	9:20	10:35	1:15	307	Corfu	Patras
17 April 1941	9:20	10:35	1:15	308	Corfu	Patras
17 April 1941	9:20	10:35	1:15	309	Corfu	Patras
17 April 1941	9:20	10:35	1:15	311	Corfu	Patras
17 April 1941	9:20	10:35	1:15	312	Corfu	Patras
17 April 1941	9:20	10:35	1:15	313	Corfu	Patras
17 April 1941	17:00	18:00	1:00	302	Patras	Salamina
17 April 1941	17:00	18:00	1:00	306	Patras	Salamina
17 April 1941	17:00	18:00	1:00	307	Patras	Salamina
17 April 1941	17:00	18:00	1:00	308	Patras	Salamina
17 April 1941	17:00	18:00	1:00	309	Patras	Salamina
17 April 1941	17:00	18:00	1:00	311	Patras	Salamina
17 April 1941	17:00	18:00	1:00	312	Patras	Salamina
17 April 1941	17:00	18:00	1:00	313	Patras	Salamina
18 April 1941				Maintenance		
19 April 1941				Maintenance		
20 April 1941	7:35	9:15	1:40	302	Salamina	Souda Bay
20 April 1941	7:35	10:15	1:40	306	Salamina	Souda Bay
20 April 1941	7:35	11:15	1:40	307	Salamina	Souda Bay
20 April 1941	7:35	12:15	1:40	308	Salamina	Souda Bay
20 April 1941	7:35	13:15	1:40	309	Salamina	Souda Bay
20 April 1941	7:35	14:15	1:40	311	Salamina	Souda Bay
20 April 1941	7:35	15:15	1:40	312	Salamina	Souda Bay
21 April 1941				Maintenance		
22 April 1941	8:30	11:50	3:20	302	Crete	Aboukir
22 April 1941	8:30	11:50	3:20	306	Crete	Aboukir
22 April 1941	8:30	11:50	3:20	307	Crete	Aboukir
22 April 1941	8:30	11:50	3:20	308	Crete	Aboukir
22 April 1941	8:30	11:50	3:20	309	Crete	Aboukir
22 April 1941	8:30	11:50	3:20	311	Crete	Aboukir
22 April 1941	8:30	11:50	3:20	312	Crete	Aboukir
23 April 1941	unknown	unknown	1:40	313	Salamina	Souda Bay
27 April 1941	unknown	unknown	3:20	313	Crete	Aboukir

*Return to base, stalled en route
**Engine overheat, landed off coast of Albania before resuming flight
***Estimate, Do 22 returned to Orahovac, Boka Kotorska

Sastav Jugoslovenskih Pomorskih Snaga (Yugoslav Naval Group) was formed eventually in Alexandria with submarine KB *Nebojša*, two torpedo boats KB *Durmitor* and *Kajmakčalan* and Dorniers. The floatplane group consisted of 29 flyers, meaning it was staffed well. The lack of technical personnel was soon apparent and so was the lack of spare parts. Do 22 306 carried most essential spare parts and tooling that could be

Rear view of the damaged 307. Waves are still pushing the floatplane further ashore. [Mario Raguž]

Following the damage on 21 January 1942, 307 was disassembled and transported back to base for repairs. Of particular interest is the random camouflage applied on the top wing as well as the new RAF style roundel. [Mario Raguž]

With floats off, crews are lifting the 307 by the struts as the flatbed trailer rolls in. [Mario Raguž]

With high tide coming in, 307 is finally ready for transport back to the base. A RAF Commer Q2 Tractor seen here is about to pull the flatbed trailer with Do 22. [Mario Raguž]

Taking a break after loading the 307 wings onto a RAF flatbed trailer. Damage to the port wing is clearly visible. [Mario Raguž]

Bomb laden 313 moments before heading out on a mission. No less than three 250 lb bombs are loaded. [Mario Raguž]

Yugoslav airmen posing during the break in between work on a Do 22. [Robert Čopec]

brought on board however that would not be enough for sustained operations. Some spare parts were also stripped from the Greek Do 22 N27 which crashed on its flight from Aboukir to Dekheila Naval Base. Ten French and British mechanics were assigned to help out with maintenance and repairs. At first Yugoslav airmen were based in Alexandria which meant they traveled daily to Aboukir to work on their floatplanes.

RAF evaluated that Yugoslav Do 22s were suitable for reconnaissance and anti-submarine warfare. As a result, they were attached to the No. 201 Group. To make communication smoother, a British liaison officer was assigned. On 7 May two Dorniers, 306 and 311 were on constant readiness at Alexandria harbor

where they were supposed to take off daily and monitor the area in the immediate vicinity of the base and the east Mediterranean Sea. The Alexandria harbor was congested with naval traffic making it extremely difficult for Do 22s to operate in such conditions. In spite of this, 311 led by pbb Stanić, pf Karić and nar Ceraj conducted the first wartime reconnaissance missions from Egypt, which lasted two hours.

In the next 12 days, Dorniers completed a total of 14 missions lasting 31 hours and 30 minutes. On 14 May, RAF handed over a mobile airplane workshop which was evacuated from the VVKJ air base at Skopje, Macedonia, to Thessaloniki, Greece and then finally to Alexandria aboard HMS *Port Halifax*. On

No. 2 Yugoslav Squadron airmen and mechanics pose for a photo in front of the Do 22 undergoing engine maintenance. [Mario Raguž]

Do 22 wing repair in field conditions. [Mario Raguž]

Do 22 undergoing engine repair. Note the roundel underneath the port wing and Yugoslav flag across the entire rudder. [Mario Raguž]

307 undergoing scheduled maintenance in the early 1942. Note severe peeling of the green paint which indicates photo was taken towards the end of its service. [Mario Raguž]

313 carefully maneuvered back into the sea. [Bojan Dimitrijević]

313 being prepared for lowering back in the water. Note the randomly applied green paint on the float surfaces is peeling. [Mario Raguž]

A Do 22 crew member carried on shore piggy back style by one of the mechanics. [Mario Raguž]

15 May a minor incident took place. While coasting in the busy harbor, 306 struck a tug and damaged its propeller. Because of poor operating conditions in the Alexandria harbor, the group relocated on May 19 to Aboukir. Crews were forced to stay at local hotels until adequate accommodation was arranged.

British No. 103 Maintenance Unit was also located in Aboukir and it supplied the Do 22s with fuel and other war materials. Friendly relations between the British and Yugoslav soldiers resulted in No. 103 Main- tenance Unit handing over several containers, which were used to tie the Do 22s to, and their interior was repurposed into workshops. Additionally, five ex RAF vehicles were handed over, which improved mobility of the Yugoslav Group.

The floatplane Group belonged to the KM until 1 June, when it was decided that it had to report to the VVKJ. The VVKJ CO visited the Group on 6 June and the very next day King Peter II paid a visit as well, inspecting the line up of floatplanes at the beach.

On 17 June floatplane Group was renamed to 2. Yugoslav Squadron Air Wing however in 230. Squadron 201. Group it was known under several different names. Within the 230. Squadron, the Yugoslav Dorniers op-

PV officers in Egypt having a chart prior to a patrol flight. Do 22 313 is visible in the background. [Aviation Museum – Belgrade]

No. 2 Yugoslav Squadron airmen lined up for inspection. [Aleksandar Ognjević]

erated jointly with the British Sunderland seaplanes. British codes were assigned to each Do 22, however these were never applied and remained only on paper.

British code	Yugoslav Escadrille No.
AX708	302
AX709	306
AX710	307
AX711	308
AX712	309
AX713	311
AX714	312
AX715	313

Pbb Ik Vladeta Petrović wrote in his combat diary:

Eight floatplanes resting in clear blue sea, freshly painted workshop containers, mobile workshop, two trucks used for pulling the floatplanes out of the sea, mast for raising the Yugoslav flag, volleyball court – all on white sand surrounded with tall palm trees and warm sun, offered comfortable and slightly romantic aspect of the No. 2 Yu-goslav Squadron. The second stage of the plan formulated prior to leaving the homeland was realized. Hard work by the entire land and air staff granted us the recognition and trust by the 201. Group. We were approved to take action in the front lines.

One of the first actions undertaken from Aboukir was a successful sea rescue of a Blenheim crew between 26 and 27 June. Yugoslav crews found the small dingy with four surviving crew members approximately 190 km offshore. They were unrecoverable on the first day, so one Do 22 returned the next day and dropped food and water packages while the other guided British rescue ships to the area. By the time they arrived, only one crew member was still alive and after recuperating at the hospital he made sure to pay a visit to his rescuers expressing his gratitude.

According to the order issued by the 201. Group CO, on 28 June Yugoslav Do 22s were used for calibrating AME (Air Ministry Experimental) radar sta-

No. 2 Yugoslav Squadron airmen in front of the 313. Shape of the covered engine indicates that the panels were removed for maintenance. Photograph was most likely taken in autumn 1941. [Mario Raguž]

No. 2 Yugoslav Squadron was handed over ex RAF vehicles which improved mobility. [Mario Raguž]

tions. These uneventful calibration flights continued through July by either one or two floatplanes flying at any given time. Most of these flights were routine and without incidents.

Do 22 312 was mistakenly shot at during calibration of anti-aircraft defenses around Alexandria. During this incident, 312 fortunately sustained no damage and returned safely to base. On 9 July, another mishap happened during calibration of anti-aircraft defenses when all gun batteries protecting the Alexandria harbor opened fire on one Do 22, this time again without major consequences. As a result of these incidents, "Kosovo Cross" insignia was removed from the wings as it resembled the German cross. On the wings, RAF style roundels were painted while the vertical stabilizer flag was reduced in size. Despite this on 14 October the entire British fleet opened fire on two occasions at the Do 22 detachment, which was tasked to protect it in the first place!

In Aboukir, the Yugoslav mechanics were able to modify the starboard float to carry fuel as well, which increased endurance during the already long Mediterranean Sea patrols. The total volume increased to 1,400 liters. Allegedly float reservoirs were donated by the Greeks, but it is not clear how they were transported from the Greek mainland. The true reason for this modification was the secret initiative by a group of KPJ supporters to fly the Do 22s over to Sevastopol, Soviet Union, since German forces did not capture Crimea yet. German advances on the front prevented any further discussion and the idea was finally abandoned. Instead, one of the Do 22s in this configuration was used to locate a Greek submarine in distress which dropped anchor 65 km east of Alexandria. Three major searches were also undertaken, for Turkish steam ship on 10 July, two shot down *Fairey* Swordfish on 16 and 17 August as well as *Bristol* Beaufighter, which crashed into sea on 26 August. During these flights, on 24 July, British journalists flew with the Yugoslav Do 22 crews recording video material from these flights, which survives to this day in British Pathé. Proudly displaying KJ flag, crews staged a typical flying day, leaving their barracks, which were adorned with the Squadron crest in English language, boarding Do 22 308, dropping anti-submarine bombs and finally landing back in Aboukir bay.

![One of the last five Do 22s, 309, seen here during maintenance]

One of the last five Do 22s , 309, seen here during maintenance. [Aleksandar Ognjević]

Do 22 313 at the end of its carrier with No. 2 Yugoslav Squadron. [Djordje Nikolić]

The only surviving part of Yugoslav Do 22 is the nameplate from 306 showing the W.Nr. 757. [Mileta Protić]

In July, Do 22 309 suffered a malfunction and on 26 August 311 (nv Ik Koroša, pf Filipović and nar Kikanović) crash landed due to engine seizure. The pilot attempted to land but struts and floats broke upon contact with the water. The crew survived and was rescued by British Sunderland shortly before midnight. 311 remained afloat and was protected by a British gunboat, however due to the damage inflicted during the crash landing, the next morning it sank without a trace. Until that incident, British records indicated that the engine in 311 had a total of 276.35 working hours.

Another loss of a Yugoslav Do 22 occurred on 20 October. Shortly before 06:00 312 took off to search for a downed *Bristol* Blenheim with pilot nar Omišl, observer pk Skopal and gunner nv Božić on board. During the flight, the engine malfunctioned and pilot attempted an emergency landing 20 nautical miles west of Alexandria and six nautical miles from the coast. Despite a hard landing, the Do 22 remained afloat and the second Do 22 306 circled above it radioing their position to base before heading back to refuel. At that time, a second pair of Do 22s, one of which was. 307, was sent up to investigate. The crew was finally picked up at 08:30 by destroyer HMS *Hotspur*. It proved impossible to tow Do 22 back to base and as a result the commander of HMS *Hotspur* ordered it to be sunk. Shortly thereafter, following hits inflicted by the machine gun fire, it sank to its final resting place.

Due to damage incurred while landing at Alexandria from Corfu in April 1941, 302 was removed from the register as late as 1 November. This process took long most likely due to the bureaucracy and because 302 was used as a source of spare parts. Once all usable parts were stripped, it became useless.

The latest two accidents proved that without adequate spare parts and tools it was difficult to keep engines serviceable and crews safe during long flights over the Mediterranean Sea no matter how hard the maintenance crews worked round the clock. The problem manifested itself especially because the Do 22 was constructed in accordance to the metric system of measurements whereas RAF used the imperial system. This meant that even simple screws or rivets could not be used from RAF stocks, so Yugoslav crews were forced to strip these parts from the shot down German airplanes found in the desert. In these meager conditions the maintenance crews managed to perform 20 minor engine repairs and 15 major repairs, which involved engine block disassembly Do 22s were used in a climate

for which they were never designed, so engine cooling proved troublesome and the lack of sand filters affected the engine operation as well. Additionally, their engines and fuselage were constantly exposed to saltwater as Do 22s were moored close to shore and seldom brought out, which induced corrosion. In September, the overhaul of 309's engine was completed with the floatplane laid down to sea on 24 September. The very next day, 308 and 313 were brought ashore and were serviced until November. The next in line for maintenance was 307, which was damaged by anti-aircraft fire from Norwegian steamship *Harbo Jensen* on 24 November at 07:00 am. Gunner nar Kikanović was seriously injured and pilot nar Pišpek immediately returned to base.

Through 1941, pilots arriving from KJ which were not entirely trained on Dornier Do 22 received necessary training and also attended English language lessons. It is interesting to note that two British pilots, Pilot-Officer Gordon and Flight-Sergeant Sumison also completed Do 22 flight training.

In 1942, No. 2 Yugoslav Squadron continued with anti-submarine reconnaissance. The frequency of flights slowed down somewhat and on average only one flight per day was conducted lasting no longer than two and half hours. No significant discoveries were made during those flights. On 21 January a major sandstorm wrecked havoc, but the Yugoslav

A scan of the Do 22 painting which to this day hangs on the No. 230 Squadron history wall. [Via Andrew Thomas]

Year	Month	No. flights	Total hours
1941	June	107	190 h 35 min
	July	61	180 h 30 min
	August	45	105 h 10 min
	September	45	96 h 10 min
	October	54	113 h 15 min
	November	72	159 h 40 min
	December	86	203 h 10 min
1942	January	45	111 h 10 min
	February	36	102 h 35 min
	March	34	74 h 45 min
	April	43	102 h 00 min

Mission	1941						
	June	July	August	September	October	November	December
Reconnaissance of Alexandria bay	58	46	31	30	31	29	48
Anti-aircraft fire calibration	4	12	3	5	4		
Pilot training	28						
Test flights	7	7	6	3	1	6	10
Sea rescue	10	4	5	6	10		
Photo-reconnaissance				1			
British war journalists		1					
Anti-submarine					3	29	28
Bombing practice						1	
General training						5	
Over flight						2	
Test flight to Edku lake					2		
Anti-aircraft recognition					1		
Reconnaissance					1		

Mission	1942			
	January	February	March	April
Reconnaissance of Alexandria bay	33	8		
Test flights			5	1
Anti-submarine	12	28	29	42

crews managed to spare all but one of their Do 22s from major damage. During that sandstorm, the ropes tying 307 broke off which caused it to come ashore damaging its port wingtip. By March the only active Do 22 was 309 with crews rotating each day. On 26 March 26 and 31 March, 313 and 307 joined the flight line-up after post-maintenance and repair test flights were successfully completed.

In April, flying was intensified with two flights taking place each day; one being in the early dawn and the other in the afternoon with the same three Do 22s used in March conducing all of the flights.

On 16 April, 309 was attacked by two Junkers Ju 88s, 55 miles from Alexandria. Following a ten minute long dogfight, Do 22 returned to base without damage.

Due to the so called Cairo Affair, which was the result of changes implemented at the top of the Yugoslav government in exile as well as within military ranks in Egypt, the last flight of Yugoslav Do 22 from No. 2 Squadron took place on 22 April with 307 (pf Ankon, por Karadžić and nar Ik Bačić) patrolling 10 miles outside of Alexandria. Crew returned after two and half hours having spotted one friendly convoy and no enemy. The very next day, No. 2 Squadron was disbanded, the men left 230. Squadron on 27 April and were sent to Abbassia camp near Cairo. The remaining five Do 22s, 306, 307, 308, 309 and 313, were pulled ashore, handed over to the No. 103 Maintenance Unit and they were thought to be scrapped. However, recent information indicates that at least one of these Dorniers survived a little while longer. According to the official RAF airplane movement cards, Do 22 307 was brought back on charge on 1 December 1942 for an unknown reason and unknown period of time. Until further information or photographs become available the exact fate of PVKJ Dorniers remains a mystery.

No. 2 Squadron performed combat flights for an entire year with high intensity. A total of 737 combat missions were accomplished with total a duration of 1,692 hours and around 204 non-combat flights lasting 74 hours. A total of three Do 22s were lost (. 302, 311 and 312). No. 2 Squadron was active for a total of 352 flying days and on average each pilot flew 180 hours.

It is important to note the No. 2 Yugoslav Squadron crews retreated from Europe with their equipment, leaving their loved ones behind. They continued to fight relentlessly against the Axis, which is a feat worthy of remembrance and gratitude.

Post War

Recent information indicates that a sole part from a PVKJ Do 22 survived the war. A nameplate from 307 was removed from the airplane by one of the PVKJ airmen, likely pbb IIk Mileta Protić and it was shown in a documentary video. After the war, the FNRJ (Federative Peoples Republic of Yugoslavia) asked for war reparations for 12 Do 22s at 2.8 Million Dinars per floatplane which was 150% increase of pre-war value. It is uncertain if this sum was ever paid out effectively, bringing closure to any official discussion concerning the Do 22 in Yugoslavia.

Top secret! Do 22 manual intended for Yugoslavia. [Airbus Corporate Heritage]

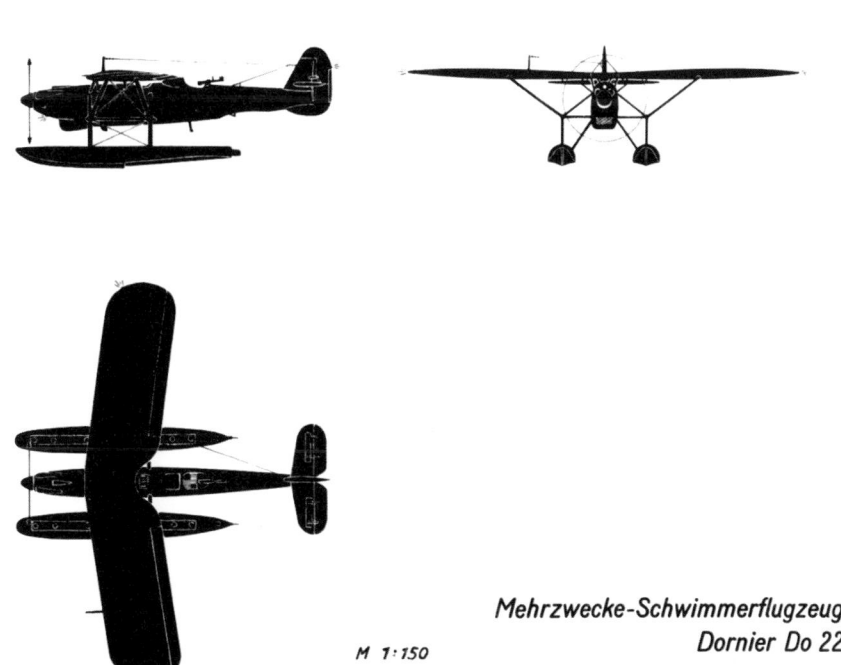

Mehrzwecke-Schwimmerflugzeug
Dornier Do 22

Original Do 22 three view drawing in scale 1:150. [Djordje Nikolić]

Observer's instrument panel was simple, featuring only basics flight instruments. The oxygen system bottles for the pilot and the observer are visible on the right hand fuselage wall. [Michel Ledet]

Do 22Kj pilot's cockpit. Instruments are labeled in Serbo-Croat language. Note the speaking tube on the right hand side fuselage wall and the float rudder pedal below the instrument panel. [Michel Ledet]

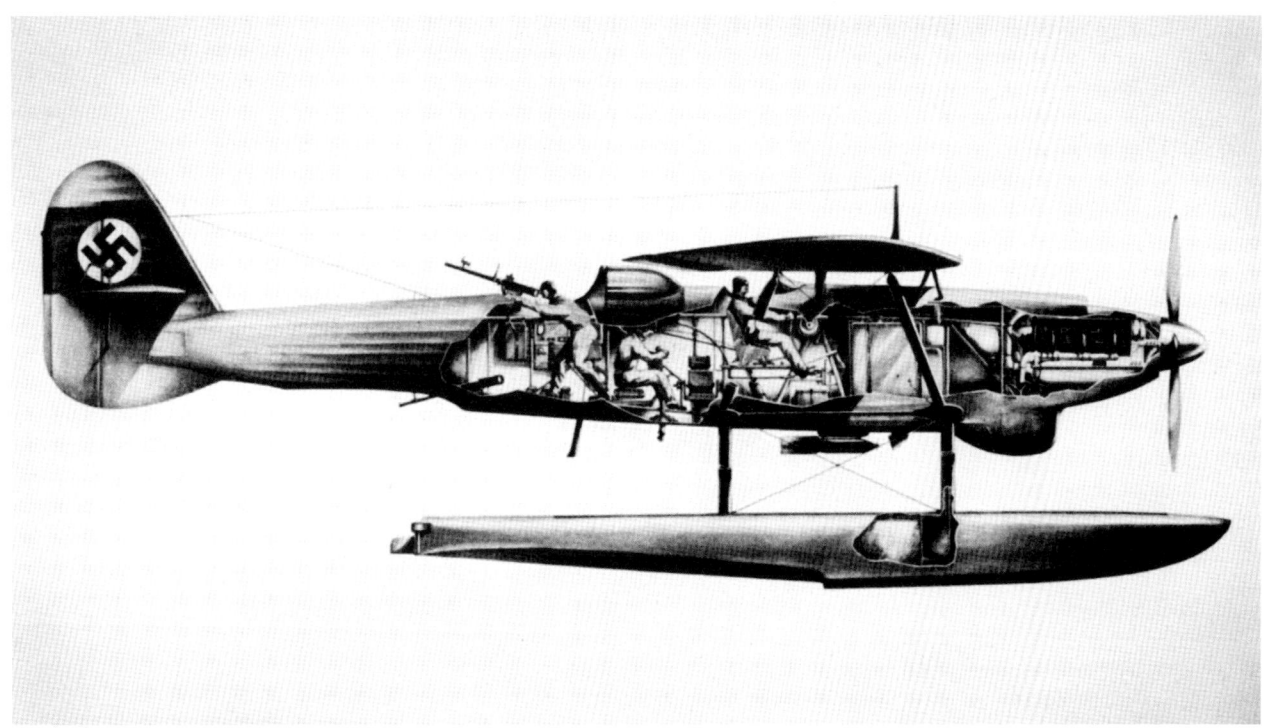

Cutaway from the original sales brochure for Do 22 dating back to 1939. Note that Do 22 has a German swastika on its tail, even though it was only sold for export. [Djordje Nikolić]

Construction Features

The cockpit offered the three man crew a spacious working environment and good visibility. The pilot, observer and the machine gunner were seated in tandem near the center of gravity. The cockpit layout was conventional, and purpose built with the instrument panel in front of the pilot's seat. The instrument panel was laid out in accordance with typical German practice of the time by grouping the navigation, flight monitoring and engine instruments in clusters. Instrument labeling was provided in the native language of the export customer to facilitative easier operation. For in flight communication each crew member had a speaking tube at the left-hand side of their station. The pilot was protected from head wind by a windshield constructed out of Plexiglas. While flying, the pilot had the ability to operate the forward-facing machine gun, which was synchronized with the propeller. On the left-hand side of the pilot's seat was an adjustment lever for raising or lowering the seat height. For steering in water, pilot coupled the control stick with the water rudder controls.

The observer sat behind the pilot and operated the camera, bomb sight and was the backup radio operator depending on the mission flown. Plexiglas hood protected the observer from the head wind. The whole hood assembly hinged to the starboard to allow for easier entry, and it could be jettisoned in case of an emergency. During reconnaissance missions, the observer used the camera to take photos through the sliding panel with Plexiglas window in the fuselage floor. During bombing missions, the observer was tasked with aiming the bomb sight through the same Plexiglas window. In case of an emergency the observer could take over flying by using the auxiliary controls.

The machine gunner sat at the very back in an open section of the cockpit and operated the top mounted rear facing machine gun and at bottom mounted machine gun for defensive purposes or ground attack respectively. The machine gunner also served the role of the main radio operator.

Do 22 was powered by a *Hispano Suiza* Type 12 Y21 12-cylinder liquid cooled engine with the maximum engine output was 800 hp at 2,100 RPM while flying at sea level and 925 hp at 2,530 RPM while flying at 3,600 m. Normal engine output was 575 hp while flying at sea level and 665 hp while flying at 3,600 m. On PVKJ Do 22s, the engine was coupled to a three-bladed variable pitch propeller manufactured by *Junkers* and licensed from *Hamilton*, type Ju H-P III. Propeller had a diameter of 3.4 m and a regulator was used to keep propeller revolutions constant by using pressurized oil to adjust propeller pitch automatically. The propeller rotated counterclockwise. The spinner was somewhat smaller in diameter than the forward fuselage to permit sufficient air flow to the engine

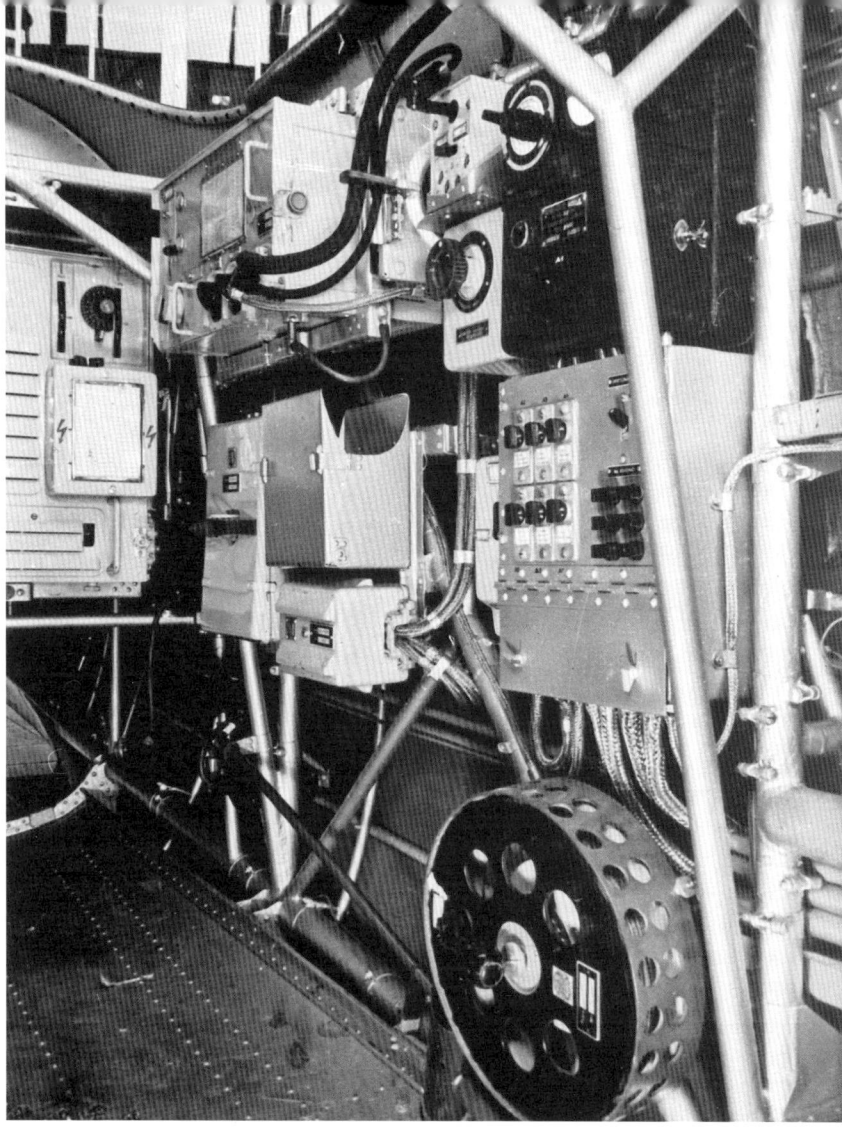

Left side view of the observer position. Drum housing the 70 meter radio antenna is clearly visible. [Michel Ledet]

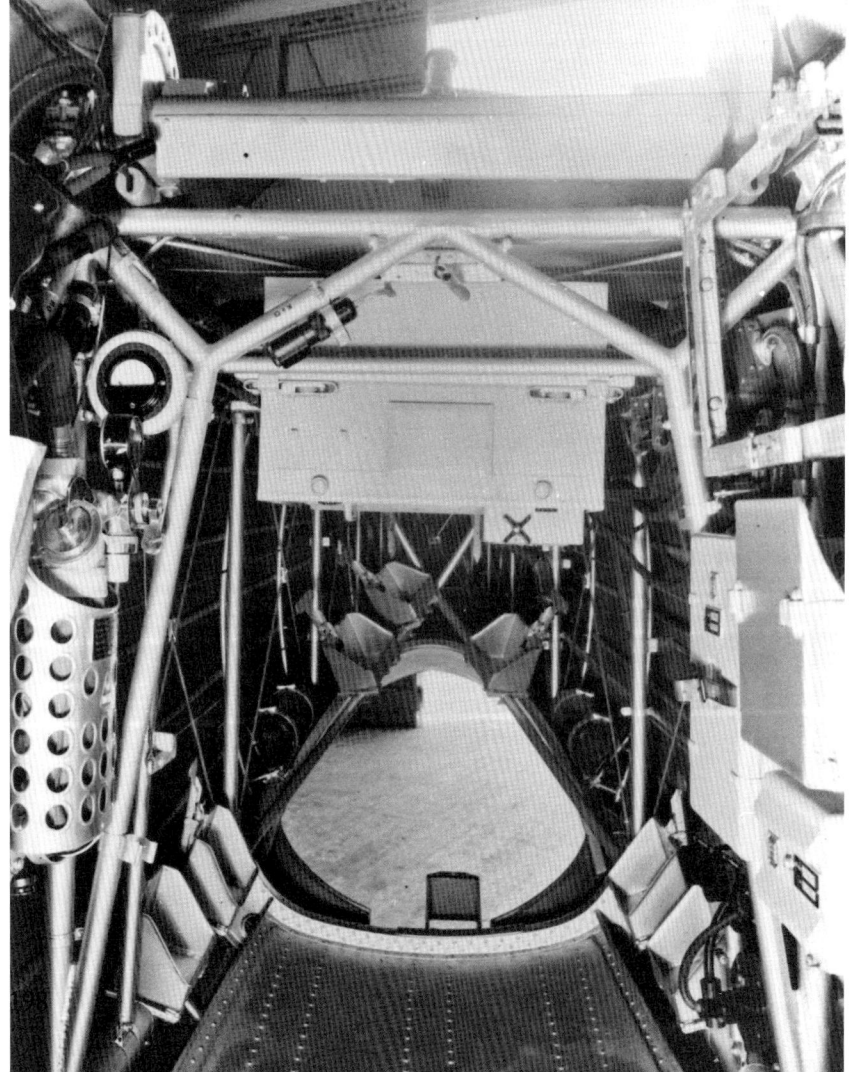

Aft view of the gunner's station. For protection from the lower rear hemisphere and for strafing ground and sea targets, the machine gunner operated the Browning FN (M.38) machine gun from a prone position. [Michel Ledet]

View forward of the gunner's station from the opening below the fuselage. Machine guns were not yet installed. [Michel Ledet]

Do 22Kj bomb rack arrangement. Ahead of the bomb racks is the emergency fuel dumping port. Behind the bomb rack is the observer's Plexiglas window used during photo reconnaissance and bombing missions. [Michel Ledet]

Top view showing the three man cockpit. Compared to the prototype, the observer's hood on the production Do 22 was more streamline and spacious. [Michel Ledet]

Hispano Suiza 12 Y21 12-cylinder liquid cooled engine. [Luftfahrt-Archiv Hafner]

for additional cooling. Engine exhausts were vented from both the port and starboard side via six exhaust tips. Yugoslav floatplanes had their exhaust manifolds removed sometime after the delivery, which is evident from the pre-war photographs and those from Egypt. It is interesting to note that the Do 22 302 had a straight manifold which followed the fuselage and vented below the machine gunner's cockpit indicating that this was

Engine access panels for servicing were well laid out. If needed, mechanics could stand on the engine panels while performing the necessary repairs. [Luftfahrt-Archiv Hafner]

The wing structure is clearly visible in this factory photo. Wing consisted of two spars stiffened by the 78 transverse beams, 41 ribs and bracing cables. [Airbus Corporate Heritage]

Browning FN (M.38) machine gun installation on a Royal Yugoslav Do 22 as evident by the starboard wing marking. Greek Do 22 N27 is seen in the background with a Do 22 wing lifted above the factory floor. Also, several Do 17 tail assemblies are visible at the busy factory. This scene was described by Flight Magazine reporter H.K. King during his visit to Friedrichshafen in 1938. [Michel Ledet]

Forward facing FN Browning (M.38) installation on a Do 22Kj. With 300 round capacity, it allowed the pilot to strafe ground and sea targets. [Airbus Corporate Heritage]

one of the early manifold adaptations. Later Do 22s had their manifolds attached to the struts and vented away from the fuselage.

Engine fuel was supplied from the main fuel tank in the fuselage and from the auxiliary fuel tank in the port float. The main fuel tank was constructed out of aluminium and was installed in the front fuselage between bulkheads 1 and 2 in front of the pilot and it had a volume of 625 liters. The reserve fuel tank in the port float was located between bulkheads 10 and 12. It had a volume

of 400 liters. Fuel was only supplied to the engine from the main fuel tank. In the Do 22 prototype only, reserve fuel tanks were not installed in the floats but in the wings. Each wing housed two small aluminium fuel tanks with 110 liters volume. The oil tank had a volume of 70 liters and was mounted directly to the engine.

The cooling system consisted of two coolant tanks and a water radiator with regulating flap. Both coolant tanks were attached to the engine, one each on the port and the starboard side having a volume of 8 liters and 6.7 liters respectively.

The main structural component of the fuselage was a welded steel tube truss with square cross section. Four longerons were connected by 13 frames, which were partially stiffened by steel tubes and adjustable span cables. Steel tubes were completely closed off to prevent any moisture ingress, which could induce corrosion. The outer frame steel tubes were lined with fabric and fuselage exterior fabric surfaces were attached directly to them. In order to maintain aerodynamic profile, the framework was constructed out of oval duralumin framing which was connected to each other with longitudinal duralumin profiles. Fuselage was conserved with layer of Nitrocellulose varnish. Fabric covered surfaces were titanized three times then covered with Nitrocellulose varnish.

The wing arrangement defined the Do 22 as a high wing monoplane. The wing consisted of three sections, middle section and two outer sections. The center section of the wing was supported against the fuselage with four short struts which were connected to the fuselage frame. The front short V shaped struts were constructed out of oval tubing. The rear short struts were constructed out of round tubing and had streamline covers which conveyed aileron control cables from the fuselage to the wings. The two outer sections were supported by the two supports struts which connected at a bolted junction with the float struts. The junctions between the center and outer wing sections were covered with bolted metal strips. Wing framework was constructed entirely out of duraluminium. The wing consisted of two spars stiffened by 78 transverse beams, 41 ribs and bracing cables to resist bending and torsion. Both wings, just like the fuselage, were conserved with layer of a Nitrocellulose varnish. Fabric covered surfaces were titanized three times and also covered with Nitrocellulose varnish.

Do 22 rested on water on top of two 4,000 liter, floats which were attached to the fuselage and wings via struts. Each float was constructed from duraluminium framework which consisted of 24 frames and was completely enclosed with duralumin panels. An anchor housed in the float could be deployed in the water to maintain position. At the very end of each float, a water rudder was installed with a pivot point being on the float to which two sets of cables from cockpit were connected. The water rudders were coupled to rudder foot pedals for use when the floatplane landed on water. Each float was connected by one forward and rear horizontal, vertical, longitudinal and diagonal strut.

For fire protection, Do 22 was equipped with a fire extinguishing system Model W2 manufactured by the *Wintrich* company. Do 22s featured a high-altitude oxygen system manufactured by *Dräger* company. There were a total of three pairs of oxygen tanks, each oxygen tank had a volume of two liters.

The radio equipment in production Do 22Kj consisted of *Telefunken* 274 aF radio stations which used medium and short waves for transmitting radio signals and long wave signals for telegraph. Just like on the prototype a 70 m long antenna was stored in a drum which could unwind the antenna cable and extended it from the fuselage during flight.

Do 22 was armed with three machine guns. The pilot operated the forward-facing machine gun which fired through the propeller. The forward facing machine gun magazine had a capacity for 300 rounds. To protect the rear hemisphere from air attacks, the machine gunner operated two rear facing machine guns; one on top of the fuselage and one below. The machine gun below the fuselage could only be operated from a prone position. Both top and bottom rear facing machine guns were equipped with four cartridges, each with 75 rounds capacity.

Pilot		Machine gunner		
Machine gun	Gun sight	Machine guns	Turret	Gun sight
Browning FN (M.38)	Colimateur Chrétien E.T.A.e modelle 1933	Browning FN (M.38)	Do 1a	E.T.A.e

PVKJ equipped its Do 22 with *Škoda* M.28 50 kg and 100 kg anti-submarine bombs with hydrostatic fuses. These bombs were already in inventory and were adapted for use when the Do 22s arrived. To supplement its inventory with new weapons, on 16 October 1939, a special commission performed a study to find suitable crutches for M. 40 50 kg and M. 38 100 kg domestic bombs manufactured by *Vistad* (Višegradska industrija inž. Stanković AD). The *Vistad* factory supplied the crutches which were compatible with Do 100/300 and GPUM 100/300 bomb racks only. Being that the Do 22 used HTM 50/100 and BV 2C 50 bomb racks, it was necessary to modify the crutches to integrate *Vis-*

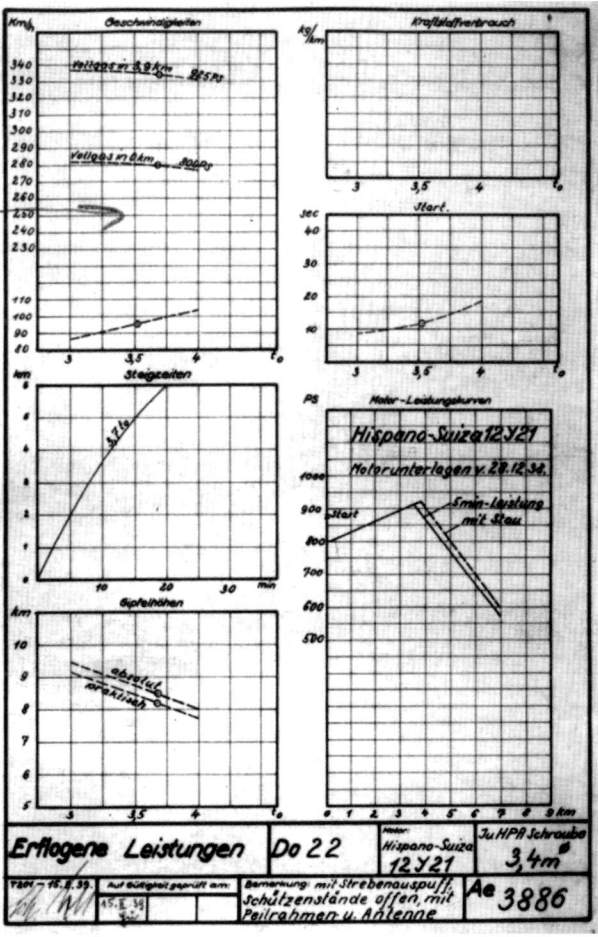

Factory chart showing maximum speed, power, climbing time and altitude of a Do 22Kj. [Dornier Museum Friedrichshafen (Airbus Group)]

tad bombs on Do 22. For bomb aiming in horizontal flight, The Do 22 used *Goerz-Boykow* bomb sight manufactured by *Akciova Společnost K.P. Goerz* from Bratislava. While using this bomb sight, light signals were displayed for the pilot to guide him to the target. For bomb aiming while diving, markings on the fuselage and floats were used to determine the dive angle. During the use in Egypt, Do 22s were adapted to carry *RAF* Mark VIII 250 lb anti-submarine bombs which were in use on *Short* Sunderland from 230. Squadron.

Do 22 prototype was equipped with 13 x 18 cm camera with a *Carl Zeiss* lens and a focal length of 70 cm. In production Do 22, F-75 or F-50 30 cm cameras were used. To operate the camera, the observer had to open the sliding cover at the bottom of the fuselage. The camera was mounted on a pivoting platform equipped with vibration dampeners. The entire assembly with the camera could be stored below the observer's seat when not in use.

The fog generator, *Šema* type, which was essentially a storage tank, could be carried by PVKJ Do 22s.

Technical Specifications

Technical specifications Dornier Do 22 prototype W.Nr. 259	
Quantity used:	1
Crew:	2 or 3 depending on a mission
Years of Service:	1935-1941
Span:	16.2 m
Length:	13.0 m
Height:	4.7 m
Wing area:	45.0 m²
Engine:	One 900 hp Hispano Suiza 12 Ydrs
Empty weight:	2,588 kg
Maximum weight:	3,350 kg
Maximum speed:	317 km/h
Cruise speed:	280 km/h
Service ceiling:	9,000 m
Maximum range:	930 km
Armament:	Unarmed. During April War received unknown defensive armament.

Technical specifications Dornier Do 22 Kj W.Nr. 753-790	
Quantity used:	12
Crew:	2 or 3 depending on a mission
Years of Service:	1938-1942
Span:	16.2 m
Length:	13.1 m
Height:	4.8 m
Wing area:	45.0 m²
Engine:	One 925 hp Hispano Suiza 12Y-21
Empty weight:	3,060 kg
Maximum weight:	3,700 kg
Maximum speed:	290 km/h
Cruise speed:	250 km/h
Service ceiling:	8,500 m
Maximum range:	950 km
Armament:	Three Browning FN (M.38) machine guns. M. 40 50 kg and M. 38 100 kg bombs or Mark VIII 250 lb anti-submarine bombs

Camouflage and Markings

Immediately following completion and during water taxiing trials, Do 22 prototype had its fabric surfaces painted in DKH Nitro enamel L40/52 *Silber* while metal panels were left bare. Shortly thereafter

DKH L40/52 color was applied which was standard for naval airplanes of the time period to the entire surface including the top of the floats. Contrary to the RLM 01 *Silber* practice, on lower float surfaces up to the high water mark, a Black anti-fouling finish was applied. Do 22 prototype was delivered to KJ without

Do 22 prototype and production airplanes were painted overall in DKH Nitro enamel L40/52 Jugograu. The rudder was entirely painted in Blue-White-Red Yugoslav flag colors. "Kosovo cross" insignia was applied on top and below the wings and Black 302 (later changed to 301 on the prototype shown here) on the fuselage. [Tomislav Aralica]

307 and 308 in Aboukir, Egypt, March 1942. RAF Dark Green was applied by brush randomly, rudder flag was painted over and replaced with a smaller one on the vertical stabilizer and the Yugoslav "Kosovo cross" insignia were painted over with RAF style roundels. [Aleksandar Ognjević]

any markings. After arrival, Blue-White-Red rudder flag and "Kosovo cross" insignia both on top and below wing surfaces were applied. The fuselage received a code number applied in Black paint.

According to the original Operations and Maintenance manual and as seen in the Appendix, Do 22Kj was painted entirely with DKH Nitro enamel L40/52 *Jugograu*. The designation *Jugograu* was adopted to distinguish this grey from the dozens of grey shades that were available and to identify the end user, even though the color was in essence identical to the DKH L40/52 *Grau* or as it became known as RLM 63. A different sheen between the metal and fabric covered surfaces is evident when looking at the available photographs. Do 22Kj fuselage codes were applied as Black numbers, 1 meter wide. The rudder was painted in Yugoslav Blue-White-Red flag and the wing topsides and undersides received the "Kosovo Cross" insignia. Neither of the Do 22s had any stencils applied. Floats and float struts were painted in DKH S 35/53 *Silber*.

Following the exile to Egypt, Do 22s had the RAF Dark Green paint applied by brush at random which

No 2 Yugoslav Squadron airmen studying a map prior to another mission. In the background are 306 and 309. Beneath the wing a new style roundel is clearly visible. [Muzej revolucije naroda Hrvatske 3907, OF-3907]

resulted in no two floatplanes being the same. To avoid confusing the Yugoslav insignia with the German cross, during September 1941 these were repainted with a RAF style Blue-White-Red or Blue-Red roundels, depending on the floatplane. It appears from the available photographs that Blue-Red roundels were used exclusively on the wing topsides. The rudder flag was painted over and instead on vertical stabilizer a smaller Yugoslav flag measuring 80 cm tall and 40 cm wide was applied.

302 seen here during factory testing. Note the difference in paint sheen between the fabric covered fuselage and metal covered tail surfaces. [The Museum of Flight]

W.Nr.	Escadrille Number
259	301
753	302
754	303
755	304
756	305
757	306
758	307
759	308
786	309
787	310
788	311
789	312
790	313

Dornier Do 22

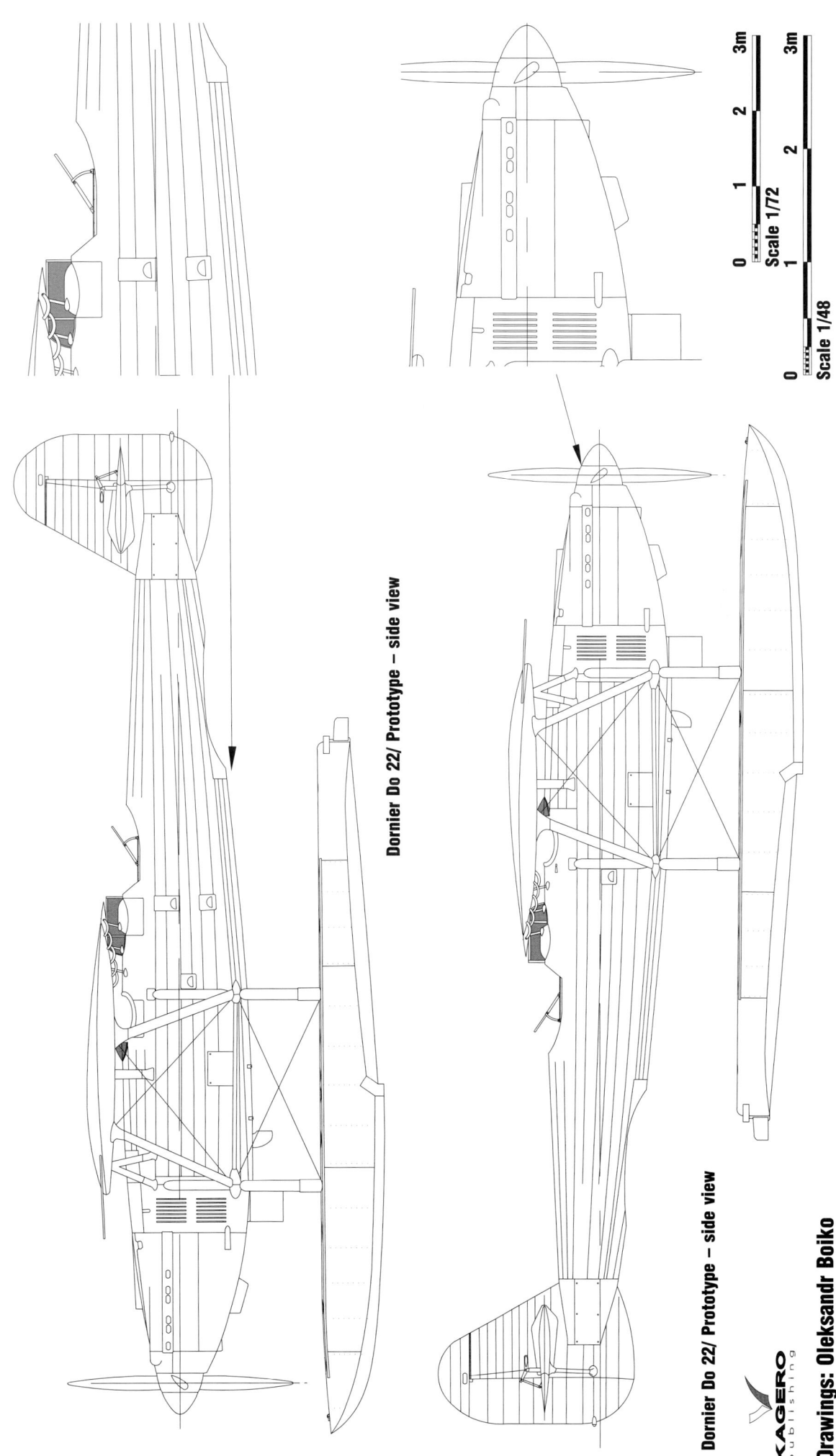

Dornier Do 22/ Prototype – side view

Dornier Do 22/ Prototype – side view

Scale 1/72

Scale 1/48

3m

3m

KAGERO
publishing

Drawings: Oleksandr Boiko

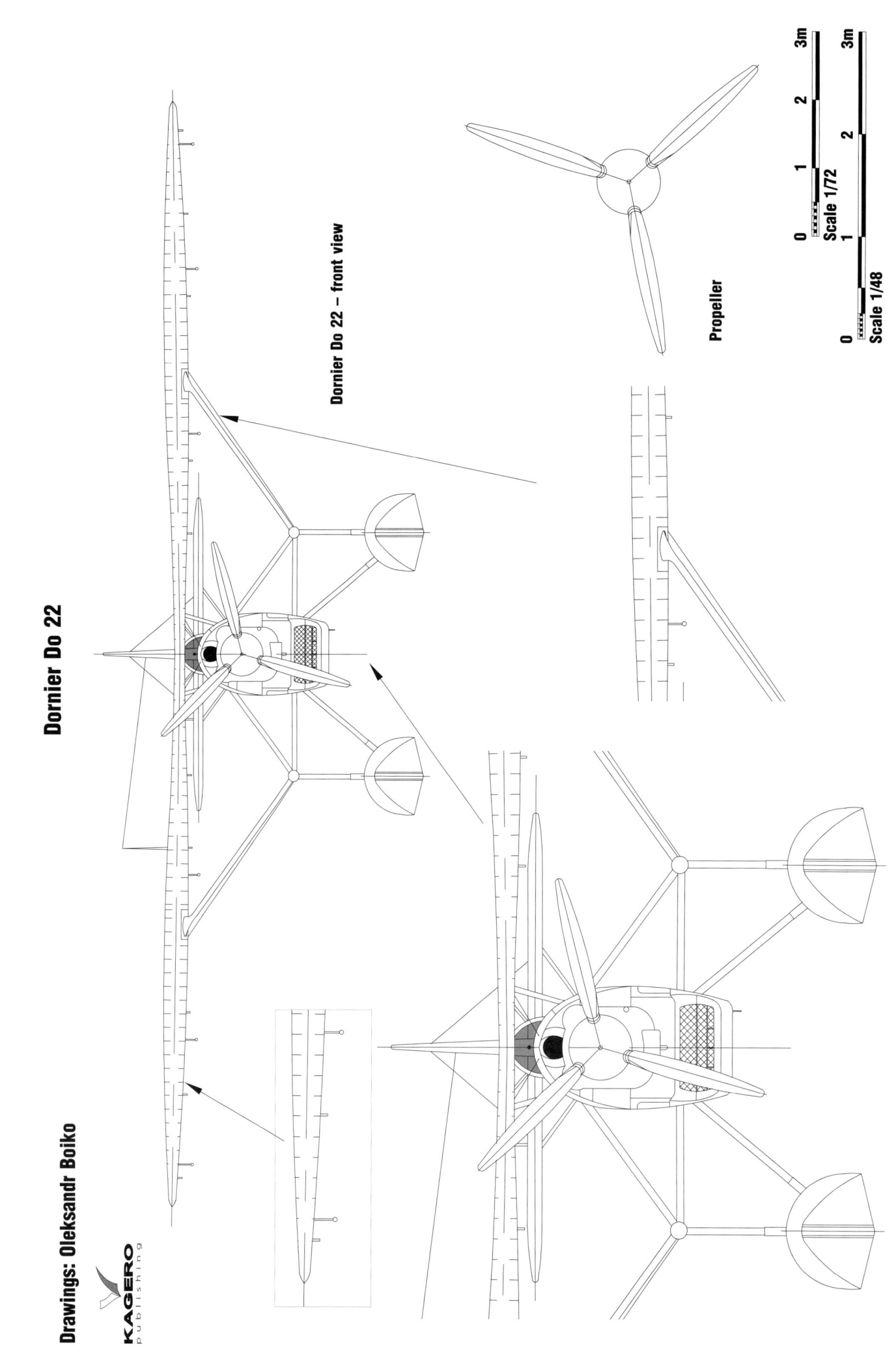

Dornier Do 22

Drawings: Oleksandr Boiko

KAGERO
publishing

Dornier Do 22 – front view

Propeller

Scale 1/72

Scale 1/48

Dornier Do 22

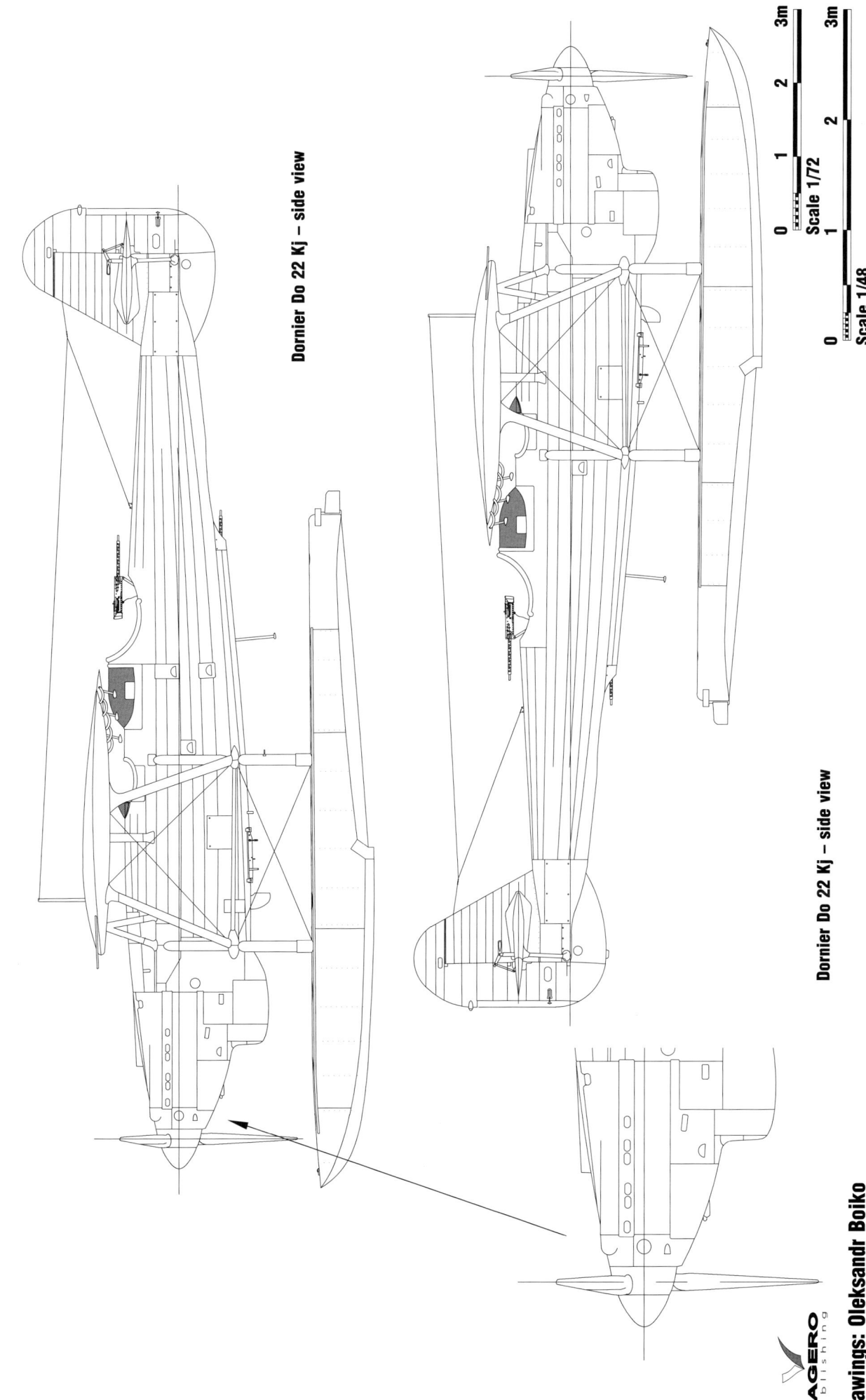

Dornier Do 22 Kj – side view

Dornier Do 22 Kj – side view

Scale 1/72

0   1   2   3m

Scale 1/48

0   1   2   3m

KAGERO publishing

Drawings: Oleksandr Boiko

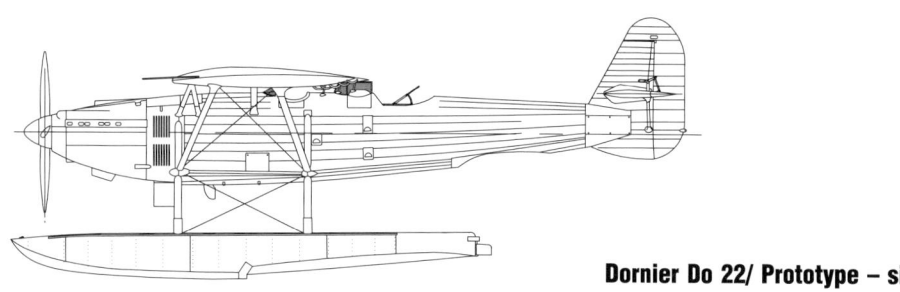

Dornier Do 22/ Prototype – side view

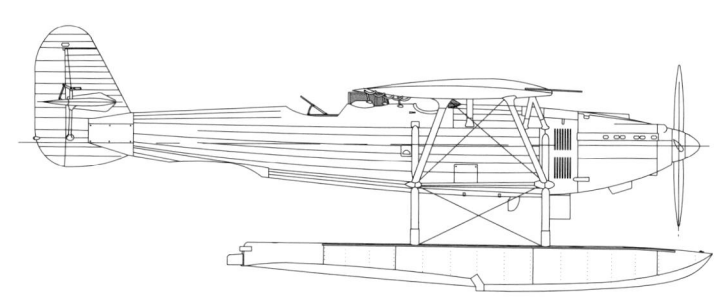

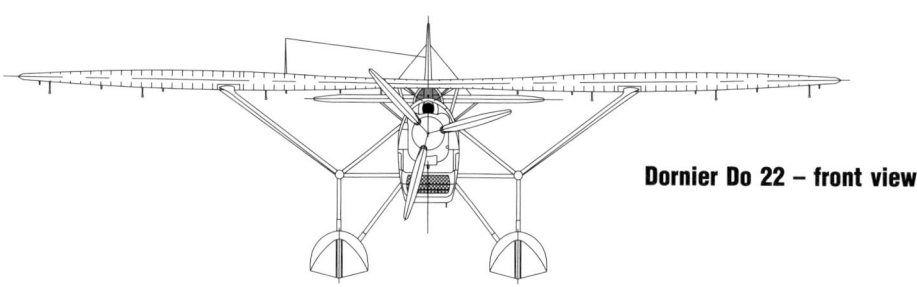

Dornier Do 22 – front view

Dornier Do 22 Kj – side view

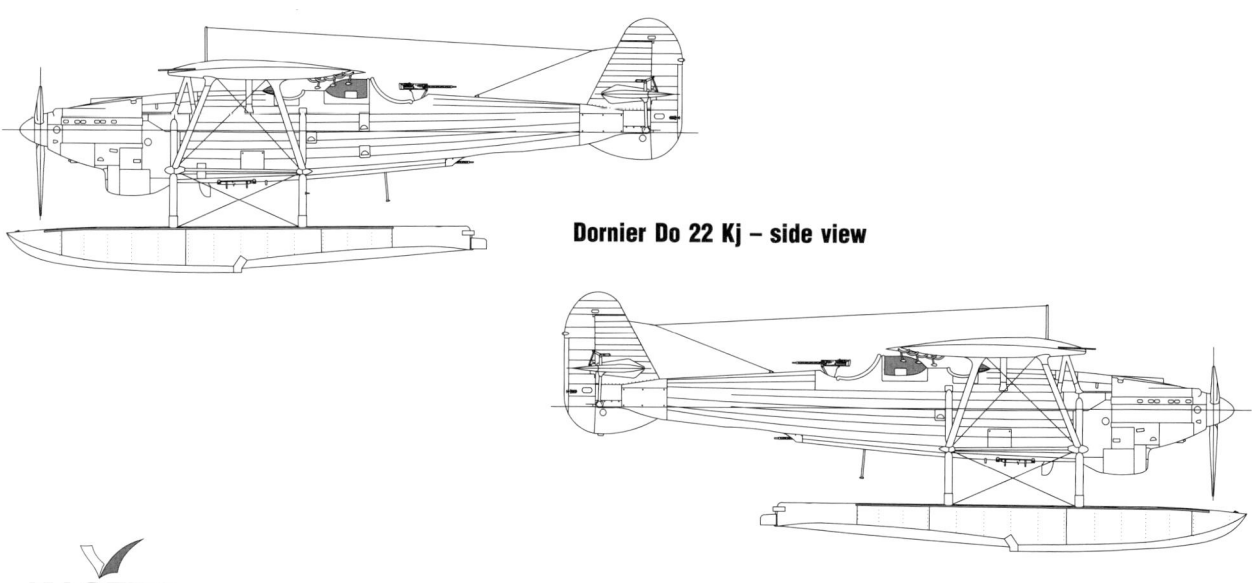

KAGERO
publishing

Drawings: Oleksandr Boiko

Scale: 1/144

Dornier Do 17K

Development

The origin of the Do 17 began in 1932. This year *Dornier Metallbauten GmbH* along with *Junkers* and *Heinkel* handed over the requirements of the Army Weapons Office (WaPr 8) for the design of a "twin-engined high-speed airplane". Following the rejection of the offer for *"two examples of a high speed airplane Do 15"*, on 17 March 1933 *Dornier* then immediately submitted a new offer for a "high-speed airliner", which

was now known under the model name Do 17, but was designed for the same demands that had already been applied for the previously rejected project. The accompanying letter to the new offer urged the earliest possible award of the contract for the work on the Do 17, since *"the cancellation of the order for the Do 15 model covers all our dispositions both in the design offices and in the workshops which have been extremely disturbed and it is now urgently necessary for us, in order to avoid further major losses, to start work on the new Do 17 airplane im-*

Do 17 V1 W.Nr. 256 fuselage during construction. The order for first two Do 17 was placed on 24 May 1933. [Airbus Corporate Heritage]

Do 17 V1 first flight took place on 23 November 1934 with Dornier chief pilot Egon Fath at the controls. [Dornier Museum Friedrichshafen (Airbus Group)]

Do 17 MV1 following completion. The nose bears high resemblance to the later to follow export K model. [Djordje Nikolić]

Do 17 MV1 (W.Nr. 691) on 26 July 1937 won the multi-seat military airplane round circuit flight competition around the Alps which took place during the 4ᵗʰ International Airshow at Zürich. [Swissair00]

mediately". However, the most important aspect of this letter is the last paragraph, which reads as follows: "*We will have the documents for the model Do 17 as a cargo airplane to follow as soon as possible. At the same time, however, we would like to point out that a certain separation of the two projects will probably be necessary.*" This letter clearly shows the true nature of this airplane, it was not for *Lufthansa* and it was not in essence going to be a "cargo airplane", rather a "medium bomber".

Although the official contract for Do 17, which the company was so reliant on, was still a long way off, work on the project, presumably on the basis of verbal agreements, started immediately with great urgency. Already on 13 April, a project meeting took place at the company headquarters at Manzell where discussion was held concerning the engines to be installed and the expected flight performance. At the same time the initial mockup inspection took place. The next mockup

inspection was conducted on 10 May while another meeting followed on 20 May, without the presence of *Lufthansa* representatives.

The last doubts about the actual purpose of the Do 17 project were cleared by the contract issued by the RLM, dated 24 May, which bears signature of none other than the newly appointed Secretary of State, Erhard Milch. In this contract only the first two airplanes, the Do 17 C and A had been ordered. Following a renewed offer from *Dornier* for a "*further (third) Do 17 high-speed airplane, equipped with a double rudder and with the installation of Hispano-Suiza 12 Ybrs*" dated 2 October 1933, RLM issued on 4 November 1933 a new contract for the delivery of this third aircraft, designated as Do 17 D. The contract from 24 May lists 15 March 1934 as the delivery date for the first two airplanes on order, for which DMB assigned W.Nr. 256 ("special airplane for RLM") and 257 ("transport service DLH").

212

German Do 17V-7 W.Nr.657 D-AQYK displayed outside of the German pavilion at the "First International Aviation Exhibition" held in Belgrade between 28 May and 13 June 1938. [Djordje Nikolić]

Special designations for the two airplanes, namely Do 17 C (W.Nr. 256) later also known as V1 (*Versuch* – Trial) and Do 17 A (W.Nr. 257), later also known as V2, were introduced to designate different engine types installed between the two.

On 20 November 1934, the head of construction supervision at *Dornier*, Dipl.-Ing. Meyer, carried out the final inspection of the airplane and three days later, on 23 November first flight of Do 17 V1 took place with *Dornier* chief pilot Egon Fath at the controls. No flight reports were published for this or the following flights. Stability problems were discovered along with weak landing gear. It was weak landing gear which caused the prototype to have an accident in mid-February 1935 while landing at Friedrichshafen-Löwenthal. The airplane was a total loss following a crash during one of the test flights on 21 December 1935.

The second prototype V2, was completed with 14 months delay but it was equipped with the twin vertical tail surfaces. It made its first flight on 18 May 1936 with pilot Egon Fath at the controls. A number of prototypes followed which were powered by a variety of engines such as BMW VI, BMW VID, *Hispano Suiza* Y, *Daimler Benz* 600, *Daimler Benz* 600C, BMW 132F and *Bramo* 323. One of these, Do 17 MV1 (W.Nr. 691), on 26 July 1937 won the multi-seat military airplane round circuit flight competition around the Alps which took place during the 4th International Airshow at Zürich. Do 17 MV1 completed the circuit in 58 minutes and 42.3 seconds while flying at an average speed of 376 km/h, a whole 7 1/2 minutes faster than the closest competitor, Belgian Air Force *Fairey* Fox. Owing to this success, which was the result of its sleek

lines and the *Daimler Benz* 600 engine, the airplane quickly gained world attention. It is important to note that this airplane had nothing to do with the M series which would follow, rather it was specially prepared for the demonstration at Zürich.

At the beginning of 1937 series production began with two series versions equipped with BMW VI engines, designated as Do 17E for bombing and Do 17F for reconnaissance. In total 536 airplanes of these two versions were produced. Do 17M version, which was used as a bomber, followed with *Bramo* 323 engines as well as the Do 17P version for photo reconnaissance. In total 200 Do 17M and 330 Do 17P were produced.

Do 17 went on to become one of the workhorses of the *Luftwaffe* bombing units during the Spanish Civil War as well as early during World War II, specifically between 1939 and 1940. It was eventually replaced in mid-1940 by more capable *Junkers* Ju 88, which had higher bomb load and range, and Do 17 was delegated to secondary duties as glider tug, for research and for training.

Do 17 for the Kingdom of Yugoslavia

In parallel with Do 17M, *Dornier* developed an export version, with designation Do 17K. The letter K could not be used for series designation early during the development phase according to Dipl.-Ing Hans Kinzler, who had worked for Dornier since 1934, due to the obvious connection of the letter K to a warplane or combat airplane (*Kriegsflugzeug* or *Kampfflugzeug*). However, this letter was free to be assigned for export

The first Do 17Ka-1 W.Nr.2381 was initially tested with the solid nose. [Airbus Corporate Heritage]

airplanes as permitted by RLM, hence it was applied to the airplanes sold to KJ. Dornier delivered 36 Do17 K which were assigned W.Nr. 2381 to 2400 and 2461 to 2467. The first 20 airplanes, designated as *Ausfuhürung A*, KA or K-1 (later also known as Ka-1 where "a" stood for *ausländish* or foreign), were produced along with 12 other *Ausfuhürung A* models, with suffix K-2 (Ka-2), based on the Do 17 E. Four airplanes, designated K-3 (Ka-3), were based on the Do 17 M version.

On 19 November 1935, during the visit to *Dornier* factory by the Yugoslav commission tasked with inspecting and accepting spare parts for Do Y, pilot Dimitrije Kneselac had the opportunity to also inspect Do 17 V3 and even be a passenger on a single flight. In his pilot logbook he briefly described this flight:

W.Nr.2381 shown here with the new, glazed nose. Note the color difference in the nose where the solid nose was removed and glazed nose was installed. [Airbus Corporate Heritage]

Starboard view of Do 17Ka-1 W.Nr.2382 (Br.2) at Friedrichshafen during a photo session by the official factory photographer. [Airbus Corporate Heritage]

Br.2 during one of the factory test flights. [Airbus Corporate Heritage]

Do 17 bomber, demonstration at Friedrichshafen airfield, Germany, 2,500 m altitude, 15 minutes, one flight, 19.XI.1935.

Following this flight and the demonstration flight by chief Dornier pilot Egon Fath, the Yugoslav commission, still impressed by what they witnessed recommended the following:

The commission is of the opinion that it would be of great use for our air force to begin closer evaluation of this airplane with respect to the use in our country considering our needs.

In accordance with the order issued by the *Ministarstvo Vojske i Mornarice* (MViM - Ministry of Army and Navy), *Komanda Vazduhoplovsta* (KV - Air Force Command) formed a seven member comission which was tasked with selecting a twin engine bomber. This comisison was provided with the required performance, which most importantly included speed and weapons, followed by payload and range of 1,500 km. The airplanes were to be purchased without engines in order to allow the use of domestic IAM K-14, which made

Aft view of Br.2 showing Dornier factory applied "Kosovo cross" insignia on both wings. [Airbus Corporate Heritage]

A close up view of Br.2 . [Airbus Corporate Heritage]

matters more difficult. From 6 May until 7 June 1936 comission visited France, Great Britain, Switzerland and Germany, evaluating five different types. These included *Breguet* 462, *Bloch* 134 and *Potez* 630 from France, *Dornier* Do 17 from Germany (Do 17E with BMW engines) and *Bristol* Blenheim Mk. I from Great Britain.

The commission recommended on 10 June that Do 17 gets purchased, but one of its members, puk Dušan Radović was of the opposite opinion. KV expanded the commission which again selected Do 17 two days later, considering it met the set performance requirements the best. Commands' recommendation was accepted on 13 July by the special MViM commission for weapons

Br.2 prior to take off from Friedrichshafen to Belgrade. [Airbus Corporate Heritage]

Generalmajor Ernst Udet visited Friedrichshafen-Löwental airfield where he met Dr. Claude Dornier. Seen standing here in front of a Do 17K from left to right are: Staff engineer Hübner, Generalmajor Udet, Dr. Dornier, Dipl.-Ing Franke, Chief Staff engineer Lucht and Oberstleutnant Junck. [Dornier Post Nr.14]

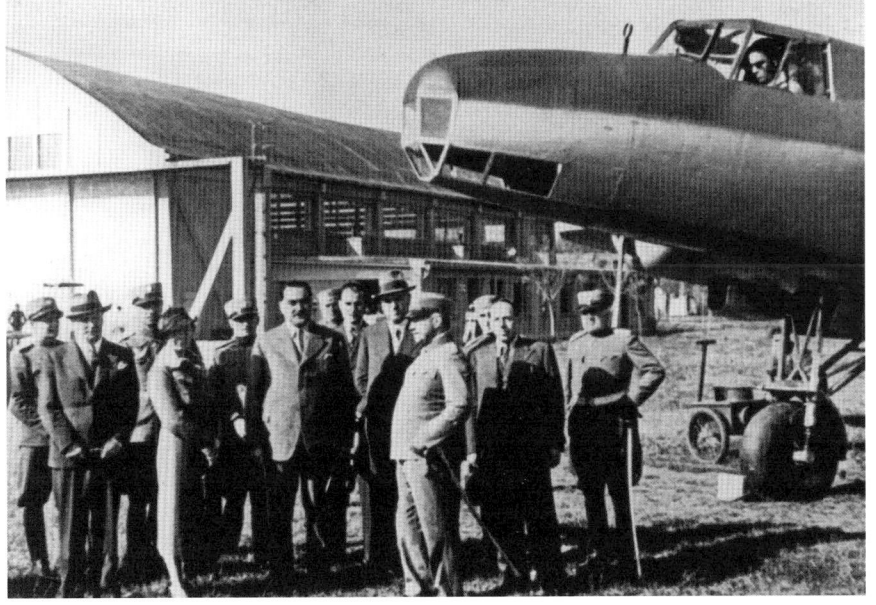

Dornier chief test pilot Egon Fath flew this machine during a ferry flight on 25 October from Friedrichshafen to Zemun. After landing the airplane was inspected by Prime Minister Dr. Milan Stojadinović along with Ministers and German envoy in Belgrade Viktor von Heeren. [Airbus Corporate Heritage]

Do 17 Ka-2 W.Nr. 2468 suffered a rough landing at an unknown location on 9 November 1938. Note partially visible registration D-AYW with the unknown last letter. W.Nr. 2471 and W.Nr. 2473 are known to have had codes D-AXWI and D-AXWM respectively. [Djordje Nikolić]

Side view of the damage to the nose. The damage appears minor and easily repairable, the main cockpit structure appears intact. [Djordje Nikolić]

Close up view of the nose. [Djordje Nikolić]

selection and a decision was made to purchase 36 Do 17 along with the license to produce this type domestically. German lobbying was very well developed within the Yugoslav politics but despite this KJ was already purchasing *Dornier* airplanes for many years and Do 17 was indeed the best choice. Not to forget, already in 1935 Germans allowed a Yugoslav pilot to fly in one of the three prototypes, which shows a high level of trust.

Do 17Ka-1 during a visit to Novi Sad, 1938.
[Aviation Museum – Belgrade]

209.E Do 17Ka-1 at Obilić airfield in the
summer of 1940. [Aleksijević family via Igor
Černiševski]

A group of VVKJ pilots in front of a Do 17Ka-1.
[Aviation Museum – Belgrade]

Do 17Ka-1 Br. 21 and in the foreground Br. 22 with 205.E airmen prior to a flight. [Vladeta Vojinović]

Kap Ilk Miodrag Petrović in front of a Do 17Ka-2 at Skoplje airfield. [Aviation Museum – Belgrade]

VVKJ airmen in front of a row of Do 17K at Skoplje airfield. [Vojislav Stankov]

Do 17Ka-1 Br.20 at the First International Aviation Exhibition in Belgrade during the Yugoslav delegation visit. [Aviation Museum – Belgrade]

The first business contacts between the Third Reich and KJ began in September 1935 when *Dornier* and MViM began negotiations to purchase Do 17 and Do 22 in a deal worth two million RM with a seven year credit. Herman Göring requested that the terms change which led to delays which resulted in KJ beginning to look elsewhere for airplanes. As soon as the word spread that France was contacted as an alternate supplier, the negotiations resumed. *Dornier* created a technical specifications document No. 1203 (*Baubeschreibung Nr. 1203*) on 14 January 1936 for "*2 engine Dornier bomber airplane Do 17 K₁ with 2 Gnome Rhône 14 Kirs with a 3:2 reduction gear or 2 Wright Cyclone SGR 1820 F-56 with 16:11 reduction gear*" which was released to KJ the next day on 15 January 1936. The contract for the delivery of 20 Do 17 with designation Pov. V.M.T. Nr. 5499/36 was signed on 9 November, without credit, and the delivery of the first airplane was set for 9 May 1937.

Not long after signing the contract, VVKJ HQ formed a commission which was sent to *Dornier Wilhelmshaven* factory to supervise the construction, acceptance and transfer of airplanes to KJ. The acceptance of the first airplane took place on 25 August 1937 with more than three months delay due to the late deliveries of *Gnome-Rhône* engines and their failure to meet marketed performance. All of the first six airplanes were completed by mid-December and the next six were completed in February 1938.

Prince Peter II visiting Dornier booth. He received a 1:20 scale model of Do 17 from Dornier staff which was subsequently displayed at the Royal Palace at Dedinje. [Aviation Museum – Belgrade]

Do 17 Br.22 at Skoplje airfield, summer 1939. Kap Ik Vlastimir Vojinović is visible through the opened cockpit sliding window. [Vladeta Vojinović]

Kap Ik Vlastimir Vojinović with family in front of a Do 17Ka-1 at a hangar at Skoplje. [Vladeta Vojinović]

A line up of Dorniers at Skoplje during the ceremony marking the arrival of the final Do 17Ka-2 airplane from Dornier. Br. 32 is on the right and Br.23 on the left. [Dinko Predoević]

The same line up with hangars, an additional Do 17K and two Breguet Bre 19 biplanes visible in the background. [Dinko Predoević]

Airplanes lined up on 2 August 1939 at Skoplje during the ceremony celebrating Sveti Ilija (Saint Elias), VVKJ patron saint. [Djordje Nikolić]

Officers in their dress uniform lined up in front of 209.E Do 17K. [Djordje Nikolić]

Potez 630 C-3 F-AREY just after delivery to VVKJ. Behind it is an Dornier Do 17Ka-1 W.Nr. 2460, the last of the Ka-1 models. [Aviation Museum – Belgrade]

The first Do 17K had a "solid" nose as the final version of the "glass" nose was not yet available at the beginning of the flight testing and was replaced by an existing nose section from the first V-type airplanes. The first flight of Do 17Ka-1 W.Nr. 2381 took place on 6 October 1937 with *Dornier* chief test pilot Egon Fath at the controls. He also flew this machine during a ferry flight on 25 October from Friedrichshafen to Zemun along with kap Ik Milutin Dostanić. The arrival of the first Do 17Ka-1 to Zemun airfield was described in detail in the article found in *Die Dornier Post* Dez./Jan. 1937/1938:

Prime Minister Dr. Milan Stojadinović visited recently at the Belgrade Zemun airfield in the company of some Ministers and prominent guests (including German envoy in Belgrade Viktor von Heeren – a.c.) the first of

VVKJ airmen in front of their Dornier. [Djordje Nikolić]

207. E Do 17 line up. Note that crews are standing in front of their airplanes. [Marko Babić]

209. E crew playing cards and smoking a cigarette. Note the wheel chocks marked with the escadrille number. [Miloš Milosavljević]

Do 17 airplanes produced by our company for Yugoslav Army. Factory friend Egon Fath flew over the machine in an outstanding time of 2 hours and 20 minutes from Friedrichshafen to Belgrade (940 km), at an average speed of 403 km/h. Dr. Stojadinović acquainted himself with the airplane from both the outside and the inside and thereafter wanted to go for a test flight with puk Zdenko Gorjup who had firsthand experience with the outstanding performance of this type. Meanwhile captain Fath had the opportunity to fly another Do 17 and better his time by 5 minutes.

Pilots in training during prequalification at Skoplje, 1939. [Ognjan Petrović]

Do 17Ka-1 at Skoplje, 18 September 1939. [Aviation Museum – Belgrade]

Do 17Ka-1 Br. 3312 from 209.E seen here at Obilić airfield preparing to take off. [Aleksijević family via Igor Černiševski]

Obilić airfield with two 209.E Do 17K visible. [Aleksijević family via Igor Černiševski]

Do 17Ka-3 Br.36 armed with forward facing machine guns at Skoplje, summer 1940. [Petar Bosnić]

To mark 550 years since the historic battle in which the Serbian army fought the Ottoman empire at Kosovo Polje on 28 June 1939, nine Do 17K from 3.BP flew in loose formation at 300 m altitude above the exact spot where the battle took place. [Milan Micevski]

The new flight time of 2 hours and 15 minutes equals the average speed of 417 km/h.

Dornier factory records show that in 1937 a total of six airplanes were delivered of which W.Nr. 2381, 2382, 2383, 2384, 2385 and 2390 are known to have been completed by the end of the year.

Next year on 9 February 1938, at 11:05, three Do 17Ka-1 took off from Friedrichshafen, Belgrade bound, entering Austrian airspace about five minutes later. In the lead airplane, coded D-AYWM, was again captain Fath:

It is 9 February 1938, 9 o'clock in the morning, blue sky dwarfs our land and the rays of the morning sun are

Ten Do 17K flying in formation. [Željko Marković]

Do 17Ka-3 at Skoplje airfield. Note the exposed Browning FN machine guns in the nose. [Aleksandar Ognjević]

Do 17Ka-3 Br.36 at Skoplje, summer 1940. [Vojislav Stankov]

Unknown 208.E Do 17K with cockpit and engine covered to protect them from the elements. [Aviation Museum – Belgrade]

eager to dissolve the dew on the grass. Today a transfer flight of three Do 17 will take place towards Belgrade. Preparations for the flight are already complete, and we are waiting for the last news of the weather in the vicinity of Yugoslavia. While the customs officer inquires about our foreign currency and the air traffic supervisor reviews the logbook, the awaited news finally arrives. It is quite favorable: the flight is possible! The decision has been made – we can go!

The engines of the three machines are already warmed up. The Yugoslav commission told us to send their regards home and all present wished us a "happy travels" and "to break a leg". The break chocks are removed and from the

outside comes the command "Free", it is time! Exactly at 11:05 we increased throttle. In a closed chain formation, the airplanes lifted off the ground and without the usual farewell loop took their bearings immediately.

After approximately five minutes we flew over the German border and said goodbye to the radio station at Stuttgart, whose range we just left. We are flying at approximately 3,500 m altitude. We have the best chance to reach our destination the fastest if we maintain this altitude during our flight.

In front of us are the Alps in their entire glory, an image, of a heavenly nature which human eyes get to see rarely. It is probably the most beautiful, but in hindsight also an

Two VVKJ airmen posing next to a tail belonging to an unknown DFA built Do 17K. [Miloš Milosavljević]

An overhead view of DFA factory at Kraljevo where Do 17Ka-3 were produced under license. [Vojislav Stankov]

eventual emergency landing undesirable landing area. We are therefore paying a particular attention to the instruments, in order to detect any anomalies as soon as possible.

We pretend to be angel flying around the area because the sky is smiling with its entire face. Above us is eternal blue, as far as the eyes can reach, below us the snow covered mountains with their lonely peaks, constantly getting closer to us, and as it were, trying to engulf us. To our left we pass the Zugspitz massif. In the background Starnberger lake is visible in the haze. Ahed of us in our course are already the Hohen Tauern, with Großglockner and Großvenediger peaks, and close ahead lies Innsbruck, which we will fly over in few minutes. I discovered the Innsbruck airport and am glad to have the ability to land for several minutes. From time to time I assure myself that both machines fly-

ing behind me, led by Schmidt-Coste and Tarnowski, keep formation. I see both planes very close to my left and right, and we made ourselves understood by waving arms that everything is in best order. Only the lead airplane has the radio set on board, so it must ensure not to lose the airplanes in formation from sight.

The engines are running perfectly and needles on the instrument panel stand ironclad in the right place. Everything on board is perfect. We receive weather report from Innsbruck, which tells us to count on encountering ground fog in Yugoslav lowlands. Belgrade is however cloudless. At an altitude of 4,000 meters we fly over the 3,700 meters high Großglockner. I am searching eagerly for skiers in the open ground, but despite "staring" I cannot succeed in finding any living being.

The Alps lie behind us and we just passed Villach. The expected ground fog is visible in the distance. As we flew over the Yugoslav border, we keep flying without ground visibility. Above us is still a clear sky and below us, as far as the eyes can reach is a white carpet above which we are flying. I received a message from Zagreb that Belgrade is still fog free. Therefore, we can count also that before Belgrade we can see the ground. I am checking the bearing in 15 minute intervals before Belgrade and am controlling my course.

Both airplanes are following me at equal distance. My mechanic Schöllkopf unpacked two dried buns which we ate together for lunch. Suddenly we were surprised by a sudden blow to the rudder. I immediately sent Schöllkopf aft to see what is going on. It was the glued insignia which loosened itself.

Two flight hours went by and we must be in the vicinity of our target at any moment. The fog is breaking up and we see Belgrade ahead of us in the haze. After 2 hours and 15 minutes we flew in tight chain formation over Zemun airfield. Then we followed in an honorable circle above Belgrade and the flight tower, which we performed very well with both machines behind me in a turn. I had to reduce the throttle somewhat for my right wingman to catch up. At 13:35 we are setting up for a landing. No one has expected that we are flying with an average speed over 400 km/h, as a result the gentlemen were very surprised that we were already there. Of course, this glorious and well ended flight was appropriately celebrated. Fath.

Two accidents are known to have occurred during ferry flights. The first with W.Nr. 2397 which crashed while landing at Zemun at the conclusion of the delivery flight on 15 March 1938. Later, this airplane will be replaced by a newly supplied W.Nr. 2460. The second was W.Nr. 2348 which force landed at an unknown location severely damaging its nose.

H. F. King, a journalist from the British *Flight Magazine*, had the opportunity to visit *Dornier* factories in April 1938. During this occasion, he witnessed a VVKJ Do 17K undergoing testing:

By the courtesy of the German Air Ministry and of Dornier Metallbauten G.m.b.H, it was my privilege to visit the three modern plants, Manzell, Allmansweiller and Löwental on the outskirts of Friedrichshafen and there to see the construction of such outstanding aircraft as the Dornier Do 17 high-speed bomber-fighter (the "Flying Pencil"). Our guide and counselor was Herr Diemer, one time Dornier test pilot…

In the shops I was immediately impressed by the completeness of jigging. It was possible to detect portions of Do 18 flying boat by the green markings, the Do 17 bomber landplane being allotted blue as its work colour…

…Though the primary object of my visit was to study the Do 17 bomber-fighter, I was able to, at the Allmansweiller works (where we were whisked by a works Mercedes, after partaking of Meersburg wine), to make a quick tour of the section devoted to the construction of Do 18. In the same shop was a small series of Do 22 torpedo bombers for the export market. This type of parasol monoplane has a welded fuselage and a Hispano Series Y engine.

I was unable to see large scale series production of the Do 17 for the Luftwaffe, as this was in progres at a number of distant factories, but i did see a fair-sized batch going through for Yugoslavia, and was subsequently shown one or two German machines which were at the adjecent Löwental for tests.

The Yugoslav machines, which were being assembled and tested under the supervision of a military mission, are powered by two Gnome-Rhône 14No fourteen cylinder two row radials and, apart from provisions for internal and external bomb stowage (the large bombs are carried on the sides of the fuselage), mounts two fixed belt fed guns in the front cowling, a free gun in the floor of the fuselage forward of the bomb compartments and a fourth weapon on a shielded mounting above the wing. At the request of the chief of the Yugoslav mission, I shall not describe the internal arrangements.

Before I left I was able to watch a Yugoslav Do 17 taking off at its gross weight. The run was protracted and the actual "unstuck" took place at a peculiar tail-down angle, but acceleration, when the undercarriage was raised, was of the rocket order.

In mid-1937 the talks were resumed concerning the purchase of further Do 17. Request for quote Pov. Broj. 200 was sent to *Dornier* which responded with an offer on 17 August 1937 for 14 Do 17 *Ausfuhürung A* (Ka-2) and two Do 17 *Ausfuhürung B* (Ka-3), although this time at much higher prices. The Ka-2 models were quoted at 189,000 RM plus 49,750 Swiss Francs and Ka-3 models at 211,200 RM plus 49,750 Swiss Francs. Finally, the contract, Pov. V.M.T. Nr.497, was signed on 14 March 1938 with 31 January 1939 as the delivery date. This time as well, the contracted delivery date was honored despite a late start. W.Nr 2461, 2462, 2463, 2465, 2466, 2467, 2468, 2469 and 2470 were delivered by 31 December 1938. W.Nr. 2464, 2471 (was made available on 15 December 1938 for delivery according to the letter from *Dornier*) and 2473 followed next. The invoice for W.Nr. 2472 was sent to MViM on 2 February 1939 for a total of 204,474.27 RM and 24,075 Swiss Francs, indicating that this airplane was ready for delivery at that time. W.Nr 2475, 2476 and 2460 were delivered in April with the last airplane arrived on 21 April 1939. The delivery of W.Nr. 2476 was delayed because the port engine, which was supposed to be delivered by 1 November, was actually delivered by the end of December 1939. *Dornier* duly informed the Yugoslav commission in Friedrichshafen on 4 January 1939 of the inability to honor the delivery schedule as a result.

According to *Dornier* factory documents, during 1938 a total of 26 Do 17Ka-1 and Ka-2 were delivered with additional six in 1939 (of which four are known to have been the improved Ka-3 version).

DFA – Državna Fabrika Aviona

All *Dornier* supplied Do 17, with exception of the first airplane, were delivered without armament and equipment. These airplanes, following their arrival to Belgrade, were flown over to Kraljevo where at VTZ they were fully equipped prior to being assigned to 3.BP.

Along with the negotiations for the second batch, a matter of license production in the country was brought up. The decision to produce bombers in the country was made due to Great Britain's reluctance to approve credit for rearmament in 1936 and 1937.

KV began negotiations with *Dornier* in 1937 and the representatives of the German company visited KJ for a meeting which took place on 27 April 1937 at Zemun with the representatives from VVKJ as well as three private factories. The original quantity of airplanes discussed was 34, which was later increased to 36. The summary report issued after the meeting lists the following:

1st Dornier produced series			2nd Dornier produced series		
Version	W.Nr.	EvBr.	Version	W.Nr.	EvBr.
Ka-1	2381	3301	Ka-1	2460	3317
Ka-1	2382	3302	Ka-2	2461	3321
Ka-1	2383	3303	Ka-2	2462	3322
Ka-1	2384	3304	Ka-2	2463	3323
Ka-1	2385	3305	Ka-2	2464	3324
Ka-1	2386	3306	Ka-2	2465	3325
Ka-1	2387	3307	Ka-2	2566	3326
Ka-1	2388	3308	Ka-2	2467	3327
Ka-1	2389	3309	Ka-2	2468	3328
Ka-1	2390	3310	Ka-2	2469	3329
Ka-1	2391	3311	Ka-2	2470	3330
Ka-1	2392	3312	Ka-3	2471	3331
Ka-1	2393	3313	Ka-3	2472	3332
Ka-1	2394	3314	Ka-2	2473	3333
Ka-1	2395	3315	Ka-3	2474	3334
Ka-1	2396	3316	Ka-2	2475	3335
Ka-1	2397	Br.17*	Ka-3	2476	3336
Ka-1	2398	3318			
Ka-1	2399	3319			
Ka-1	2400	3320			

*EvBr. Not assigned as airplane was destroyed on landing on 15 March 1938, before EvBr. system was introduced

The Yugoslav government is looking for advice concerning the feasibility for the start of production of Do 17 by the private industry, respectively state factory at Kraljevo.

The private airplane industry consists of 3 companies:
1. Ikarus A.D. factory in Zemun, Director is Mr. Konjević
2. Zmaj, Zemun, Owners are Mr. Petrović and Sterić
3. Rogožarski, Belgrade, Director is Mr. Nikolić

Both of the first companies are capable of building mixed constructions and are at the moment producing English Hawkers under license per orders from Zemun

HQ and each company is building a complete machine, Ikarus 24 and Zmaj 14. Besides these new builds both companies have repair contracts for Breguet and Potez. Rogožarski is the smallest of the three and is solely capable of building wooden structures. This factory is at the moment producing one airplane (of domestic construction), as well as repairing Breguet and Potez airplanes.

The state factory at Kraljevo was set up in 1927 or 1928 to produce Breguet airplanes and is with respect to the workshop, space and installations superior to the private companies. The factory has all the facilities to take on the construction of metal airplanes successfully.

After visiting four factories, there was a meeting at the Command which was presided by Oberstleutnant Stanojević, in which a number of questions were raised concerning the basics of license production. Mr. Oberstleutnant Stanojević shared with us that the intent would be to at first produce 34 machines, 24 through private industry and 10 through the state factory at Kraljevo. This split was founded on the fact that the Yugoslav government is anxious to get a strong private industry, on the other hand by building the same airplanes at the state factory the control of the private companies could be possible. The question was also asked how would the production of 24 airplanes be divided to the three private companies and which installations should be made by these companies and secondly, how well is the state factory set up to produce Dornier airplanes on its own. These questions, as well as many others, which are supplied in the provided questionnaire, were answered in the provided Exposé.

After submission of the Exposé, the private industry was invited for a meeting at the Headquarters which was presided by Maj Popović from Kraljevo. During this meeting a protocol was prepared which is included in this report and from which it is possible to deduce that the opinion of the Headquarters to produce complete airplanes at two sites including the private industry was abandoned and that it is leaning towards producing airplanes only at Kraljevo. In the above meeting, Oberstleutnant Stanojević was not present, but private industry was, but at a special meeting he spoke with him and advised him to accept this proposal, according to which all the airplanes would be produced in Kraljevo and that from Kraljevo the private companies may be provided with sub-assembly work. Oberstleutnant Stanojević welcomed this proposal, but expressed concerns with respect to administrative difficulties while producing the machines in Kraljevo if the circumstantial management for the procurement of materials, machines, furnishings leads to the delays in deliveries. I asked thereon the question if it was not possible to eliminate the existing administration at the start and to deal with it first on a private sector

basis. He thought that this could eventually be possible, however he noted that the in order to justify the production at Kraljevo he needed the costs of the necessary investments and start up times:

1. When all 34 machines are produced at Kraljevo

2. When all 34 machines are produced by private industry as an alternative

3. When 24 machines are produced by the private industry and 10 by the state factory

Oberstleutnant Stanojević hinted that the Command, opposite of the earlier statement, always expected that the state factory would be expanded and that the required conditions for that are already present. Only the necessary justification to do so was missing at the time. This would be justified by the start of Do 17 production, by full production of all parts at Kraljevo, as much as possible.

On 28.4 (Wednesday), Oberstleutnant Stanojević spoke to General Simović, whom he proposed to begin production of Do 17 at first at Kraljevo and that the Command is looking for means to establish the existing administration to acquire the materials, equipment and so on in a reasonable amount of time. Oberstleutnant Stanojević agreed that factory manager at Kraljevo, Mr. Zeluška, should come in approximately 14 days to Friedrichshafen, to inform himself about the production of Do 17.

The representatives from the three private companies argued against the decision made by KV, most vocally indicating that DFA should only be employed in the case of war and that it was unsuitable for such complex airplane production. KV's decision, made rightfully so, was final and production at DFA was given a go ahead.

The report further states that production of three to five airplanes at DFA is possible with approximately 500 to 680 men (depending on labor efficiency drop off) and a minimum of 11,000 m² factory floor space and that one Do 17 requires at total of 45,000 man hours to produce.

It is interesting to note that Dornier sent a number of engineers Ikarus, Zmaj, Rogožarski on 21 and 22 April 1937 and DFA on 23 and 24 April, before the above meeting took place, to assess the abilities for domestic production, either at one location or split between several factories. This provided the German side with unprecedented access and insight into the airplane production in Yugoslavia, which significantly aided preparing future plans for utilizing these factories in the German war effort after the war was won in KJ.

In December 1937 KV sent a group of airmen, engineers and technicians to Friedricshafen to "get *acquainted with the production of Do 17, in accordance with the contract for license production and other tasks related to production of these airplanes at Kraljevo".* By 24 December *Dornier* delivered technical drawings to KJ for review and to prepare necessary tools and jigs.

Negotiations were concluded on 27 June 1938 with a contract for the delivery of material for the completion of 36 license airplanes. The material needed for production was received in three batches, the first batch arrived in October 1938, the second in January 1939 while the third batch arrived by August of the same year. The first batch included major sub-assemblies such as complete vertical stabilizers and landing gear while the subsequent batches were pre-assembled, placed in storage and pulled from storage for final assembly.

The license production was at first assigned to airplane workshop belonging to VTZ which began assembling Do 17 towards the end of 1938. However due to complicated administrative regulations, the state founded on 21 March 1939 an autonomous state entity which was named *Državna fabrika aviona*. Factory began work on 15 May 1939 when the work on the initial series of 16 airplanes began. German engineers, which were assisting production, also conducted regular quality control checks which were to say the least very strict. The workers, used in manufacturing Breguet biplanes, had difficult time adjusting to the quality requirements of such complex and modern airplane such as the Do 17. German records indicate that on 6 April 1941 a total of 1,019 men and 25 women were employed at DFA.

DFA consisted of the following buildings which were used directly or indirectly in the production of Do 17:

Building	Area
Hangar A	11,750 m²
Hangar B	8,340 m²
Warehouse	4,700 m²
Fire station	430 m²
Power station	1,300 m²
Garage	875 m²

The first domestically produced Do 17Ka-3 took off in mid-November. Following completion, the airplanes had their weapons and equipment installed as well as compasses compensated prior to hand off to 3.BP (*Bombarderski puk* - Bomber Regiment). The first 26 Do 17Ka-3 were produced identical to those airplanes from *Dornier* with short engine nacelles, while the last 10 were modified with longer engine nacelles, strengthened fuselage to prevent damage during belly landing and asymmetrical windshields with an additional 7.9 mm *Browning* FN heavy machine gun on the starboard side.

These modifications were implemented without the knowledge of *Dornier* engineers (due to their restricted access) and were first implemented under the leadership of engineer Bora Petrović during 1940 at a Hangar B belonging to VTZ. The first 14 license produced Do 17Ka-3 had French built engines while the remainder used those produced in the country.

The production was slow and the contracted deadlines were exceeded. DFA delivered only seven airplanes by March 1940. On 1 April 1940 VVKJ had 41 operational Do 17 of which 34 were produced by *Dornier* and seven domestically. Nine airplanes were supposed to be delivered by 1 July and 20 by the end of 1940. Until April 1941 only 32 airplanes were delivered.

During the second, and final, annual regular meeting of *Udruženje Vazduhoplovne Industrije Kraljevine Jugoslavije* (Yugoslav Aviaton Industry Association) held on 23 March 1941, DFA received favorable praise:

The results achieved by our youngest aviation industry manufacturer during the year and a half since its founding, were such that one can conclude absolute justification and hope it was given to it.

Immediately after the acquisition of the workshop belonging to Vazduhoplovno-tehnički Zavod, the factory Management with few of the inherited personnel, started working full of energy to attain the assigned goals. At first it needed to create an entirely new organization, technical and commercial, to organize the necessary manpower, to implement manufacturing processes and to apply modern principles necessary for quick and rational manufacturing.

The Management encountered unforeseen problems along the way. Mastering such a technologically complex machine such as Do 17, training personnel, beginning with the most basic fabrication processes, working against large problems which were the result of war situation in the world, was a immense undertaking which could only be overcome with maximum effort by the employees.

Today, after working for year and a half, we can concluded that Državna fabrika aviona accomplished all those tasks despite unfavorable work conditions and yielded results which in their entirety justify the expectations set forth by those in charge.

Right before the April war, in the middle of March, most of the Germans engineers left Kraljevo and those which remained were threated roughly following coup d'état on 27 March due to the growing anti-German sentiment. DFA factory was of strategic importance, therefore on 4 April 1941 due to imminent war danger it was placed under military control. On the same day one Do 17Ka-3 EvBr. 3366 lacking some of the instruments was delivered to 210.E. On 6 April work went on as usual where workers were assembling a batch of Do 17Ka-3 which were slated for hand off at the end of next week. As soon as the war started, work pace slowed down since a number of workers did not show up. On 10 April 1941, the fourth day of the war, an order was received to prepare the factory for demolition to prevent it from falling in the enemy hands. Significant amount of explosives were placed in all fabrication areas to ensure machines and tools were destroyed beyond use. The airplane assembly area was loaded with barrels full of gasoline, but not before two airplanes were completed and delivered. When on 11 April the news spread that Germans may be close, an order was issues to detonate the explosive charges however the engineering units were at another location in Kraljevo, unable to complete this task. Therefore, the airplanes undergoing assembly were doused in gasoline and torched. The fire extinguished itself as quickly as the news arrived that Germans were still far away. The last airplane to be delivered was fortunately in front of the assembly building and was not subjected to torching. Finally, the fate of DFA was sealed on 14 April when the factory was blown up right before Germans entered Kraljevo.

During the war DFA delivered two additional airplanes: EvBr. 3368 on 6 April to 210.E and 3365 on 12 April with mixed crews consisting of 209.E pilots and 5.LP (*Lovački puk* – Fighter Regiment) copilots. Photographs of DFA factory show four almost complete Do 17Ka-3, which are suspected to be EvBr. 3369, 3370, 3371 and 3372 as well as a fuselage of another machine which had what appears to be a belly turret, similar to that on Do 215, which could have been an entirely ingenious design or could have been replicated from drawings provided by *Dornier*. This was likely a factory prototype which would have served as a pattern airplane for the second domestic production series, which due to the April war, never materialized.

According to the secret protocol from the 12th meeting of the German and Yugoslav governmental economic committee which took place on 19 October 1940, another delivery of materials for 40 license built Do 17Ka-3 was approved. However, the deliveries began as late as February 1941, due to intentional delays from the German side. As a result, the 40 airplanes ordered from *Dornier* and 90 domestic IAM K-14NO engines remained undelivered.

Despite the extensive damage, which in the available photographs appears to have put DFA effectively out of service, Germans recognized the importance of this factory and invested significant amount of time and money to make it operational so that it can become part of their war effort.

DFA produced Do 17Ka-3			
EvBr.			
3337	3346	3355	3364
3338	3347	3356	3365
3339	3348	3357	3366
3340	3349	3358	3367
3341	3350	3359	3368
3342	3351	3360	3369*
3343	3352	3361	3370*
3344	3353	3362	3371*
3345	3354	3363	3372*

*Captured incomplete and damaged at DFA

Dornier sent a letter on 15 August 1941 to the German commission which visited DFA to survey the damage. Of interest to *Dornier* was could the factory be repaired so that it can begin supplying parts for Do 215, which used similar components as Do 17. The extensive 26 page report details surveyed damage in each of the factory buildings, what was salvageable and what not as well as the state of repairs on the day of the submittal of the report. Both the north and the south wing of Hangar A had some 1,050 m² damage each and the concrete pillars collapsed and damaged the equipment. In Hangar B some two thirds of the reinforced concrete roof caved in and the power plant and the boiler room were significantly damaged amongst others.

Still, by the end of October the damage was expected to be repaired in time for the harsh winter months. At the time of the report the majority of the concrete structures, flooring and glass were repaired while the machining area as well as the forge and metal hardening areas were fully operational. The work has already begun on tail sub-assemblies and spars for Do 17 as well as fabrication of components for Do 215. Tools were desperately needed though in order to improve production of Do 215 rudders and ailerons and the arrival of these was expected from Germany.

In the northern part of the factory, the work was nearing completion on Do 17 which were captured during assembly. Herr Hofmann wrote about the status of the Yugoslav Do 17 repairs as of 15 August 1941:

The work on completing the 11 Yugoslav built Do 17 was recorded on 15.8. The engines for the said airplanes must be sent to Rakovica engine factory for testing and changes to the compressors. Apart from that many instruments are missing, which have been requested. The completion of the airplanes depends on the arrival of the engines and the equipment. Another difficulty for the airplanes destined for Germany is that they must get German equipment such as Telefunken 1001 bF, 118 N and 119 which may require certain modifications. If the modifications are required this will affect the delivery timeline. It would be necessary to establish soon, who will get six Do 17 airplanes brought to the airfield, so that the correct equipment can be installed accordingly.

German records indicate that on 16 August 1941 a total of 865 men and 22 women were still employed at DFA.

By 28 August 1941 a total of nine airplanes of various types were delivered to ZNDH while others were undergoing evaluation and repair. The commission was of the opinion that despite 56 airplanes of various types are destined for ZNDH, it may not be possible to make all of them operational as some of the more damaged airplanes would be used as sourced of spare parts.

Following an extensive report, on behalf of the commission Herr Hofmann sent a reply back to *Dornier* on 11 September 1941 which concludes that:

By the way, the overall conditions here are very unfavorable. Wheatear it will be possible to get a certain benefit out of the work here will be shown in the future. Daily sabotages in Serbia and the associated traffic congestion, which often lasts for a long time, does not allow for optimistic hopes.

Peace Time Use

Upon arrival to KJ, all Do 17 were based at Zemun. This allowed for the core of the newly trained crews to be trained on and familiarize with this type. One of the setbacks which affected the training of new crews was that, at the time of the Dornier purchase, no adequate dual control trainer airplanes were ordered, therefore it was only possible to train new crews on the actual type.

Starting with the spring of 1938 Do 17 began to be delivered to 3.BP based at Skoplje, Macedonia, which was tasked with protecting southern Serbia and Macedonia. 3.BP consisted of 63.BG with its 205.E and 206.E as well as 64.BG with its 208E. and 209.E. On 25 June of the same year 12 Caproni Ca.310 were purchased for use as trainers which made the training and conversion process at 3.BP less complex.

The first public appearance of the newest bomber in the country took place during *Prva medjunarodna izložba aviona* (First International Aviation Exhibition) held in Belgrade between 28 May and 13 June 1938. Two Do 17s were displayed, one was a VVKJ Br. 20 displayed in the Yugoslav pavilion and the other a German Do 17V-7 W.Nr. 657 D-AQYK displayed outside of the German pavilion. Also, part of the German air echelon which arrived at Belgrade was 9./LG 1 with 12 Do 17E. Towards the end of the Exhibition, his highness Prince

Peter II visited *Dornier* booth. He received a 1:20 scale model of Do 17 from *Dornier* staff. This was subsequently displayed at the Royal Palace at Dedinje.

On 1 April 1939, 63.BG and 64.BG were assigned a third escadrille each, becoming the only VVKJ units to have such formation, when 207.E and 210.E were added respectively. To mark the completion of deliveries of all Do 17Ka-2 to 3.BP escadrilles, a ceremony was held in 1939 at Skoplje with neatly lined up airplanes. The occasion was used both for official purpose as well as for propaganda. Airplanes were lined up for other ceremonies and celebrations such as the *Sveti Ilija* (Saint Elias) day which takes place every 2 August, to commemorate the VVKJ patron saint.

To mark 550 years since the historic battle in which the Serbian army fought the Ottoman empire at Kosovo Polje on 28 June 1939, nine Do 17K from 3.BP flew in loose formation at 300 m altitude above the exact spot where the battle took place.

On 15 May 1940 209.E relocated to Obilić auxiliary airfield for a period of two months to practice operation from remote airfields, a measure which will prove useful in the upcoming war.

On 6 March 1941 63.BG relocated from its home base at Skoplje to Petrovac auxiliary airfield, and a week later reservist began to arrive and by 3 April it had full strength. 12 hangars were relocated from Kraljevo to Petrovac, the base of 63.BP. Hangars were used during minor repairs on Do 17 as well as for storage of spare parts. Along with the hangars a number of workers were sent over to assist with the repairs on site.

Por Rudolf Papist from 8.AČ (*Aerodromska Četa* – Airfield Company) assigned to 64.BG noted:

Peace time organization within 64.BG, which consisted of one company and three escadrilles, was founded on old principles and was very weak. The entire operation was burdened with the administration, and there was very little time for actions which would yield true results. Combat readiness was poorly organized. The airplane equipment was very good and active officers were skilled and well trained in using the airplanes and weapons, but the anti-aircraft defenses were weak. For the entire 64.BG, there were only 12 machine guns, of which the majority were obsolete and experienced frequent jamming, which proved ineffective in combat. All units were evacuated to auxiliary airfields at the beginning February and March. On 12 March 64.BG flew over to an auxiliary airfield at Kosovo Polje (Stubol and Obilić – p.a) near Priština. The ammunition and bombs were transported to auxiliary airfields, but only to those where the units were based. This proved to be a mistake because at the time of war, the ammunition

Do 17Ka-1 Br.17 crashed on 15 March 1938 at Zemun airfield. Crew escaped unharmed while VVKJ CO brig đen Borivoje Mirković, who was the passanger, suffered multiple arm and leg fractures. This machine was replaced by Germans free of charge as a gesture of good will. [Šime Oštrić]

Do 17Ka-2 Br.25 from 206.E crashed when engines failed on take off from Skoplje, fortunately without casualties. [Dinko Predoević]

should have been stored at other auxiliary airfields because airplanes had to constantly relocate to avoid discovery of their airfields. Because this was not done before the war, the airplanes were forced to relocate to auxiliary airfields with insufficient ammunition stocks, which later had to be transported from other airfields.

To practice its offensive missions which were to take place against enemy positions in Bulgaria, during March practice bombing missions with concrete bombs were conducted at Romanovci military proving range. Not long thereafter, on 1 April, the alert status was raised to its highest level and 3.BP was waiting for the inevitable enemy attack and was prepared to undertake its assigned offensive missions.

Accidents

Intensive use and flying led to a number of accidents with two total losses and five damaged airplanes. The first total loss took place on 15 March 1938 when Br.17, which was just delivered from Germany, stalled and burst into flames on a take-off during a demonstration flight at Zemun. Crew escaped unharmed while VVKJ CO brig đen Borivoje Mirković, who was the passanger, suffered multiple arm and leg fractures. This machine was replaced by Germans free of charge as a gesture of good will.

On 14 September 1938 Do 17Ka-1 Br.6 collided with Breguet XIX Br.1338 while landing. Br.6 suffered heavy damage but was later repaired at DFA and returned to service.

Mechanic vod Milivoje Ilić was involved in two peacetime accidents, one in 1938 and the other in 1939:

I remember it as it was yesterday. I took-off with my Dornier with pilot Zdenko Gorjup. During landing the pilot committed a crude error. He negligently cut the throttle and the airplane stalled near the ground. I acted as fast as lightning. I switched off the fuel line and retracted the landing gear. Thereafter the airplane coasted and belly-landed into some marsh. Gorjup was petrified with fear. I remember being commended for saving the lives of the crew and sparing the expensive airplane.

In July 1939 (3 July 1939 – a.c.) I took off with pilot narednik-vodnik IIk Radovan Milinić and when we were few hundred meters above the ground, shortly after getting airborne, there was a loud bang and a jolt. The right engine ceased. The left one would not even move. We are in a dive! We toppled down right on the surrounding buildings. I am screaming at the pilot to turn in the direction of forest and Vardar river at least. I quickly closed the fuel lines, retracted the landing gear... We crashed into Vardar river. Still breathing... The whole wing was brought to the scene to learn how I made it, and for that I was awarded with a months leave.

The first accident resulting in a fatality took place on 12 June 1939 when two soldiers snuck in the fuselage disturbing the center of gravity and interfering with the control cables in the fuselage, eventually causing the crash at Kitka mountain in which 205.E CO kap Ik Miodrag Pavlović, observer por Dimitrije Jovanović and mechanic nar Milivoje Janičijević perished. To make matters worse, the airplane crashed into a flock of sheep killing the shepheard and most of his flock.

A brand new Dornier was severely damaged in the summer of 1940 at Kraljevo airfield when a pilot in training, Landratović, during his first solo flight in a Fizir FP-2 biplane came in low and crashed into a Do 17 on a compass swinging platform, ending between its wheels, but escaping uninjured.

The last known accident took place on 11 September 1940 when Do 17 Ka-1 EvBr.3318 from 210.E landed with landing gear up two kilometers north of Skoplje due to engine trouble. This airplane was dismantled on site and taken to DFA where it remained until capitulation.

1941 April War

On 6 April 1941 there were 63 Do 17 on strength, of which 60 were operational. All but two airplanes were assigned to 3.BP, while two were at *Vazduhoplovna škola bombardovanja* (VŠB – Air bombardment school). DFA delivered two additional airplanes during the war.

3.Mešovita Vazduhoplovna Brigada (3.MVB – Mixed Air Brigade)
3.BP - CO puk Zdenko Gorjup
63.BG – based at Petrovac
CO maj Branislav Djordjević
205.E CO maj Mato Čulinović
206.E CO kap IIk Mihajlo Djonlić
207.E CO kap IIk Miodrag Nikolić
7.AČ CO kap Ik Borivoje Nikolić
64.BG - based at Stubol
CO maj Branko Fanedl
208.E CO kap Ik Sima Mijušković
209.E CO kap Ik Dušan Milojević
210.E acting CO kap Ik Vojislav Grujić (kap Ik Živojin Lazić was jailed prior to the war)
8.AČ CO kap Ik Čedomir Jovanović

3.BP was tasked with offensive actions against the western parts of Bulgaria as well as the support of the armed forces along the border regions. Two days after the secret mobilization order from 4 March 1941 by the VVKJ HQ, 63.BG relocated from Skoplje to Petrovac auxiliary airfield. On 12 March 3.BP ordered in secret its reservists to report in for exercise.

The same day 64.BG flew over towards Stubol and Obilić auxiliary airfields. A critical mistake was made at the time by transporting the ammunition only to those auxiliary airfields where the airplanes were based, ignorning completely alternate loca-

As evident from this photograph, the airplane came to a more or less gentle landing in a forest close to Vardar river. [Dinko Predoević]

Do 17 from 207.E destroyed at Petrovac during the well organized attack. [Jan van den Heuvel via Aleksandar Ognjević]

207.E Do 17K remains at Petrovac. [Djordje Nikolić]

tions where airplanes may need to relocate to avoid detection.

Until the coup d'état on 27 March, 3.BP spent intensively preparing for offensive actions, with airplane maintenance taking place as well as practice bomb runs with concrete bombs over Romanovci range.

In the afternoon of 5 April 1941 28 armed Do 17K from 63.BG were prepared to take off from Petrovac towards Uroševac auxuliary airfield, however a sudden an unexpected order was received to lower the alert status. It is unclear who sent the order, however due to its timing and trecherous content, it is likely that it was a deliberate act of sabotage. Having heard the news, puk Zdenko Gorjup ordred bombs removed from the airplanes and he sent the men from 205.E and 206.E on leave for the evening. Puk Gorjup issued the same order to 64.BG CO maj Branko Fenedl, who rightfully ignored it. By the evening, specifically 21:00, VVKJ HQ sent a coded message reversing the prior order however puk Gorjup brushed it off with an excuse that there is

time to relocate the airplanes next morning. This was a fatal, and likely intentional, mistake as the war began the next morning on 6 April.

207.E pilot kap IIk Drago Krivokapić recalls the early morning attack on 6 April:

I went towards the window of the office where i slept and observed four airplanes coming from the direction of Pčinja, from the east side of the airfield. I thought that those were our fighters based in the vicinity of Kumanovo and that they were flying so early in order to defend the border areas with Bulgaria. The highest alert status was called off by puk Zdenko Gorjup. 206.E and 205.E officers went on leave to the city last night and they did not return until eight in the morning. Mato Čulinovic, who was promoted to major, took them to Skoplje to celebrate his promotion. My 207.E remained on alert at the airfield. As I mentioned, I thought that those airplanes which were coming from the direction of Bulgaria belonged to us, but I immediately noticed German markings. They were Stukas. They flew in the Latin letter V formation at an

DFA produced Do 17Ka-3 EvBr. 3357 at Petrovac with German soldier posing on one of the starboard engine. Note other destroyed Do 17K which were placed close together. [Jan van den Heuvel via Aleksandar Ognjević]

Another destroyed Do 17 from 207.E at Petrovac. [Jan van den Heuvel via Aleksandar Ognjević]

The same airplane from another angle. Intact 607.TE biplanes are visible in the background. [Jan van den Heuvel via Aleksandar Ognjević]

altitude of 2,000 m, and when they approached close to our airfield, they lined up one after the other at a distance of 500 m. When their leader arrived above the airfield he turned left in the direction of the wind. He flew towards a machine gun nest which was located along the airplanes. Bomb fell close by causing no damage. The second Stuka bombed another machine gun nest on the opposite side, also without effect. The third Stuka bombed a machine gun nest on the opposite side while the fourth dropped bombs on a group of parked Dorniers from 205.E, which were lined up very close to one another, almost grouped. Following the bombing, one by one, flying low above the hills, Stukas

239

Dornier produced Do 17K remains at Petrovac. [Jan van den Heuvel via Aleksandar Ognjević]

Do 17Ka-1 EvBr. 3319 captured at Petrovac and later on pressed into Bulgarian service. [Peter Petrick via Denés Bernád]

The tail of Do 17K-1 EvBr. 3319. Note light damage to the vertical stabilizer. [Peter Petrick via Denés Bernád]

Do 17K-1 EvBr. 3319 along with another Do 17K at Petrovac. [Jan van den Heuvel via Aleksandar Ognjević]

turned in the direction of Bulgaria and disappeared in the horizon. Together with several mechanics, officer on duty and ppor Zečević (nvtč Ljubomir Zečević – a.c.) I ran to the airfield immediately after the attack. When we almost reached the edge of the airfield, we saw tracer rounds flying above and around us. I fell in a ditch by the road, turned around and saw an entire column of airplanes circling to the left of the Karadžica mountain peak, leaving formation one by one and diving slowly at the airfield. The first German airplane flamed a Breguet XIX and one Do 17. One after the other Messerschmitts attacked and after each burst some of our Do 17 would catch on fire. There were 17 Messerschmitts. Following the low level attack, they went into a left turn while climbing and repeating the attack. They actually closed the circle and kept attacking for exactly 21 minutes. As soon as one exhausted its ammunition, it would move away from the airfield and circle at low level. As soon as two airplanes freed up, they would fly in pair towards the border.

The well organized and coordinated attack by *Luftwaffe* caused unimaginable destruction with around 15 Do 17K destroyed while the remaining ones were damaged. 206.E CO and 207.E CO, kap IIk Mihajlo Djonlić and kap IIk Miodrag Nikolić

DFA produced Do 17Ka-3 EvBr. 3367 was one of the last delivered airplanes. Port wing has small diameter "Kosovo Cross" insignia. [Jan van den Heuvel via Aleksandar Ognjević]

Port engine from a captured EvBr. 3319 as seen from the cockpit. In the background are destroyed Do 17K and 607.TE biplanes. [Jan van den Heuvel via Aleksandar Ognjević]

EvBr.3367 as seen from another angle. Note that port wing does not have "Kosovo Cross" insignia on the underside. [Jan van den Heuvel via Aleksandar Ognjević]

206.E Do 17K at Petrovac. Note that the port tire is flat. [Jan van den Heuvel via Aleksandar Ognjević]

Do 17K at Petrovac inspected by German soldiers. The machine in the background belongs to 206.E. [Jan van den Heuvel via Aleksandar Ognjević]

respectively, ordered their men to immediately begin repairs to the damaged airplanes and to remove any usable equipment from those which were destroyed. Crews were able to repair only two airplanes from each escadrille preparing them to fly over to Uroševac auxiliary airfield. Mato Švarc ordered nvtč Stanko

Kovačević to immediately take charge of puk Gorjup's Do 17 at Skoplje and fly it over to Uroševac awaiting further orders. The first airplane to attempt the flyover crashed not long after takeoff, due to apparent damage to its controls which was not detected on the ground. This airplane, which belonged to 206.E, crashed on

Do 17Ka-1 EvBr. 3319 with German soldier standing through the open emergency hatch. [Jan van den Heuvel via Aleksandar Ognjević]

German soldiers inspecting EvBr. 3319. [Peter Petrick]

A line up of Do 17K captured by Bulgarian forces at Petrovac airfield. [Jan van den Heuvel via Aleksandar Ognjević]

DFA produced Do 17Ka-3 EvBr. 3364 belonging to 206.E. This machine was captured at Preljina. [Djordje Nikolić]

Do 17Ka-3 EvBr. 3366 attempted a flyover to the Soviet Union with nine on board, but it crashed as a result of stormy weather in the vicinity of Cândeşti village in Romania. [Denés Bernád]

Do 17Ka-3 EvBr. 3333 piloted by nvtč Ivan Pavelić while attempting to fly over to the Soviet Union force landed at around 15:30 3 km from Nagyskomkút and ended on its nose [Gyorgy Punka via Milan Micevski]

the side of the hill, fortunately not catching fire but crew sustained serious injuries. Nar Aleksandar Cucić was aboard and gave his acoount of what took place during an interview:

The airplanes from our escadrille were positioned little to the corner of the airfield, so they did not fare as bad as the others. After the attack, kapetan Djonlić ordered the transfer of remaining airplanes to Uroševac. I took off as a mechanic-gunner in an airplane piloted by nar Radomir Nedeljković 'Džigeran', with mechanics vod Milivoje Ilić and nar Vladimir Kolarić. We did not even reach 300 meters, when all of a sudden we fell. It seemed that horizontal rud-

Do 17Ka-3 EvBr. 3362 from 210.E was captured at Kapino Polje. This airplane was deliberately sabotaged but was later repaired by the Italians. [Giancarlo Garello]

Do 17Ka-3 EvBr. 3362 at Kapino Polje. Note the row of captured Bücker Bü 131 trainers. [Giancarlo Garello]

Do 17K EvBr. 3362 at Kapino Polje. The airplane remained in pristine condition. [Aviation Museum – Belgrade]

Two Do 17K were destroyed at Mostar, of which one was EvBr. 3326. Here Italian soldiers are inspecting the wreckage. [Statto Magiore Aeronautica]

Italian soldiers inspecting what used to be nose mounted Browning FN machine guns. [Statto Magiore Aeronautica]

An Italian soldier inspecting bomb crates. In the background is the second destroyed Do 17K at Mostar airfield. [Statto Magiore Aeronautica]

der cable was cut by some shrapnel without anyone noticing it. Nedeljković gave him the left foot (pushed left rudder pedal – a.c.) and turned the airplane so as not to fall straight on the nose. Instead, we crashed on the side of a small hill, the engines tore off but fortunately we did not catch fire. Nedeljković, with broken arcade and thinking his eye has leaked, got out of the wreck and ran away. Ilić broke a thigh bone and had a sprained shoulder, Kolarić had his left leg crushed. I spit eighteen teeth after striking the machine gun and since then I am toothless! I also had cuts on throat and head. I was pressed by turret and barely got out. After helping Kolarić and Ilić out and giving them first aid, I walked in the direction of the airfield and fell. Next thing I knew was hearing someone telling 'Sir, this one is alive'. They placed us on a truck and Djonlić threw his leather coat over my body. We were taken to a military hospital in Skoplje and there I met Nedeljković in a cellar during an air raid.

Do 17Ka-3 EvBr. 3363 arrived to Paramythia on 14 April 1941. [Ronald Dudman via Ian Carter]

Kap Ik Dušan Milojević and a British soldier in front of EvBr. 3363 at Paramythia, 14 April 1941. [Ronald Dudman via Ian Carter]

EvBr. 3363 belonged to 209.E but the lack of the white circle on the fuselage is noticeable. [Ronald Dudman via Ian Carter]

On 19 April 1941 EvBr. 3363 and 3348 along with two Savoia-Marchetti Sm.79 bombers arriver at Heliopolis. [Len Cooper via Ian Carter]

On 15 April Italian Macchi C.200 fighters swept in low and strafed Yugoslav bombers at Paramythia, destroying five and damaging Do 17Ka-3 EvBr. 3348. [Giancarlo Garello]

German technical commission during a visit to DFA. Despite heavy damage, the commission concluded that rebuilding the factory is feasible, as it was needed to support the war effort of the Third Reich. [Aleksandar Ognjević]

Exterior view of the damage to the DFA building. [Aleksandar Ognjević]

Interior view of DFA also shows the scale of destruction at the factory. [Aleksandar Ognjević]

A large amount of collapsed materials as a result of the demolition. [Aleksandar Ognjević]

Looking at the level of damage at DFA, it is unclear how German commission concluded that rebuilding the factory was profitable? [Aleksandar Ognjević]

One of the destroyed machining sections at DFA. Special tools manufactured by famous Schuler company are visible. This company still exists today. [Aleksandar Ognjević]

Part of the roof at the factory collapsed. [Aleksandar Ognjević]

A member of German commission observing one of the machining departments at DFA, which was left in complete chaos as a result of deliberate explosions. [Aleksandar Ognjević]

Besides the obvious evidence, Royal Yugoslav intelligence officer responsible for destruction of the factory was after the war blamed by the communist that DFA was "intact" when it was captured by the enemy!!! [Aleksandar Ognjević]

On 11 April, Army engineering units were at another location in Kraljevo, unable to detonate the explosive charges at DFA. Therefore, the airplanes undergoing assembly were doused in gasoline and torched. Note "war flags" and still unpainted rudders. [Jan van den Heuvel via Aleksandar Ognjević]

A total of four unfinished Do 17Ka-3 were captured at DFA in final stages of assembly. [Jan van den Heuvel via Aleksandar Ognjević]

Two of the captured Do 17Ka-3 without glazed nose sections. [Jan van den Heuvel via Aleksan-dar Ognjević]

Overhead view of the same Do 17Ka-3. [Jan van den Heuvel via Aleksandar Ognjević]

Behind the railing is a large crate with brand new Gnome-Rhône engine. [Jan van den Heuvel via Aleksandar Ognjević]

A very interesting photograph showing an unknown Do 17 with a belly turret, similar to that on Do 215, which could have been an entirely ingenious design or could have been replicated from drawings provided by Dornier. [Jan van den Heuvel via Aleksandar Ognjević]

The remaining airplanes flew over to Uroševac successfully, the airplane piloted by nvtč Ilija Petričić was unable to retract its landing gear, likely again due to damage from the strafing attack. Still, the flight was condluded sucessfully but on landing the airplane got stuck in mud and remained there for the rest of the war where it was captured. Once at Uroševac, as a result of the critical mistake not to transport ammunition to all auxiliary airfields, no offensive actions could take place from this airfield for the remainder of the day. 3.BP CO puk Gorjup's passive actions led to his deputy maj Dragomir Žikić issuing orders to make good of what was left of their airplanes. He intended to join 64.BG in offensive operations.

At 7:00 3.AR (*Aeroplanska radionica* – Airplane Workshop) with its mobile workshops arrived at Petro-

The only surviving artifact from VVKJ Do 17K is this nameplate from W.Nr. 2474 which was captured and sold to NDH. [Jan van den Heuvel via Aleksandar Ognjević]

A well known photo of Do 17Ka-3 EvBr. 3363 at Heliopolis. [Aviation Museum – Belgrade]

Do 17Ka-3 EvBr. 3363 in Egypt still wearing the Yugoslav "war flag" on the rudder. White DFA logo is also visible. [Djordje Nikolić]

Do 17Ka-3 EvBr. 3348 was handed over to RAF where it received new code AX 707. [Djordje Nikolić]

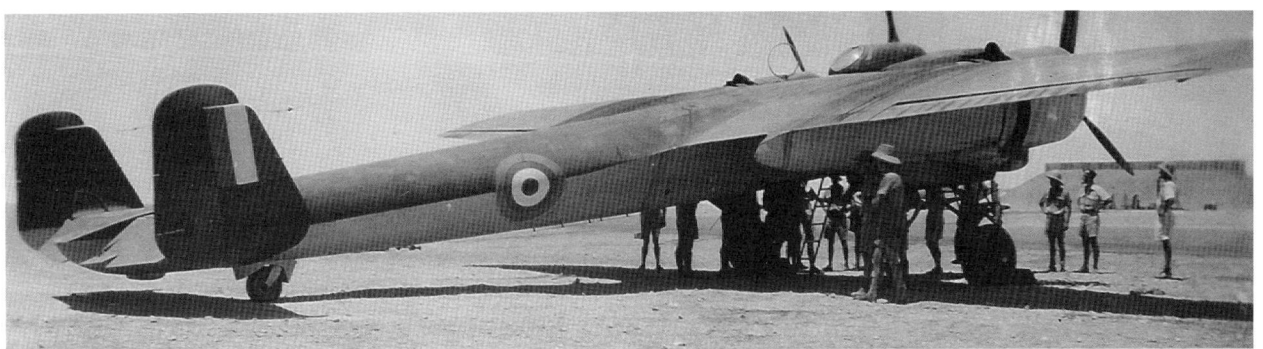

102. Maintenance Unit took charge of Yugoslav Dorniers. EvBr. 3362, now known as AX 706, inspected by curious British soldiers. [Aviation Museum – Belgrade]

The same crowd around ex EvBr. 3362. [Peter Petrick]

vac to assist with the repairs. Three Do 17 were repaired as a result but another group of Messerschmitts arrived at 13:10 destroying another five Dorniers. Do 17K EvBr. 3314 was the least damaged, while 3305, 3319, 3343 and 3350 were beyond repair. The next day 3.AR managed to complete repairs on another Do 17Ka-1, 3314, fol-

lowing which this airplane flew over to Uroševac and joined the three already Do 17 there.

The men from 205.E were passive during the first two days of the war, mostly due to their CO maj Mato Čulinović's lack of action to assign orders to repair damaged airplanes. Failing to heed to the orders to reaterat

AX 706 and 707 had British roundels applied in place of the "Kosovo cross" insignia. [Peter Petrick]

AX706 warming up its engines prior to a flight. [Robert Čopec]

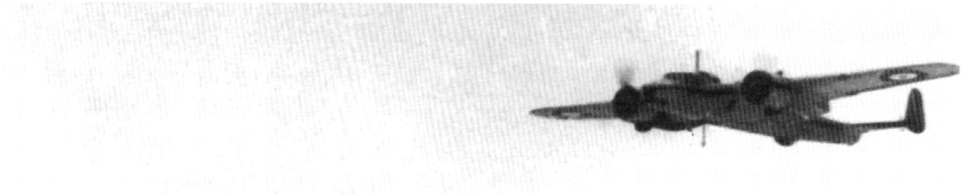

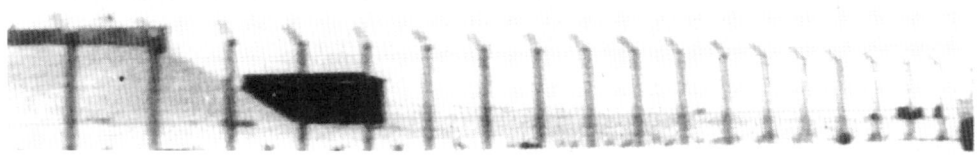

AX 706 in low level flight above the Egyptian landscape. [Peter Petrick]

Aft view showing lack of insignia on the wing topsides. [Peter Petrick]

Head on view of AX 706. Both airplanes were damaged on 27 August 1941 during an aerial attack and were struck off charge on 12 September. [Peter Petrick]

on 7 April, Čulinović relinquished his command and left his men to fend for themselves. The majority of 205.E men were captured but some managed to reach and join 64.BG.

3.BP began its retreat imminently before Germans captured the airfield heading southwards. Some were captured but those who avoided capture arrived at Athens on 15 April and boarded a British merchant ship two days later. It is interesting to note that 3.AR managed to retreat with three of its mobile workshops which found their way to Egypt where they were assigned to No 2 Yugoslav Squadron flying Do 22s.

Two Do 17K which remained at Skoplje airplane workshop, EvBr. 3339 and 3341, were not repaired in time and were captured by German forces the next day, 7 April.

64.BG managed to avoid the German surprise attacks during the first day of the war but were hit hard by the news of disarray coming from 63.BG. Having ignored the order to disarm the airplanes and call for a night's leave, maj Branko Fenedl was ready to call

Air Chief Marshall Sir Arthur Longmore inspecting one of ex Yugoslav Dorniers in Egypt. [Flight magazine]

After repairs at Kapino Polje, Do 17Ka-3 EvBr. 3362 was flown over to Tirana, Albania. [Luigi Gorena via Giancarlo Garello]

EvBr. 3362 received Italian markings before being transferred to Guidonia test center. [Luigi Gorena via Giancarlo Garello]

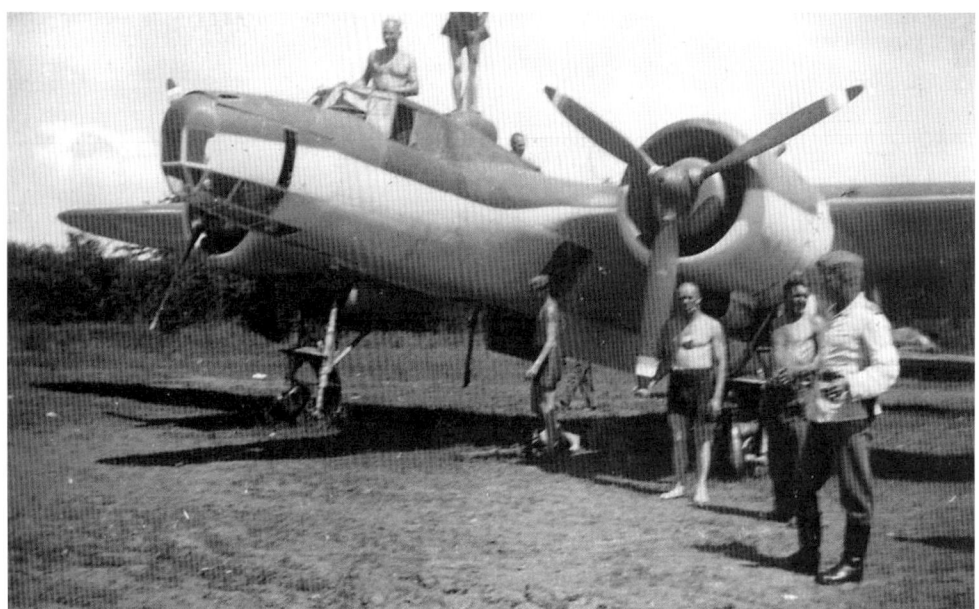

Bulgarian soldiers posing with one of the captured Do 17K at Petrovac. [Jan van den Heuvel via Aleksandar Ognjević]

Airplanes were neatly lined up awaiting repairs prior to being put back into use. [Jan van den Heuvel via Aleksandar Ognjević]

Bulgarian soldiers posing in front of a Do 17K which received a new livery. [Jan van den Heuvel via Aleksandar Ognjević]

his men to arms and issue the orders for first offensive actions of the war. This action was described by nvtč Milan Srdanović, who took off in the lead Do 17K from a total of three machines belonging to 208.E:

Our section was ordered to head towards the Bulgarian border and bomb enemy troops which were advancing along the road from Ćustendil to Kriva Palanka, if they were there and if not, to attack Radomir airfield in the vicinity of Sofia. It was cold, early spring morning. The villages beneath us were waking up, likely unaware of what was happening at our borders. Sharp sound of our engines tore through the fresh air. The visibility was clear. Only a few white clouds whose shadows quietly pass above the scenery of the hazy horizon. We immediately observed the enemy. An endless column of motor vehicles was traveling from within Bulgaria towards our country, and the head of it reached our border already. Those were our old acquaintances, acquaintances of our fathers and brothers, from Cer and Rudnik, from Dobro Polje and Kajmakčalan (places from World War I battles – a.c.)…

The game of death has begun. We jettisoned our bombs from the appropriate altitude. The effects were satisfactory. We had no losses. Once we returned, we discovered numerous bullet holes on our airplane.

Another nine airplanes, this time from 210.E, took off at 06:10, followed some five minutes later by four more from 208.E. Nine airplanes from 209.E were the last to take off, some one hour later, losing the element of surprise. One airplane from 210.E and one from 209.E had to return due to malfunction. 64.BG was ordered to attack a long column of enemy vehicles traveling on the road to Kriva Palanka. This well organized attack, with well armed airplanes and motivated crews caused havoc amongst the enemy

with numerous hits scored on the vehicles and men on the ground. Three airplanes from 210.E bombed the main street in Ćustendil as the German troops passed through the city resulting in dozens of civilians dead and only two German soldiers killed.

All airplanes successfully returned to their airfields. Immediately after their return, crews began repairing any damage, refueling and rearming them for further action. When the necessary preparations were complete, a number of airplanes belonging to 209.E took off at 09:30 and attacked the same enemy column at Kriva Palanka again inflicting damage. No further flights took place until 11:00 when nine airplanes, likely alerted by German troops under attack, appeared at low altitude. Bf 109E from II.JG/27 strafed Dornier bombers destroying EvBr. 3336 from 64.BG, four from 208.E, one from 209.E and four from 210.E. 208.E had one airplane damaged, 209.E three and 210.E two. One from each escadrille was repaired by the next morning.

The surviving airplanes were immediately manned and their crews used radio direction finder to locate the airfield where these fighters originated from and attacked it. They caught the enemy by surprise damaging several airplanes while three Bf 109E-7s allegedly crashed while trying to take off and intercept them. The account of kap IIk Dimitrije Naumović who flew as the observer in the lead Dornier is as follows:

Before we arrived at the enemy airfield at Radomir (Vrba – a.c.), we arrived over the crossroads with numerous enemy vehicles to the west of the airfield and bombed it with 100 kg bombs. Bombs were released on target from the altitude of 600 meters. There were several enemy airplanes at the airfield which were not moved to a shelter. Bombs fell on the airfield. After the bombing we flew

Three ex VVKJ Do 17K at an airfield in Bulgaria. [Jan van den Heuvel via Aleksandar Ognjević]

over Radomir where we were shot at by enemy AA, but without success. Immediately thereafter we were attacked by two enemy fighters. Only one of them opened fire at us, hitting our airplane with one burst in the nose, another by the machine gun turret, but he did not manage to cause any serious damage. The mechanic-gunner nar Ratković Čedomir returned fire at the enemy airplane with a machine gun from the turret and likely hit it, considering that it suddenly turned and left. The second enemy airplane did not even attempt to attack us, because we in the meantime increased speed and altitude.

Maj Fanedl asked kap IIk Naumović how come he dared to follow and attack the enemy to which he responded confidently:

Major Sir, they exhausted their ammunition, while I had four ready machine guns and 800 kilograms of bombs. Therefore, they were unable to attack me. Even if they were to have noticed me, they would have thought I was one of theirs, because they have the same type of airplane. It was most important to me that they lead me to their airfield, so I don't have to look for it. That is exactly what happened.

The second Do 17K, with pilot ppor Mileko Tošić observer por Zlatko Kudrna, lost the lead airplane and turned towards the capital of Bulgaria, Sofia, where the crew jettisoned eight 100kg bombs in the suburbs around 14:05 killing six civilians.

Further attacks continued in the afternoon when four Do 17K from 210.E attacked a German column along Kriva Palanka – Stracin road. One of the attackers with pilot nar Radoslav Nestrović was heavily damaged and force landed at Samodreža auxiliary airfield. No other actions took place for the remainder of the day but two of the escadrilles changed their locations with 208.E relocating to Plemetina and 210.E to Samodreža

where a brand new Do 17Ka-3 EvBr. 3368 was delivered from DFA. 209.E remained at Stubol. The evening was used to service the airplanes and prepare plans for the offensive operations for the next day.

Before dawn five machines from 208.E led by kap Ik Miljušković took off and at around 05:00 and attacked a large formation of parked enemy vehicles in an open field. During a low level bombing run they dropped their bombs precisely. Several airplanes were damaged during the attack but they managed to land at the auxiliary airfield at Peć where they were joined by another airplane from their escadrille which did not take off during the early morning attack. 210.E with four machines attacked enemy in the same area, while three joined the 208.E airplanes at Peć and one returned to its originating airfield at Samodreža. Do 17Ka-3 EvBr. 3366 and 3368 were ordered to fly to Samodreža.

Two airplanes from 209.E took off from Stubol and one from Laplje Selo (aborted soon after takeoff and returned due to faulty bomb jettisoning mechanism) heading towards Ihtiman airfield in Bulgaria. As this airfield was abandoned, crews led by kap Ik Dušan Milojević attacked Radomir railway station. During their return a Ju 87 Stuka attempted to attack the lead Dornier but it faced return fire and was allegedly shot down. Two 209.E machines landed at Laplje Selo while the remainder of escadrille's airplane arrived there early in the morning the next day.

During the afternoon, five Dorniers from 208.E, several airplanes from 209.E and three from 210.E (of which one was EvBr. 3366), attacked German troops between Stracin and Kumanovo. Three surviving Do 17K from 63.BG flew over to Peć and joined 208.E as an independent section. This group flew over to Uroševac as bombs were stored at that location. Along with EvBr. 3314 which arrived from Petrovac they attacked at 17:30 a bridge spanning river Pčinja in the vicinity of Stracin, missing their target but scoring hits on enemy vehicles located there. Due to the heavy anti-aircraft fire some of the airplanes sustained serious damage.

On 7 April at 04:50, seven airplanes form 208.E led by maj Dragomir Žikić took off to attack an armored column near Djeneral Janović railway station. This target was well defended and several airplanes were damaged, of which that flown by nvtč Stanko Kovačević, who flew over to Plementina the last evening to join 208.E, had to force land at Peć. At 05:00 two detachments from 209.E each with three airplanes, took off to bomb vehicle column at the entrance to Kačanik gorge from an extremely low altitude of only 150 m causing havoc on the ground. At 05:15 another two detachment from 210.E each with three airplanes followed but having lost the elements of surprise they faces stronger defenses which included a pair of Me 110 from II./ZG26 which shot down the airplane flown by nar Josip Telar, who survived by parachute while the rest of the crew, nar Životije Kojadinović and nar Dušan Mihailović perished. Despite the losses and knowledge that Germans are standing by for another air

attack, two more detachments from 209.E bombed the same target. Nar Ivan Oreškovic, who was the machine gunner in the airplane flown by por Desmir Smiljković described this engagement:

When the observer dropped the bombs, I machine gunned the column and observed direct bomb hits which was the proof of the precise and correct flight direction. All eight bombs fell from the altitude of 150 meters. The enemy opened strong machine gun fire and our airplane was heavily damaged. The rudder controls were not responding, the fuel tank was punctured and landing flaps were inoperable. Thanks to the engines and ailerons, the pilot managed to return to Stubol.

During the same engagementnvtč Petar Simović landed his Do 17Ka-1 EvBr. 3313 whose tail caught fire at Stubol. Knowing of the importance of halting the advance of the German column which had the limited ability to maneuver in the Kačanik gorge, another three machines from 209.E were sent to attack, two returned to Stubol while one landed at Samodreža.

13 airplanes which were gathered at Peć around 10:00 in the morning were preparing for repeated attacks against Kačanik gorge when an enemy airplane circled above. To avoid detection and destruction, kap Miljušković ordered them immediately to take off, however not all airplanes were able to leave on short notice. Still, this was proven to be a wise decision as this group of Dorniers observed enemy airplanes in

Do 17K with 1./5.BO emblem below the cockpit. [Djordje Nikolić]

259

The only Do 17K used in Hungary was EvBr. 3333 which was modified with the installation of four GKM 8 mm machine guns in the nose and two reconnaissance cameras. [Aviation Museum – Belgrade]

the distance which were heading towards the airfield they originated from. German airplanes did attack Peć airfield destroying Do 17Ka-1 EvBr. 3303 from 210.E and damaging Do 17Ka-2 EvBr. 3334 and a number of others. Two serviceable airplanes from 210.E left Peć towards Tatojevica.

EvBr.3310, 3344 and 3313 belonging to 209.E were destroyed at Stubol during a strafing attack by Bf-110 from II./ZG26 while the surviving airplanes flew from Stubol and Samodreža towards Kraljevo and Tatojevica in the afternoon.

Due to worsened weather on 8 April, 3.BP did not conduct any offensive operations but a sole Do 17Ka-1 EvBr. 3318 took off from DFA at Kraljevo under the commands of kap Ik Djordje Jankovski and attacked German troops in the vicinity of Niš. After refueling and rearming the pilot flew the airplane this time towards Kačanik gorge where alert German troops heavily damaged the attacking Dornier which managed to return to base but suffered a collapsed landing gear leg.

As a result of poor weather also on 10 April, only two flights took place. EvBr. 3346 with maj Žikić and 3364 with nar Esad Halibegović at the controls transferred to Preljina and Kraljevo respectively. When the weather improved, these airplanes flew over to Tatojevica.

Crews began discussing how to escape to safety, options being Greece or Soviet Union. The first airplane to set on this risky voyage piloted by ppor Milenko Tošić was Do 17Ka-3 EvBr. 3366, which attempted a flyover to the Soviet Union with nine on board, but it crashed as a result of stormy weather in the vicinity of Cândeşti Deal village in Romania. According to the reports from the witnesses in the village, the airplane was in flames before it struck the ground, which does not exclude that it was hit by anti-aircraft defences. The crew was buried there.

The second airplane Do 17Ka-2 EvBr. 3333 piloted by nvtč Ivan Pavelić force landed at around 15:30 three km from Nagyskomkút and ended on its nose. It was later repaired and pressed into Hungarian service. The last airplane to attempt fly over towards Soviet Union was Do 17Ka-2 EvBr. 3326 which was flown over by kap Ik Momčilo Petrović to Ortješ airfield near Mostar but his airplane was disabled by Ustaša sympathizers, where it was destroyed during an attack by the Italian fighters and bombers on 13 April.

The weather on 11 April also prohibited any flying activities, which did not prevent German advance deeper into the Yugoslav territory. The next day, after the weather improved, three airplanes led by kap Ik Mihajlo Djonlić attempted an attack against a bridge across Morava river near Ćuprija, however their attack was unsuccessful and the bridge was captured intact. Three additional, brand new DFA made Do 17Ka-3 were available at Kraljevo, and ppuk Leonid Bajdak, 5.LP CO, wanted to make good of them being there but lacked pilots. He managed to find volunteers from 209.E troops in retreat and along with three pilots from 5.LP formed crews which attacked German vehicle column. The account of one of the 5.LP pilots, kap Ik Velimir Veličković is as follows:

Since two brand new Do 17 were already loaded with 100 kg bombs, they took off on the mission first. Captain Todorović and I had problems starting the engines. I cannot exclude sabotage by the factory workers, therefore we took off somewhat later. By then the first two airplanes were already returning from the mission and were flying in the direction of Užička Požega (Gorobilje – Tatojevica auxiliary airfield – a.c.) for landing. We flew extremely low, at 300 meters when we left Velika Morava valley. I observed German mechanized columns along Ćuprija – Jagodina road. Then I ordered the pilot to turn, so I was able to begin bombing of the entire column from the end to start. Pilot Todorović immediately followed the order. As soon as the enemy saw us, they immediately stopped. German officers were likely convinced that our Do 17 was armed with 100 kg bombs like the previous two, because there were no visible external differences between our three airplanes, they ordered vehicles to be abandoned and to disperse in the fields by the road. That was an ideal target for us! Bombing was now completely safe, because all German soldiers were actually unarmed, which is proven by the fact that we did not get a single bullet strike in our airplane during the entire bombing. Lying on my stomach above the bombsight and to the right of the pilot (Todorović was sitting somewhat above me), I observed that he did not have a good view of the target below us. Accordingly, he could not zig zag the road enabling me to drop bombs in groups of two or four. That was the reason why I had to grab hold of the column with my left hand and zig-zag the airplane. The pilot in the meantime assisted with the rudders and held the column. At the same time, I was turning the lever with my right hand and dropped bombs. Such "double Niš command" functioned perfectly all the way to the last bomb and we bombed Germans from Stalać almost until Jagodina.

Following this successful attack, they landed at Tatojevica where the other two Dorniers joined them after refueling at Kraljevo. They took off then along with another four Dorniers to bomb the same column. The first three airplanes landed at Rajlovac airfield near Sarajevo following the conclusion of their mission and then flew over to Butmir. The four Dorniers also completed their mission. Another four unknown Dorniers completed the final group mission of the day at around 16:00, however one of them was damaged and barely made it to Tatojevica where it was captured. The last solo mission by ppor Ratomir Manojlović took place in the late afternoon bombing German armored vehicles in the vicinity of Kragujevac all while evading German fighters.

On 13 April at 04:30 seven Do 17s in two groups took off from Tatojevica. Airplanes flown by maj Branko Fenedl and nar Ivan Babogredac led the formation towards Jagodina – Ćuprija road where they attacked a vehicle column from low level. Only two airplanes managed to drop bombs successfully, two were unable to get on target so they dropped their bombs in a forest and landed at Butmir. The remaining two airplanes split up above Kragujevac due to strong anti-aircraft fire, receiving numerous hits and returning back to Tatojevica. Nar Babogredac managed to fly his airplane back to Tatojevica where his airplane was captured along with the other two which returned there. The lead airplane belonging to maj Fanedl was severely damaged by anti-aircraft fire. Two of the crew including maj Fanedl and kap Ik Teofilov attempted unsuccessfully to bail out while gunner nar Franjo Ribič crashed along with the airplane.

As the retreat of Yugoslav forces continued, it was up to VVKJ to ensure that the King, Prince Peter II, and other government officials reach the relative safety of an allied country not yet overrun by Germans. As a result, early on 14 April kap Ik Dušan Milojević flew from Butmir to Kapino Polje near Nikšic and then to Paramythia in Greece where he transported a Greek liaison officer who was to establish contact with the Greek and British forces to ensure a safe escape evacuation. Kapino Polje is where a total of seven available Do 17 from 3.BP flew over to fulfill this mission. They relocated shortly thereafter to Grab auxiliary airfield and then back to Kapino Polje where a group of fighter pilots embarked on them to be transported to Paramythia on 15 April. Do 17Ka-3 EvBr. 3362 was sabotaged likely by one of the mechanics who switched its magnetos due to the spite that he was unable to escape with the higher ranking officers. This airplane was as a result captured by the Italians. Six Do 17 which managed to take off arrived at Paramythia around 16:00 but some ten minutes after their landing Italian Macchi C.200 fighters swept in low and strafed Yugoslav bombers, destroying five and damaging

Do 17Ka-3 EvBr. 3348. This airplane was repaired the next morning.

The last flight from KJ by VVKJ Do 17K was with kap Ik Milojević on board Do 17Ka-3 EvBr.3363 which flew towards Agrinion and then Athens with 60 kg of gold. Two days later, this airplane flew over to Heliopolis in Egypt.

Two Do 17K from 3.BP, of which one was Do 17Ka-2 EvBr. 3326, were assigned to VŠB at Jasenica near Mostar. When the Italians attacked in the early hours of 6 April, they caught the airfield defenses by surprise and destroyed numerous airplanes on the ground. Two Dorniers managed to avoid destruction but were damaged on purpose by own crews. On 13 April, there were plans to repair them so they could join the rest of the airplanes at Kapino Polje prior to the evacuation to Greece however the Italian Stukas and fighters repeated the attack at the airfield that day, destroying both of them.

3.BP had 63 Do 17K on strength, of which 60 were serviceable and 23 airplanes survived German attacks. On the first day of the war 3.BP had over 30 combat flights without losses. Following the fiasco at Petrovac, 3.BP used the correct tactics, where airplanes constantly changed their locations which most certainly saved them from certain and early destruction.

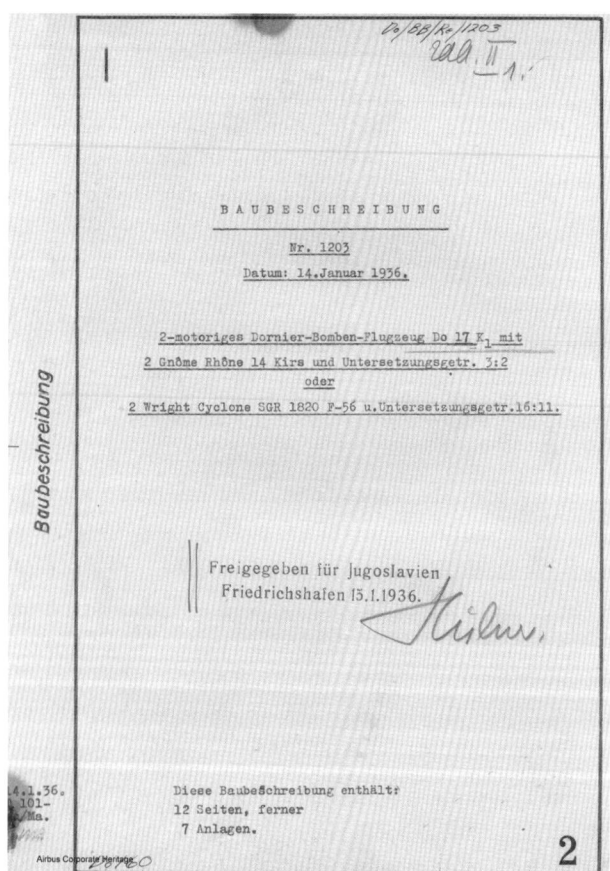

Technical specifications document No. 1203 for Do 17Ka-1 was released to KJ on 15 January 1936. [Airbus Corporate Heritage]

The sole surviving part from a VVKJ Do 17 is a nameplate from Do 17Ka-3 W.Nr. 2474 (Br. 3334) which was captured at Peć on 17 April and was later sold to ZNDH.

VVKJ Do-17s in Foreign Service

Only two airplanes managed to escape KJ and reach the Allies. Do 17Ka-3 EvBr. 3348 and 3363 reached Egypt where the crews were ordered to hand them over to the RAF on 6 May 1941. Airplanes were assigned to 102. Maintenance Unit where they received new code numbers AX 706 (EvBr. 3363) and AX 707 (EvBr. 3348). Due to RAF's lack of interest and need for these airplanes, they were used only sporadically until they were damaged in the German attack on 27 August at Ismalia airfield and subsequently struck off charge on 12 September.

Germans captured a total of 18 Do 17 of various versions. EvBr. 3305, 3319, 3343 and 3350 were captured at Petrovac while 3339 and 3341 were captured at Skoplje. EvBr. 3323, 3308 and 3314 were captured at Tatojevica while 3318, 3370, 3371 and 3372 at DFA. EvBr. 3364 was captured at Preljina, 3313 and 3344 at Stubol and 3334 at Peć. These airplanes were handed over to ZNDH.

Do 17K following repairs at DFA were delivered to ZNDH where they flew throughout the war. One of them defected in April 1945 to the Partisans. In total only three survived until the end of the war.

Do 17Ka-3 EvBr. 3362 captured at Kapino Polje by the Italians had its magnetos repaired and it was transferred to *1° Centro Sperimentale di Guidonia* (1st Experimental Center of Guidonia) for testing where it remained until September 1943.

EvBr. 3305, 3319, 3343, 3350, 3339, 3341 entered Bulgarian service. There they served within 1./5 BO (*Bombardirovičen orlijak* – Bomber Group) throughout the war. Towards the end of the war EvBr. 3339 was captured at Skopje by the Partisans while 3343 was destroyed in a RAF bombing raid. After the war Bulgaria returned four Do 17 in a dismantled state via rail but due to their obsolescence these were not put back into service.

The sole Do 17Ka-3 EvBr. 3333 which landed in Hungarian territory was repaired. During repairs it received a modification which consisted of four GKM (*Gebauer Motorgéppuska*) 8 mm machine guns in the nose and two reconnaissance cameras. With designation J101 it remained in service until 1944.

Construction Features

Do 17 was a high wing monoplane of all metal design consisting of four major assemblies: main fuselage, cockpit, wings and tail assembly.

The cockpit housed the crew of three including the pilot, bombardier/navigator and a radio operator/machine gunner. Pilot sat at the left side of the cockpit while behind him offset to the center sat the bombardier/navigator. Do 17 was equipped with only a single control column while throttle, flaps and other controls were to the left of the pilot. The navigator's seat was collapsible forward which permitted him to aim and release bombs during the attack run. Controls used during the bombing run were mounted on the right side wall towards the base of the cockpit. Behind the cockpit was an open rear facing Plexiglas covered machine gun position, which was longer in shape on Ka-3 than previous versions. The crew entered the cockpit through a hinged hatch with a window between frames 10 and 11 on the fuselage side, while the top of the cockpit had an emergency exit hatch. There was another fuselage access hatch between frames 14 and 15, directly behind the bomb bay.

The narrow, stretched fuselage was a monocoque design with metal skin. Do 17Ka-1 W.Nr. 2382 (Br.2) had a Plexiglas nose cone while all other airplanes had duraluminium ones with additional glazing. The fuselage starts off more or less oval but quickly changes to a trapezoid structure, returning once more to oval section as it nears the tail. Built up frames and intermediate stiffeners were employed in the fuselage construction being notched to receive the channel section stringers, the lips of which are flattened where required, at the intersection with the frame. Fuselage was covered with large formed metal panels.

The cantilever trapezoid wing with rounded tips had sheet metal skin, apart from the fabric skin in the center part of the wing underside. Two spars spanned the entire length of the wing while the main ribs were built up of duraluminium channel sections where intermediate ribs have tubular bracing. Mechanically operated ailerons with trim tabs and flaps were installed on the trailing edge of both wings. On Ka-1 and Ka-2 versions, fabric covering extended between the spars on the lower surface of the wing, while the remainder of the wing had smooth metal covering with flush rivets, while on Ka-3 version the wing was entirely covered with duraluminium panels. The leading edge of the wings was heated by the engine exhaust gases. Navigation lights were housed at the wing tip of each wing while port wing had a landing reflector and the pitot tube at the leading edge.

Tail assembly consisted of metal horizontal stabilizers which had dual metal vertical stabilizers, one installed on each end. Horizontal stabilizers were adjustable from -1° to +13° degrees. Rudders were covered with fabric. Trim tabs and mass balances were incorpo-

Original performance document for Dornier produced Do 17Ka-1 along with engine power as well as dimensional and cutaway drawings. [Airbus Corporate Heritage]

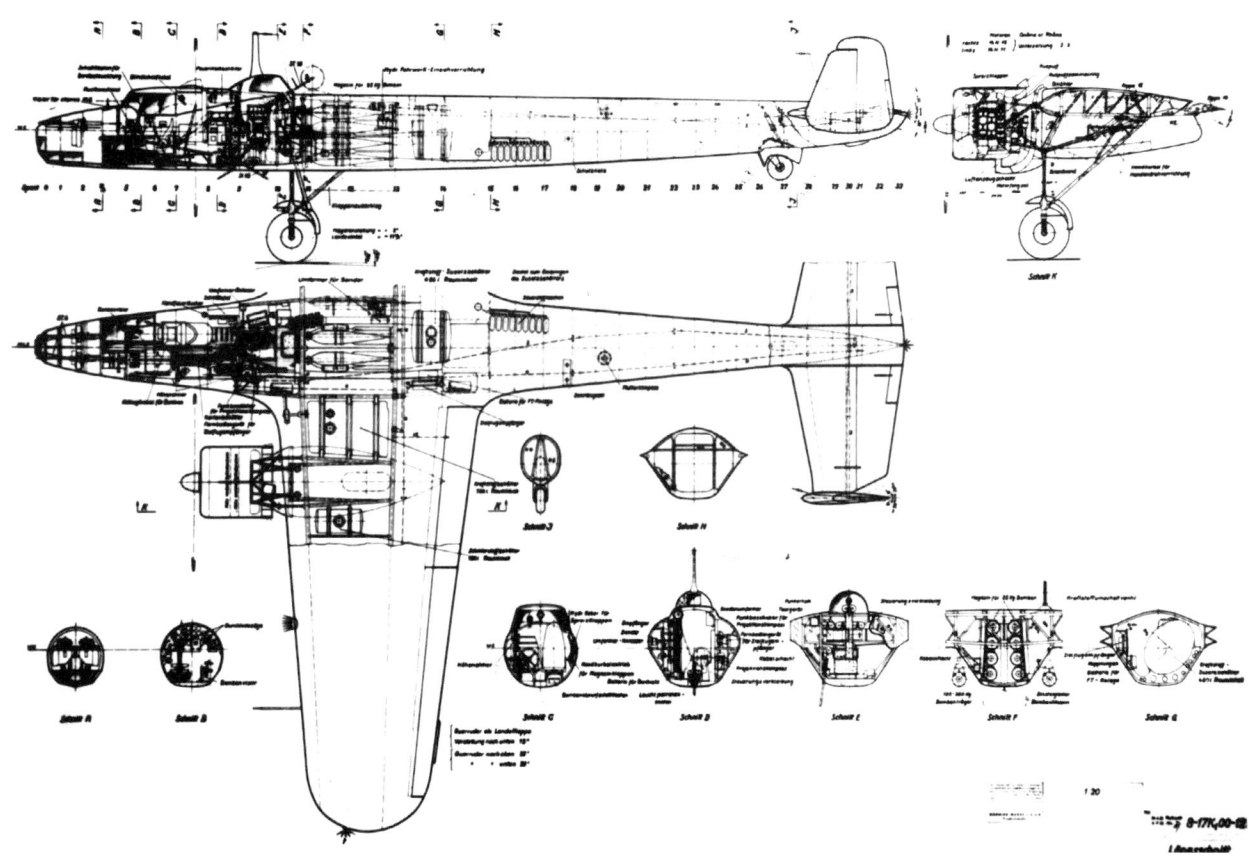

Do 17Ka-1 cutaway drawing dated 25 August 1938 . [Airbus Corporate Heritage]

A detail view of the nose of a captured Do 17K shows. Plexiglas nose is where the bombardier aimed from at the targets below. [Jan van den Heuvel via Aleksandar Ognjević]

Cockpit of DFA produced Do 17Ka-3, likely one of the machines handed over to RAF. [Ronald Dudman via Ian Carter]

rated in all movable tail surfaces. A sole position light was installed at the very end of the tail.

Two engines were installed in nacelles extending from the wing leading edge. Dornier produced Do 17K used a 850 hp *Gnome-Rhône* 14K while those produced at *DFA* used domestically license produced IAM K-14NO engines. Exhaust collector ring removed gases from all cylinders and it terminated on top of the wings. The engines were coupled to three blade VDM (*Vereinigte Deutsche Metallwerke* – United

Do 17Ka-1 interior fuselage structure showing the location where the two spars would be located, above the bomb bay. [Airbus Corporate Heritage]

German Metal works) variable pitch propellers with 3.3 m diameter and the propellers rotated in the opposite direction to counter the inertia forces. A single Do 17 W.Nr. 2383 (EvBr. 3303) was equipped with *Ratier* propeller and required minor modification to the NACA engine cowlings. NACA cowlings of long chord with trailing edge cooling flaps were prominent features. Two Dornier manufactured Do 17Ka-3 machines were used as pattern machines for domestic production and they incorporated longer engine nacelles than the Ka-1 and Ka-2 versions and the spacing between the engines was allegedly some 50 mm longer.

87 Octane fuel was stored in two wing mounted 700 l tanks in between the spars on the inboard sections of the engine and in one fuselage mounted 480 l tank behind the bomb bay. Oil was stored outboard of each engine in the wings in 90 l tanks.

The landing gear was housed in the engine nacelles and was retractable both hydraulically and mechanically. The break equipped wheels folded forward into the engine nacelle and two automatic clam shell doors closed the wheel bay. The tail wheel was also completely retractable. Main wheels had dimensions of 950 mm x 350 mm and the tail wheel 380 mm x 150 mm.

Oxygen apparatus consisted of nine two liter *Dräger* oxygen bottles installed in the fuselage at the trailing edge of the starboard wing and each crew member had an oxygen mask at his position. Radio equipment consisted of FuG III *Telefunken* 274 af radio set, direction finder was *Telefunken* 128H and a radio compass P 63 uN. Radio antenna was installed on top of the rear facing machine gun position cover.

Do 17Ka-1 and Ka-2 produced by *Dornier* were delivered with internal *Maga* 4/85 bomb racks and P.V. 125 external bomb racks. Bombs were carried horizontally between frames 11 and 13, between two

External bomb rack was located immediately behind the entry hatch. [Airbus Corporate Heritage]

Port Gnôme-Rhône 14K engine mated to the engine mount at Dornier factory. [Airbus Corporate Heritage]

Detail view of the Gnôme-Rhône 14K engine with the exhaust manifold installed. [Airbus Corporate Heritage]

Retractable tail wheel detail. [Djordje Nikolić]

Br.2 showing its three tone camouflage scheme. Note the factory logo, airplane designation and number on the vertical stabilizer. [Airbus Corporate Heritage]

wing spars. In total eight 100 kg bombs could be loaded in the fuselage while and a total of two 200 kg bombs could be carried externally. DFA produced Do 17Ka-3 had varying internal bomb arrangements, with either eight 50kg bombs or two D32 bomb racks for a total of 64 12 kg bombs used to attack troop concentrations and vehicles. Several months prior to the war a bomb rack which could carry 12 90 or 100 kg bombs was installed for trials. Tests were conducted in the assembly building with wooden bomb models loaded with lead which was followed by flight testing and live drops of red colored bombs across the Kraljevo airfield. *Goertz* ET 219c bomb sight was installed in the nose in all the models.

Ka-1 and Ka-2 versions were armed with two forward facing machine guns, one fuselage and one belly mounted rear facing machine gun. All machine guns were Belgian 7.9 mm *Browning* FN. All Ka-1 airplanes were modified to house forward facing 7.9 mm FN Browning machine guns in the nose which were accessible through a large removable hatch. A single machine gun was installed in the belly between frames 9 and 10. At DFA, machine guns were bore sighted by the workers by installing a light source on the machine gun mounts which projected a red dot on a grid wall. The machine guns were adjusted so that the red dot intersected with the appropriate grid lines. Following the adjustment of the machine gun mounts, VTZ armorers installed the actual machine guns in their place.

Löschner A.I.47 and A.II cameras could be carried in a photo reconnaissance role although thus was never used in VVKJ service.

Technical specifications

Technical specifications Dornier Do 17Ka-1/2/3	
Quantity used:	68 (4 undelivered)
Crew:	3
Years of Service:	1938-1941
Span:	18.0 m
Length:	16.6 m
Height:	4.5 m
Wing area:	55.0 m²
Engine:	Two 630[1]/850[2] hp Gnome-Rhône 14K
Empty weight:	4,645 kg
Maximum weight:	7,000 kg (7,800 kg Ka-3)
Maximum speed:	420 km/h at 3,480 m
Cruise speed:	374 km/h at 3,200 m
Service ceiling:	9,000 m
Maximum range:	1,030 km
Armament:	Ka-1: Eight 100 kg bombs in the fuselage and two 200 kg bombs beneath wings. Two fixed 7.7 mm Browning FN machine guns in the nose, and one rear facing 7.7 mm Browning FN machine gun in the fuselage turret.

[1] Nominal power
[2] Take off power

Camouflage and Markings

Do 17 was the first Yugoslav twin engine military airplane to be painted in the new three tone camouflage scheme. It was a sole example where color patterns were mostly followed in accordance with the regulations, owing mostly to the existence of an official pattern guide. The machines produced at Dornier factory followed the

Crew conducting maintenance to one of 209.E Do 17Ka-2 Br. 3330. Note the newly applied "war flag" on the rudder of the first airplane while the second airplane, Br. 3312, retains large Yugoslav flag. [Aleksijević family via Igor Černiševski]

same template using *IG-Farbe Industrie* brand colors. The underside of Dornier made airplanes was Silver while the underside of DFA made airplanes was light Blue-Gray. The topsides on all airplanes consisted of Dark Brown, Dark Green and Ochre Yellow. This pattern was sprayed with fine, soft edges between colors. The total weight of paints applied on Do 17K was 39 kg.

The "Kosovo cross" insignia were applied on wing topsides and undersides and were defined in overall size as ¾ of the wing chord. Dornier produced machines' rudders carried large Yugoslav Blue-White-Red state flag on the outer surfaces only. Vertical stabilizers had large Dornier logo, airplane name and designation "Dornier Do 17" as well as Cyrillic airplane serial num-

German soldier sitting on the nose of Do 17Ka-3 EvBr. 3364. Note the asymmetrically applied "Kosovo cross" insignia, present only under the starboard wing and on top of the port wing. [Jan van den Heuvel via Aleksandar Ognjević]

This machine was captured at Preljina belonged to 206.E CO kap Ilk Mihajlo Djonlić as evidenced by double White triangles on the fuselage. To differentiate machines belonging to Escadrille CO, fuselage markings were doubled up. [Dinko Predoević]

ber starting with Бр. (*Broj* – number). On the starboard side of the fuselage, just ahead of the tail were airplane masses written in Black letters. DFA produced machines had White DFA logo (ДФА) on the vertical stabilizer along with airplane designation "Do 17" in Cyrillic letters as well as the newly introduced EvBr. starting with digits 33 while the original serial number prefix Бр. remained ahead of it. Service inscriptions both in Black and White letters were located at a number of locations on the airplane.

In accordance with the new rule POV.V.B.Br. 1191 from 26 January 1940, the "Kosovo Cross" insignia size and number used was reduced, retaining one marking on the underside of the starboard and one on the upper side of the port wing, in an asymmetric arrangement. Standard VVKJ tail stripes (the so-called "war flags") were applied in place of a large state flag. This rule was gradually applied on all 3.BP Dorniers by the autumn of 1940.

Do 17K had distinct markings on the fuselage where 63.BG airplanes had triangles while those belonging to 64.BG had circles, applied in Blue, White or Red color analogous to those on the KJ flag. To differentiate machines belonging to Escadrille CO, fuselage markings were doubled up, of which twin red circles and white triangles are known.

Escadrille	Fuselage markings
205.E	Blue triangle
206.E	White triangle
207.E	Red Triangle
208.E	Blue circle
209.E	White circle
210.E	Red circle

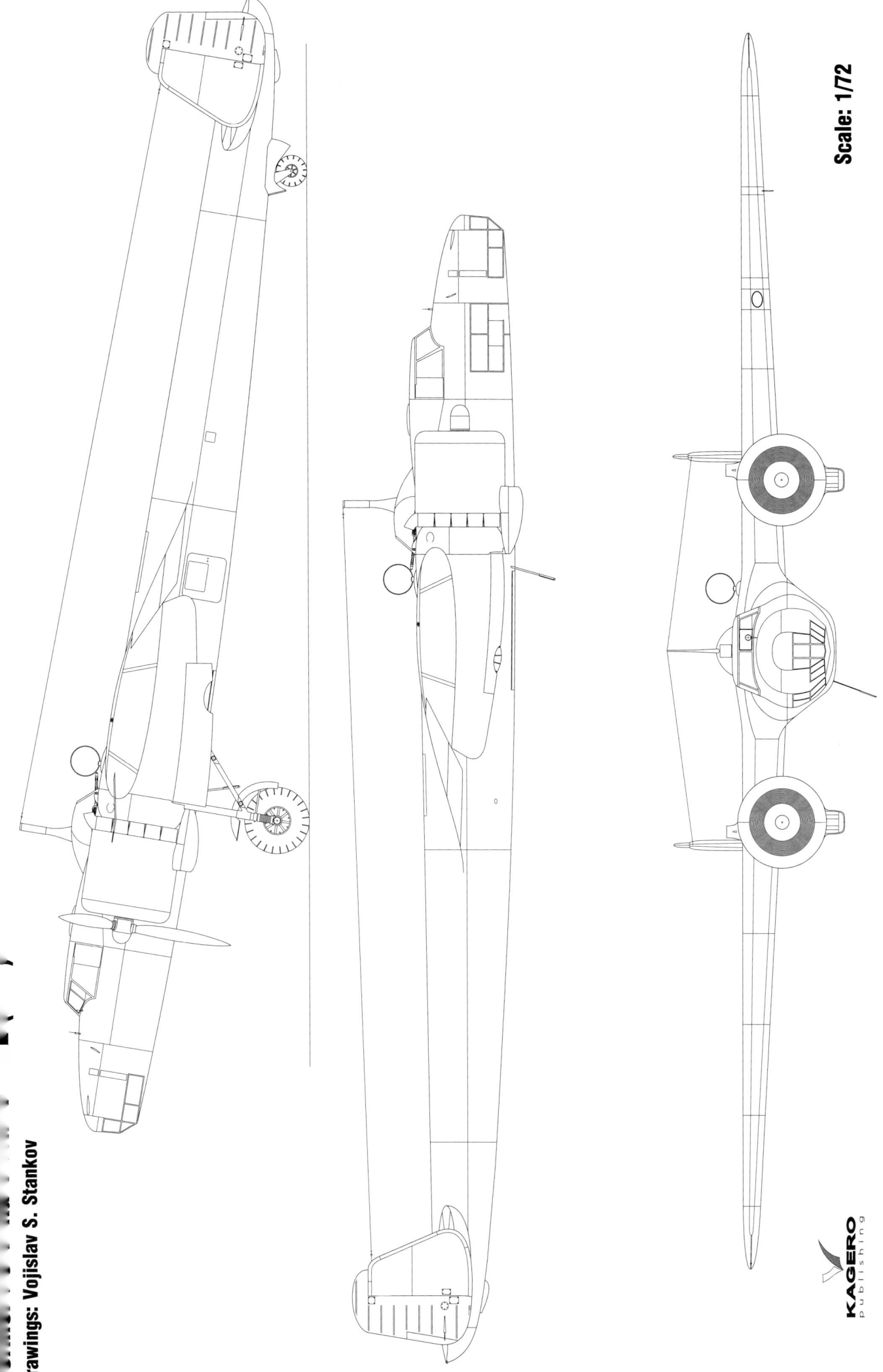

Drawings: Vojislav S. Stankov

Scale: 1/72

KAGERO
publishing

Dornier Do-17Ka-1 W.Nr. 2382 (Br.2)
Drawings: Vojislav S. Stankov

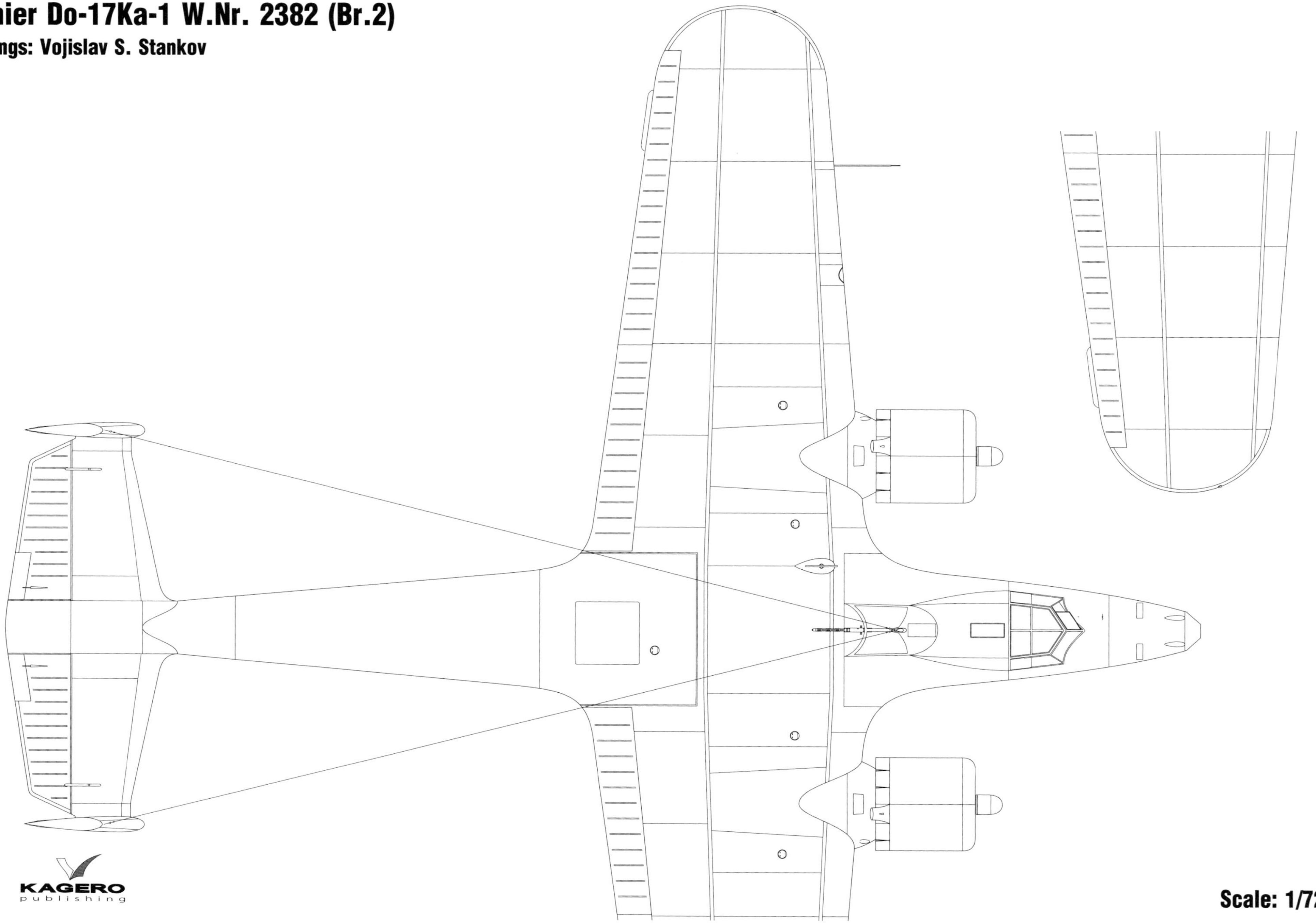

Scale: 1/72

KAGERO
publishing

Drawings: Vojislav S. Stankov

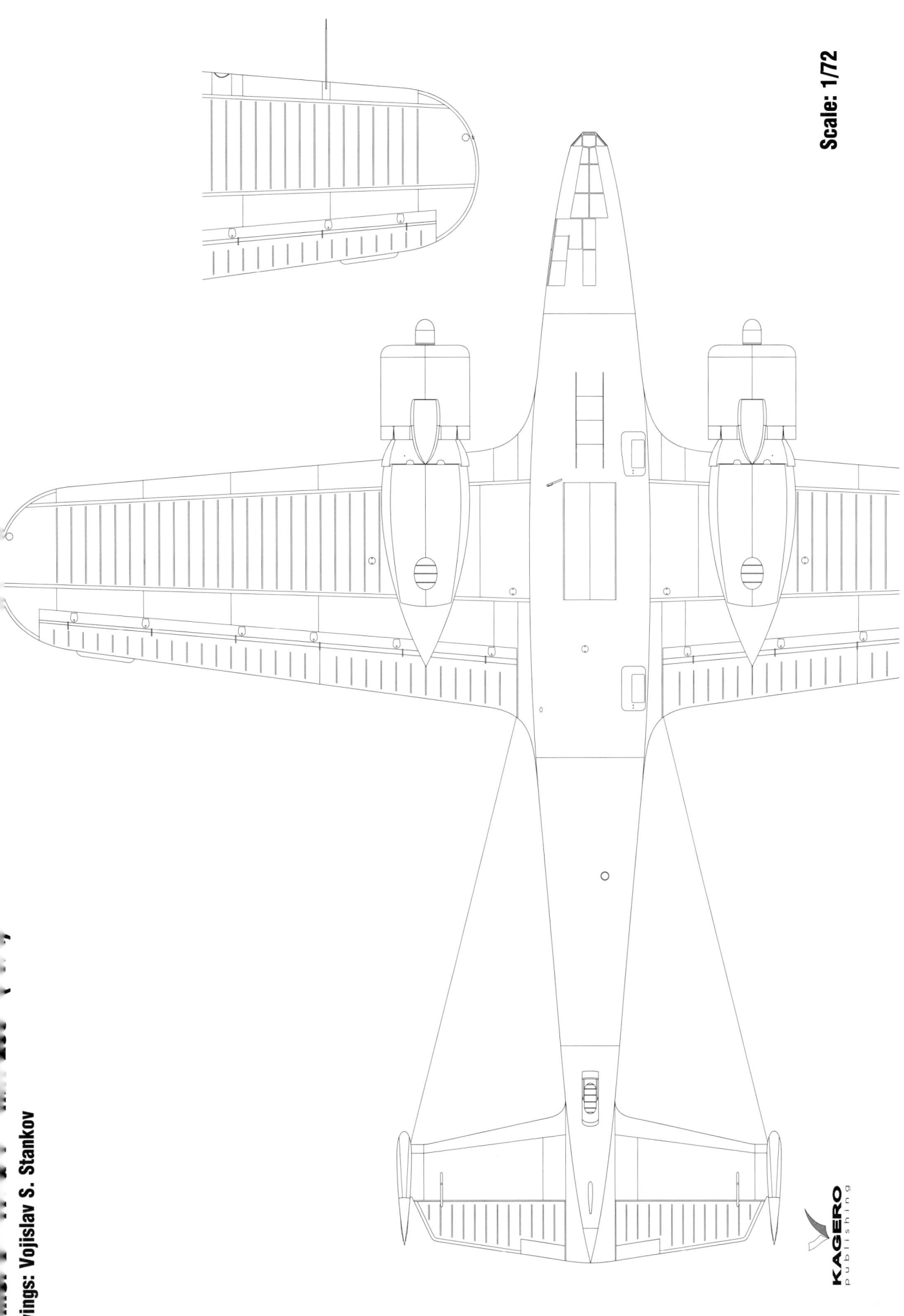

Scale: 1/72

KAGERO
publishing

Dornier Do-17Ka-2 (Br.22)
Drawings: Vojislav S. Stankov

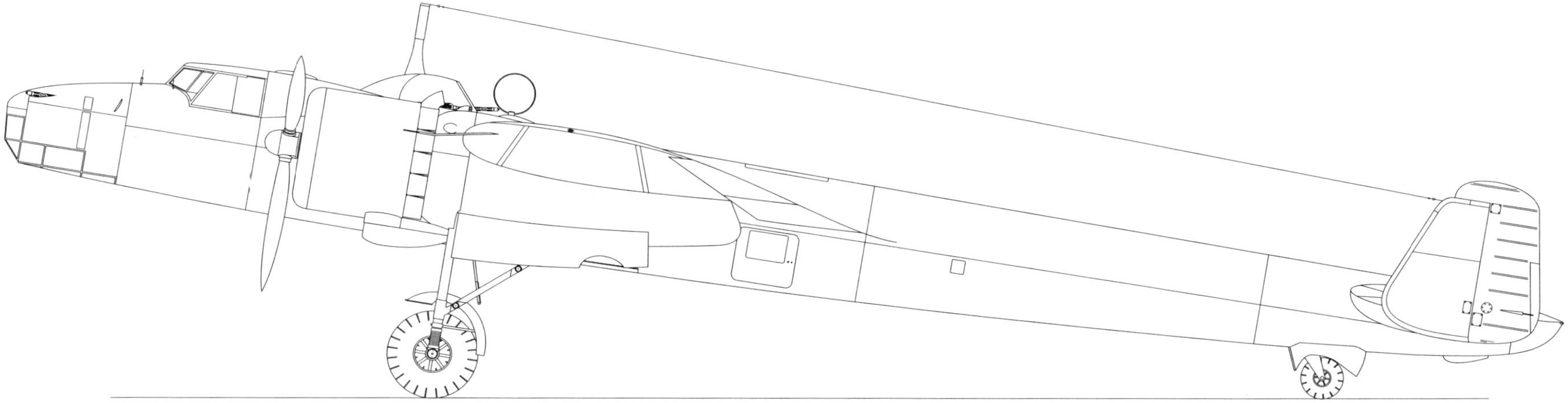

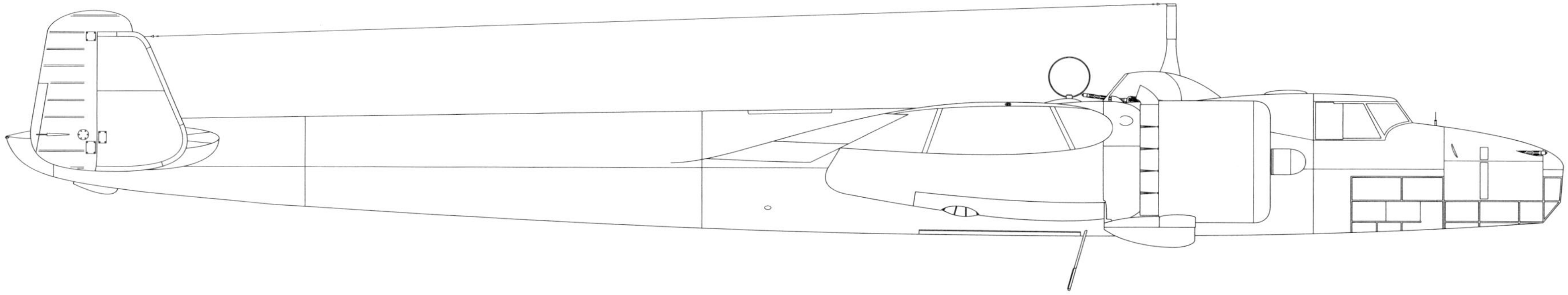

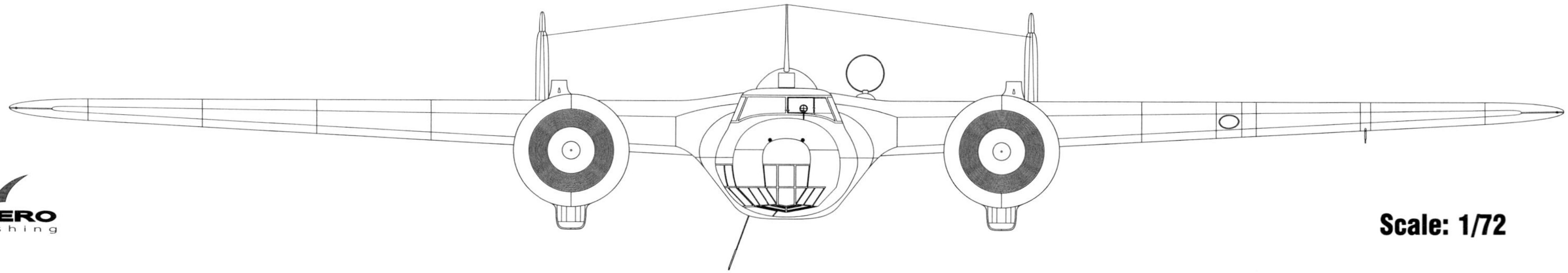

Scale: 1/72

KAGERO
publishing

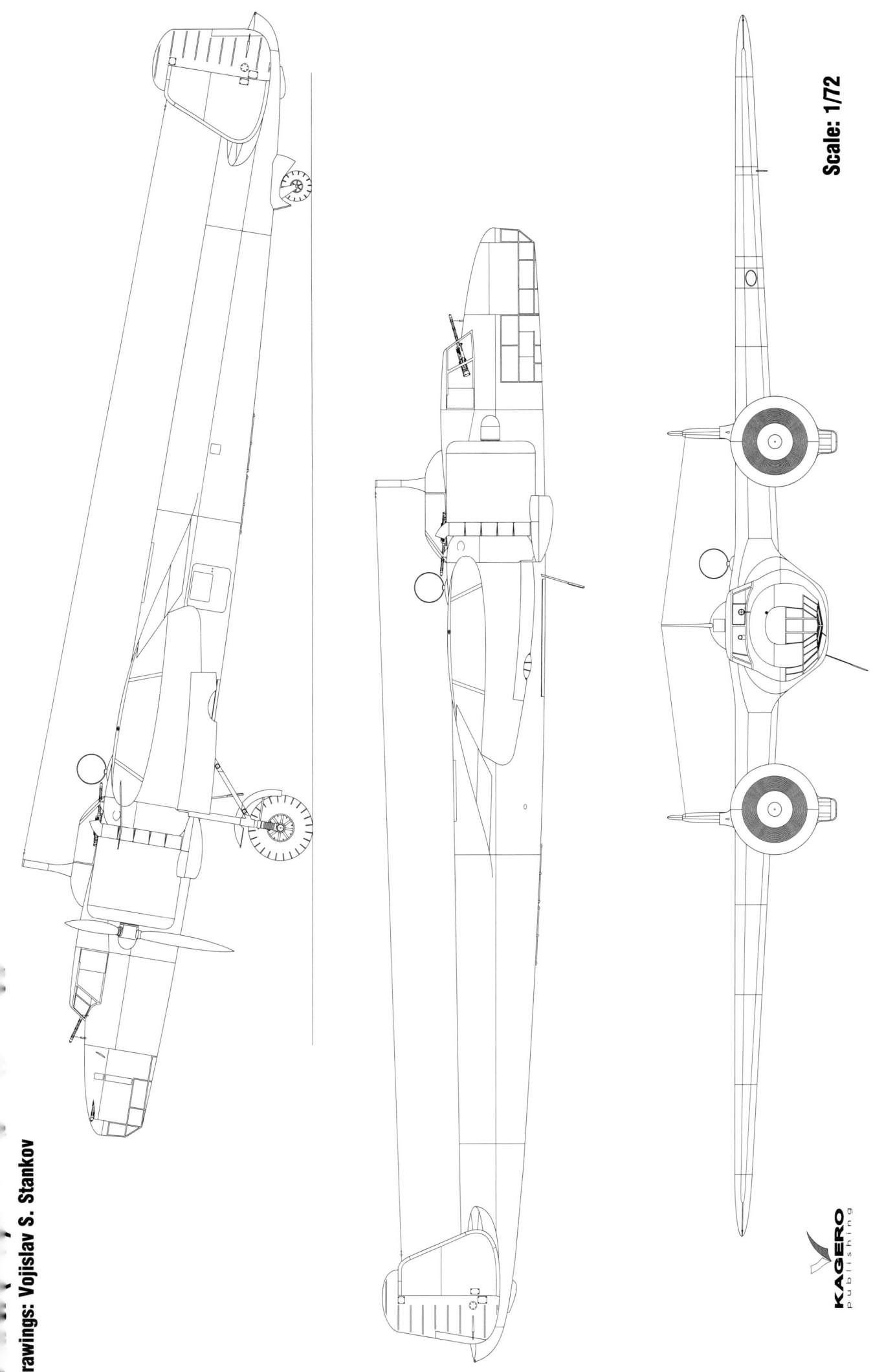

Drawings: Vojislav S. Stankov

Dornier Do-17Ka-3 (Br.36)
Drawings: Vojislav S. Stankov

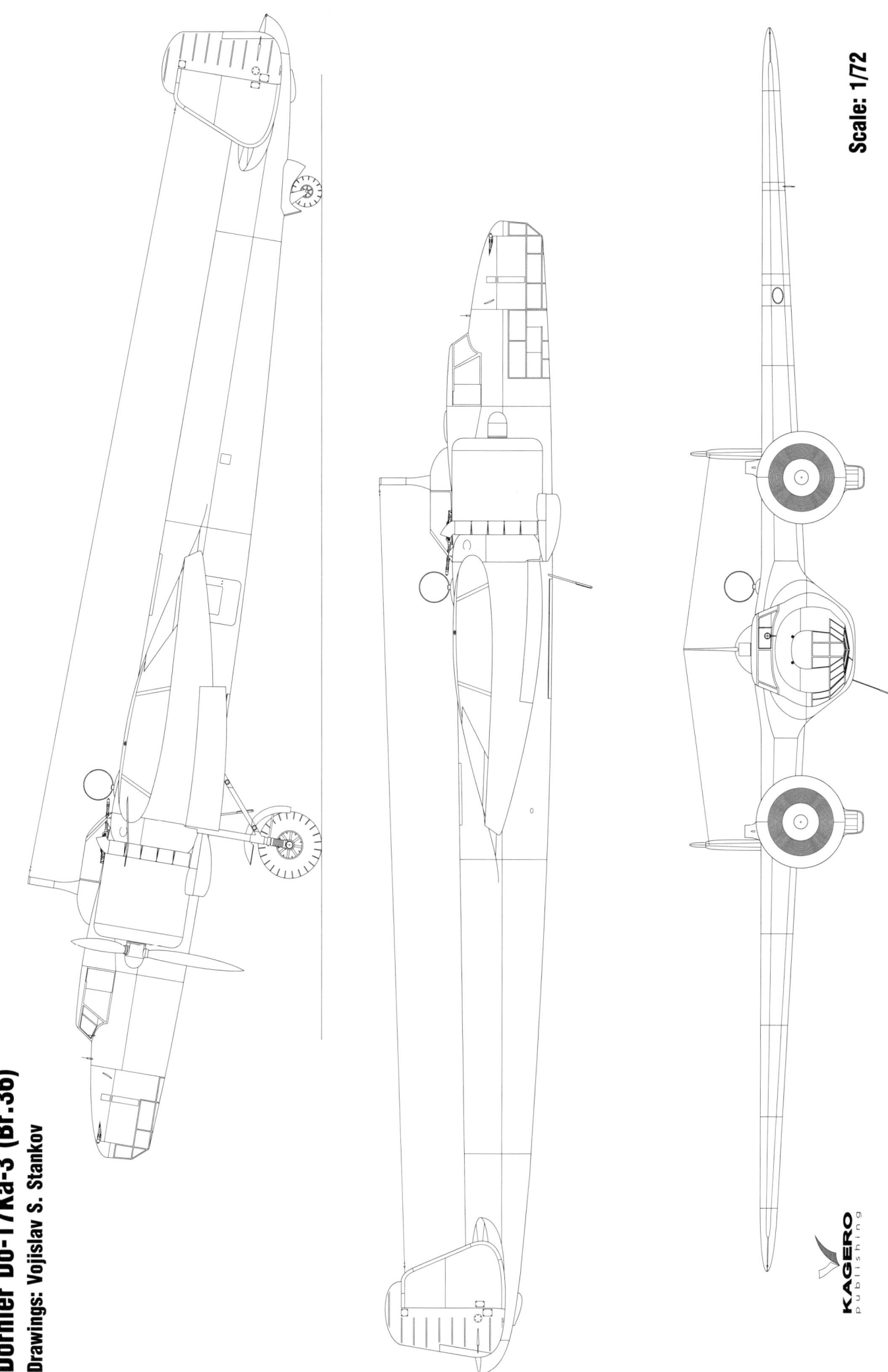

Scale: 1/72

KAGERO
publishing

Scale: 1/72

Dornier Do-17Ka-1 W.Nr. 2382 (Br.2)

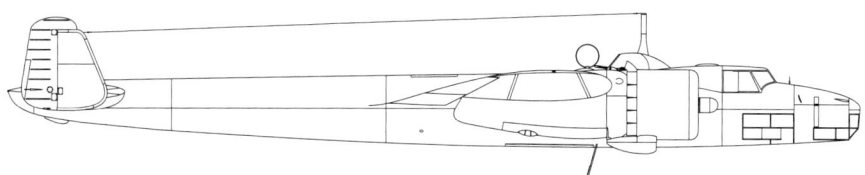

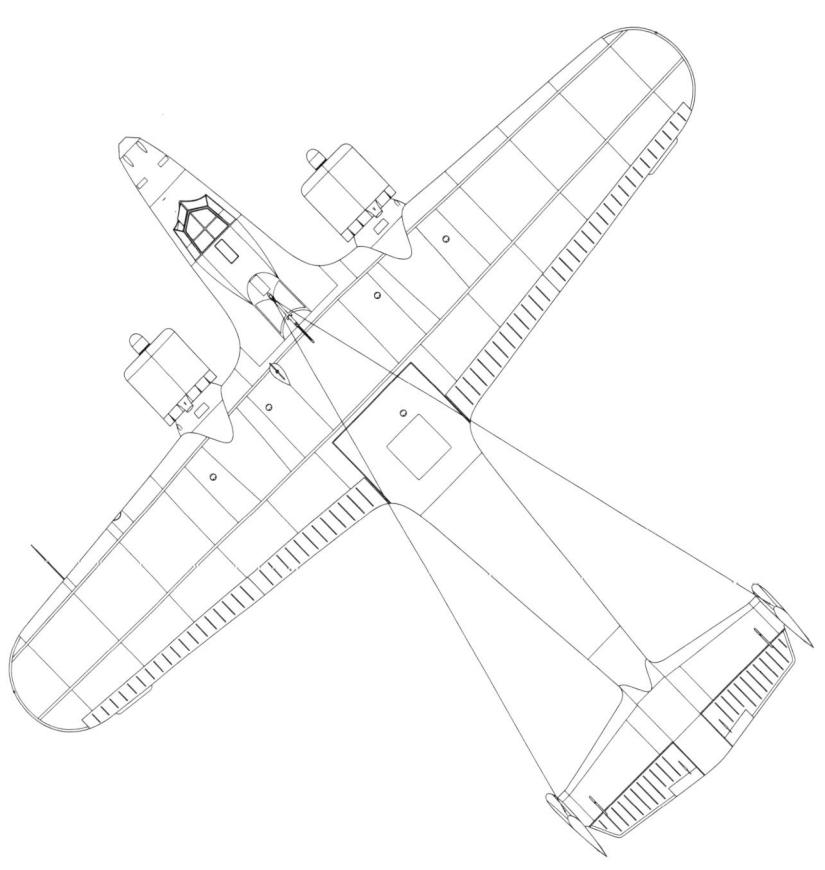

Drawings: Vojislav S. Stankov

Scale 1/144

Dornier Do-17Ka-1 W.Nr. 2382 (Br.2)

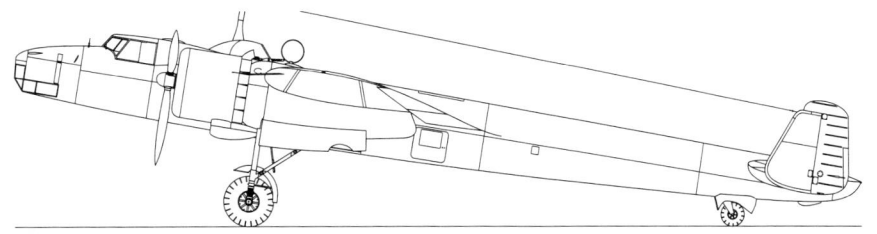

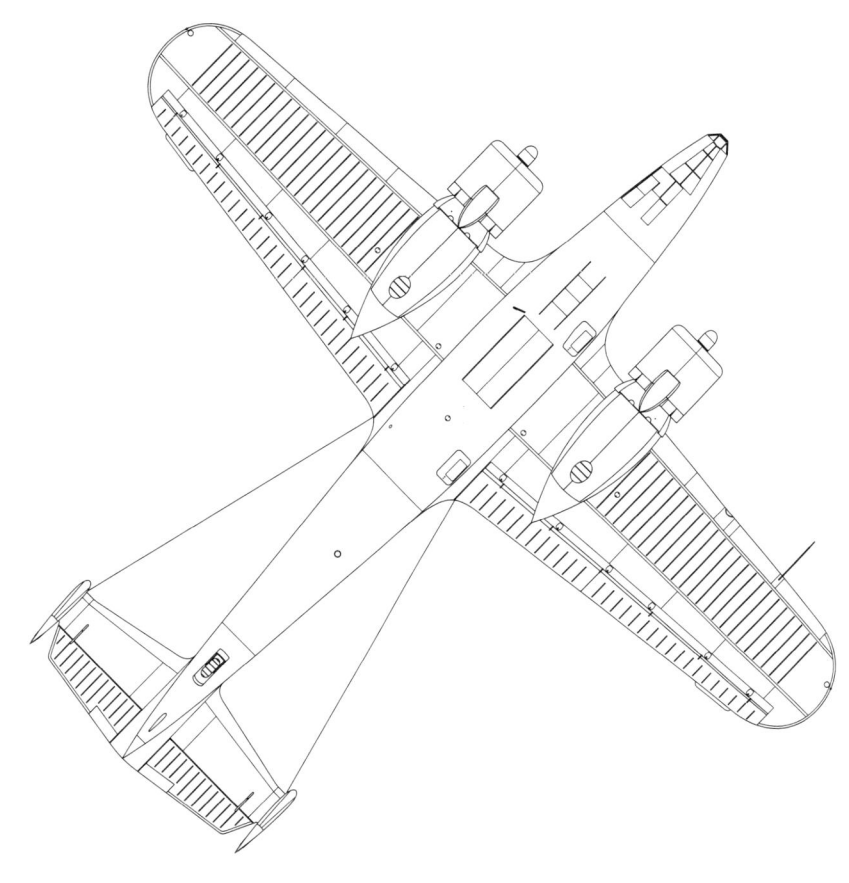

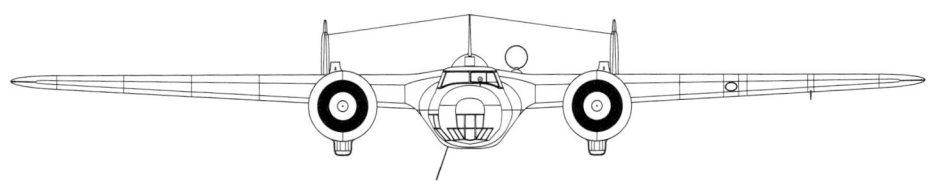

Drawings: Vojislav S. Stankov

Scale 1/144

Dornier Do-17Ka-2 (Br.22)

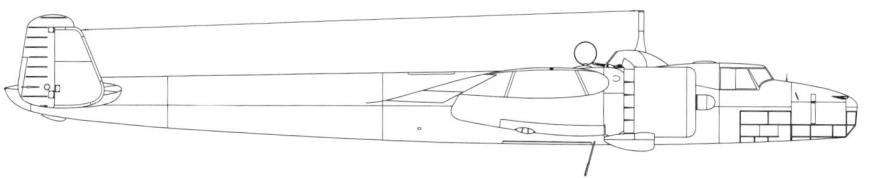

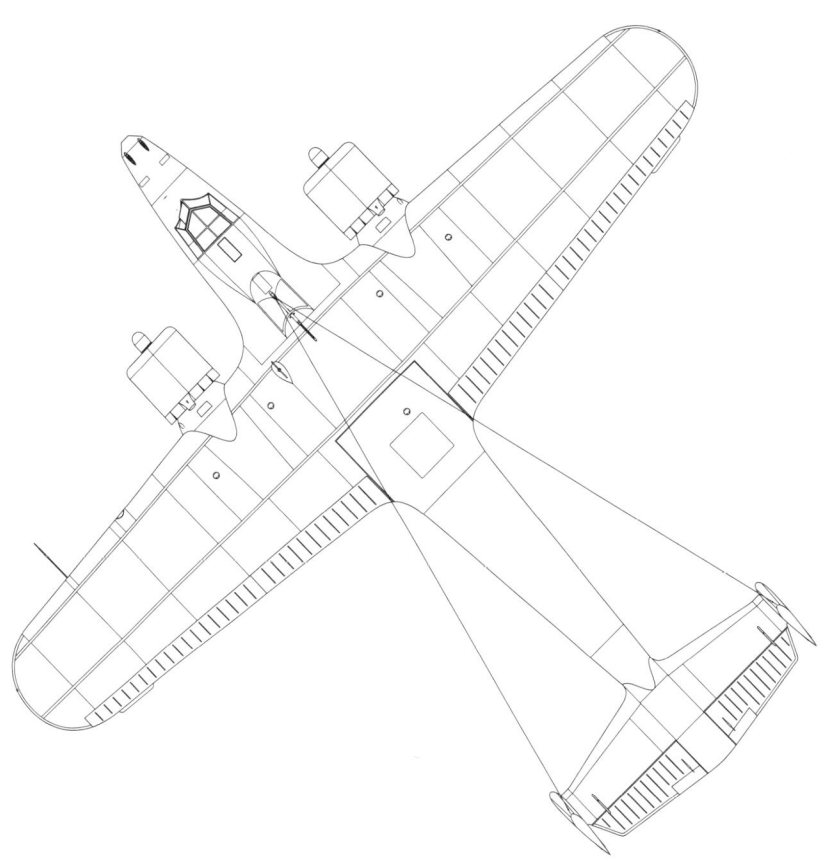

Drawings: Vojislav S. Stankov

Scale 1/144

Dornier Do-17Ka-2 (Br.22)

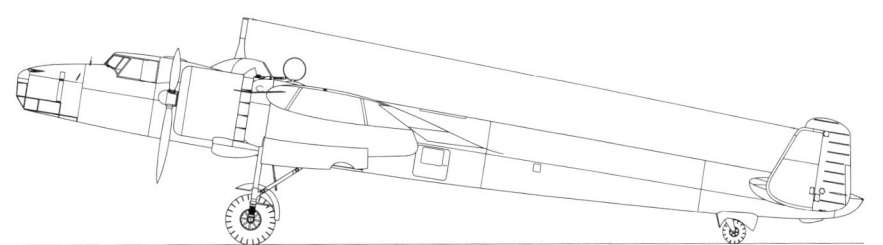

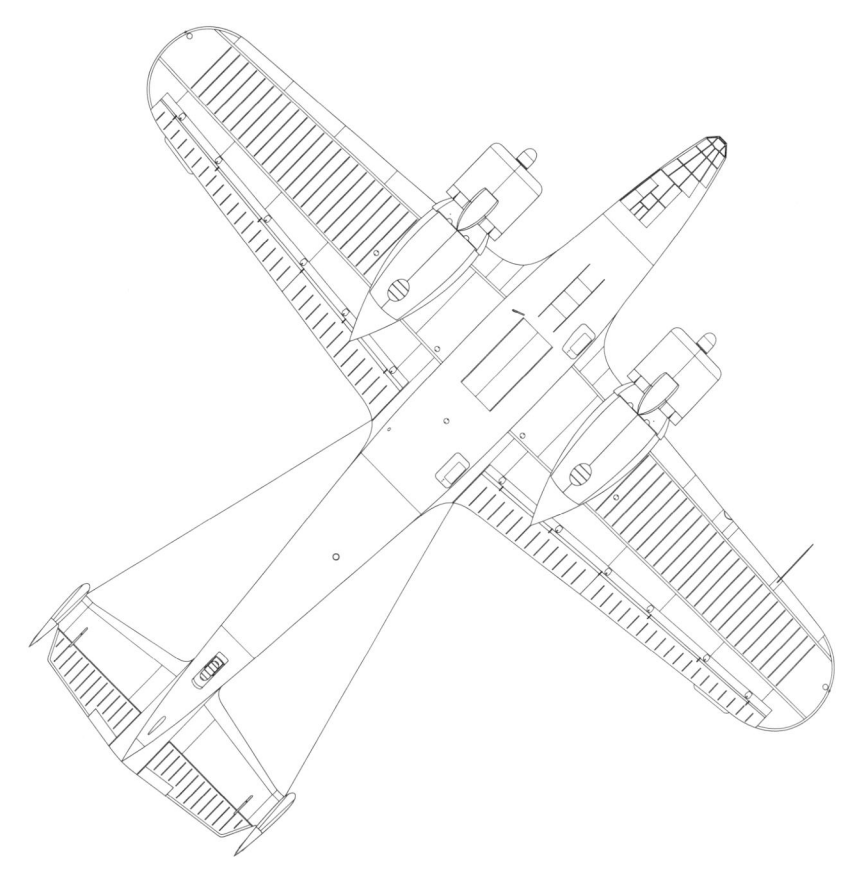

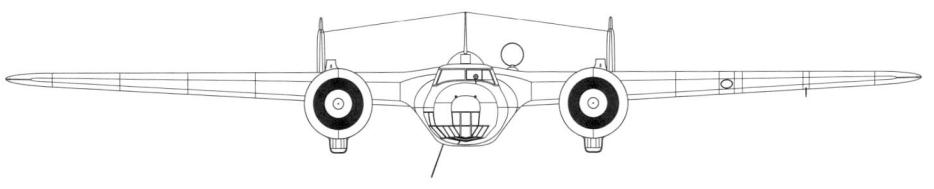

Drawings: Vojislav S. Stankov

Scale 1/144

Dornier Do-17Ka-3 (Br.36)

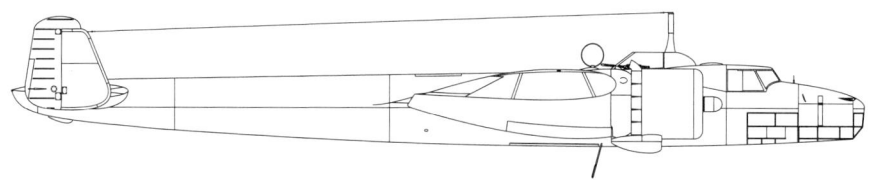

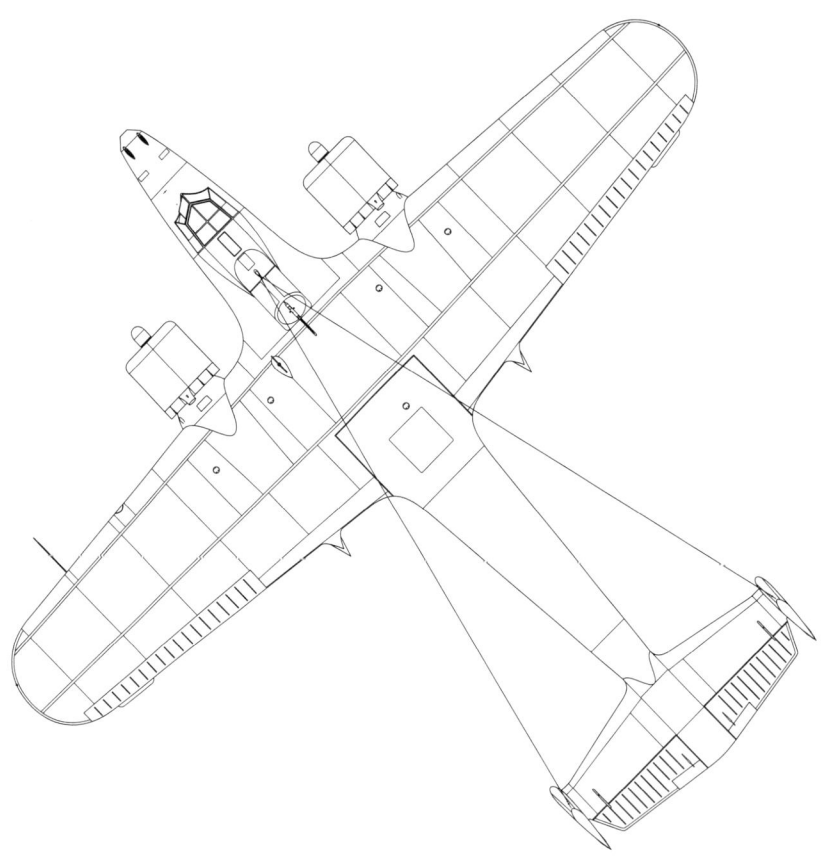

Drawings: Vojislav S. Stankov

Scale 1/144

Dornier Do-17Ka-3 (Br.36)

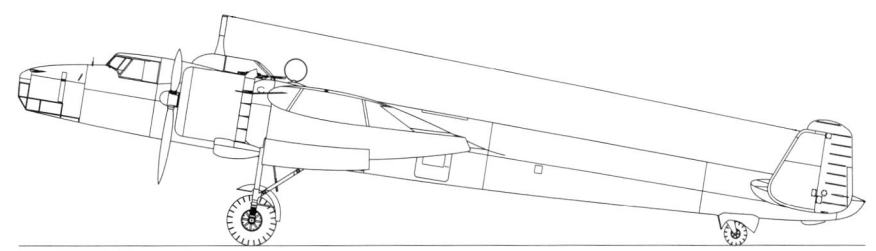

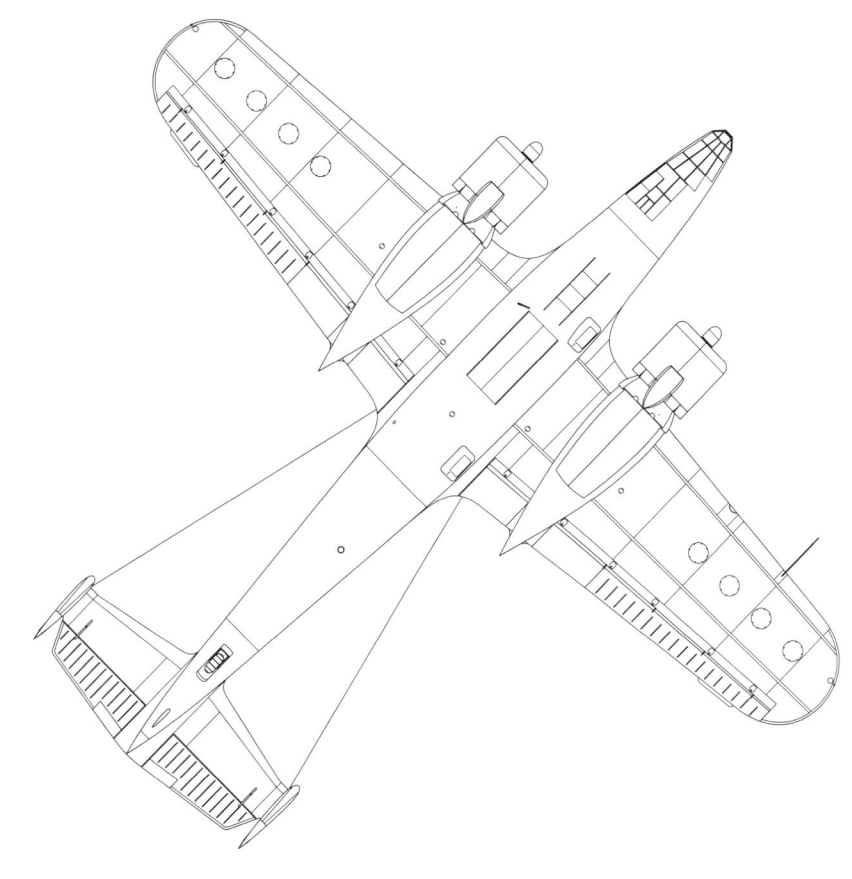

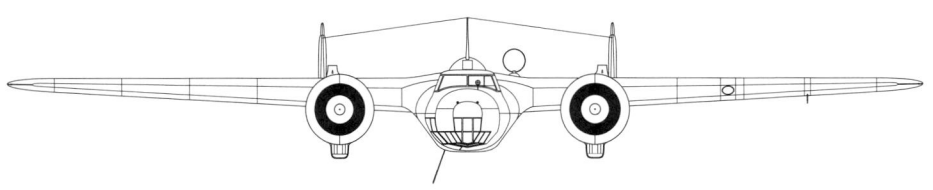

Drawings: Vojislav S. Stankov

Scale 1/144

Dornier (DFA) Do-17Ka-3 EvBr. 3362

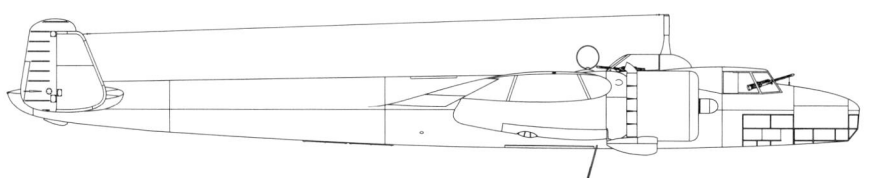

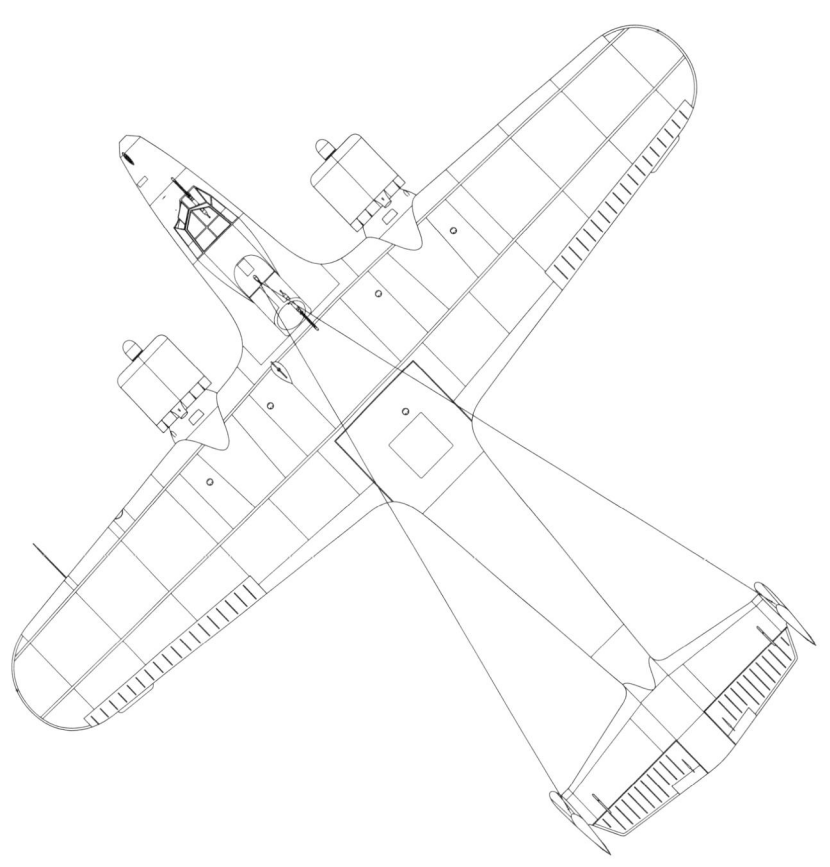

KAGERO
publishing

Drawings: Vojislav S. Stankov

Scale 1/144

Dornier (DFA) Do-17Ka-3 EvBr. 3362

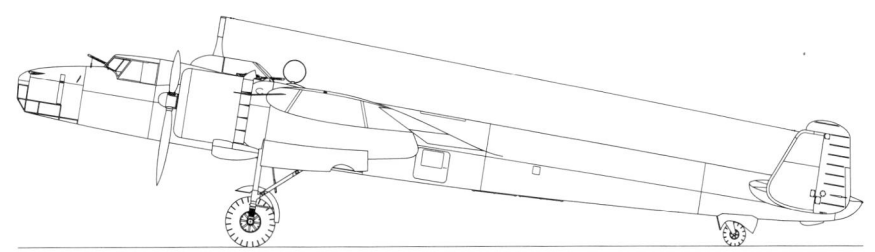

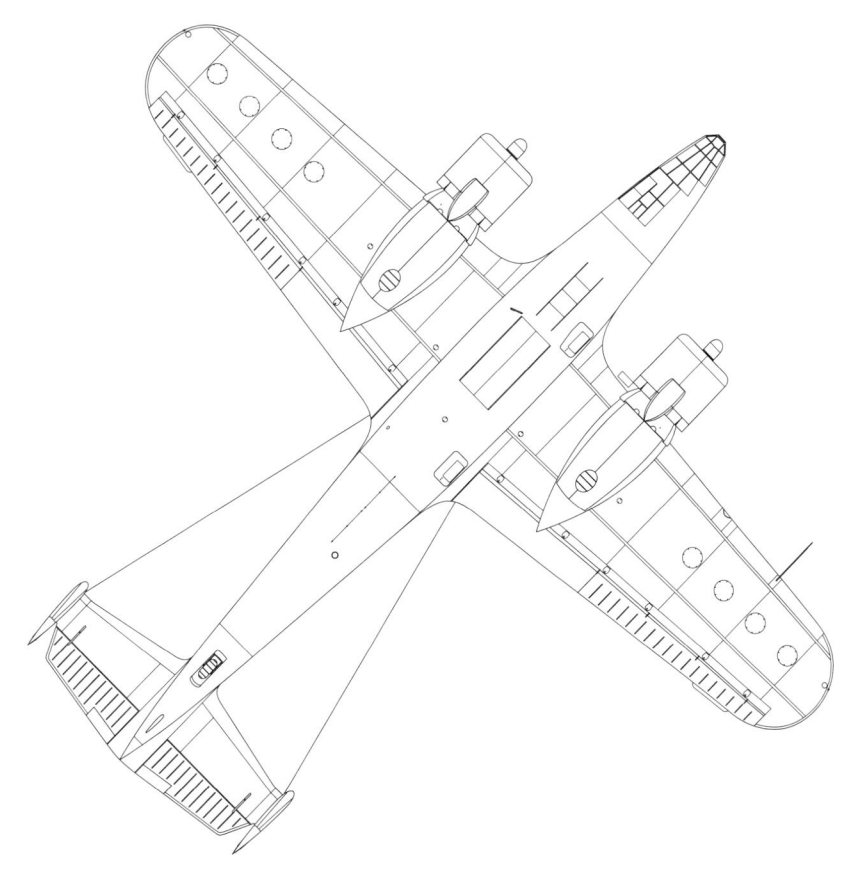

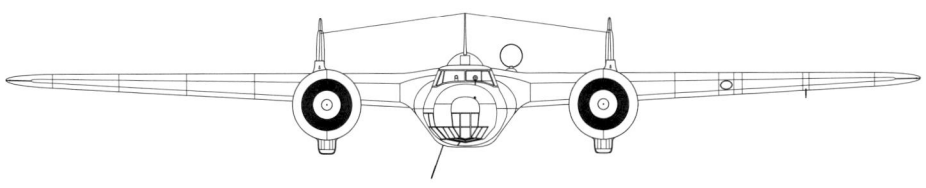

Drawings: Vojislav S. Stankov

Scale 1/144

Dornier (DFA) Do-17Ka-3 EvBr. 3363

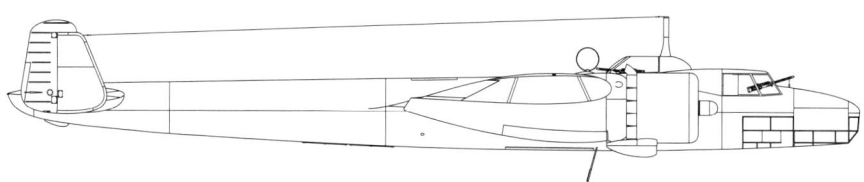

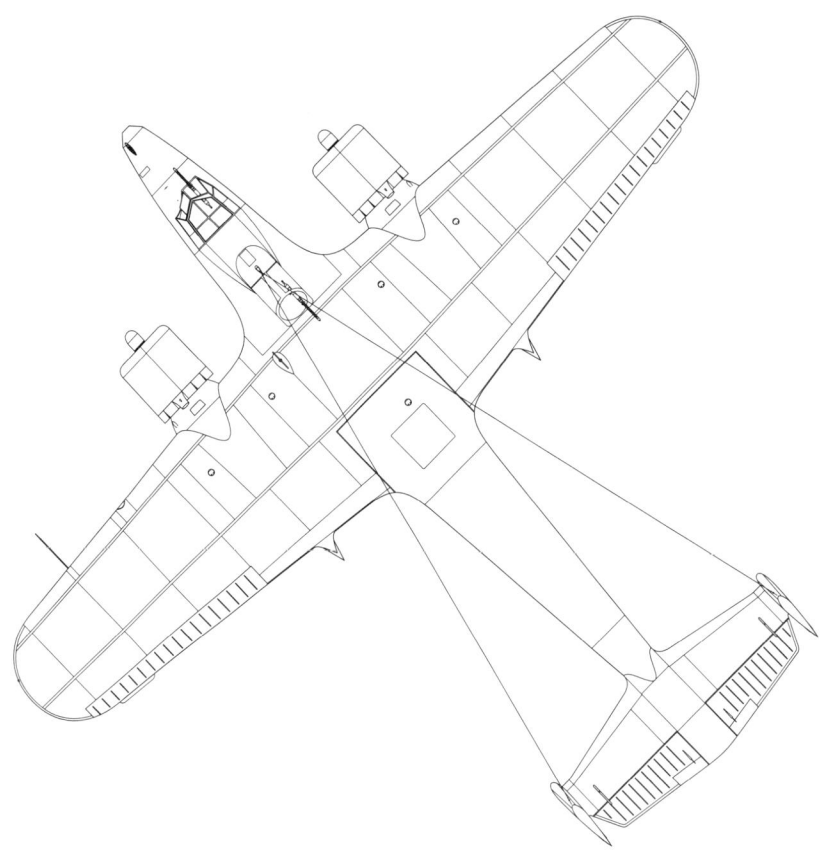

KAGERO
publishing

Drawings: Vojislav S. Stankov

Scale 1/144

Dornier (DFA) Do-17Ka-3 EvBr. 3363

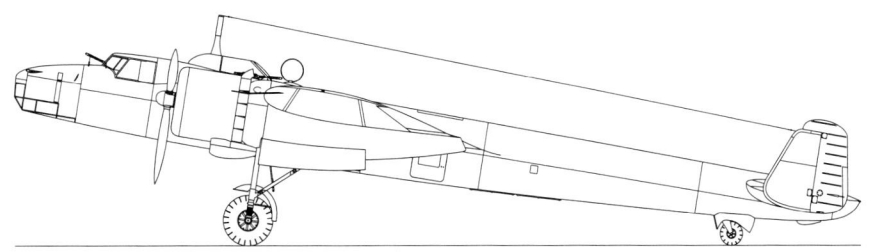

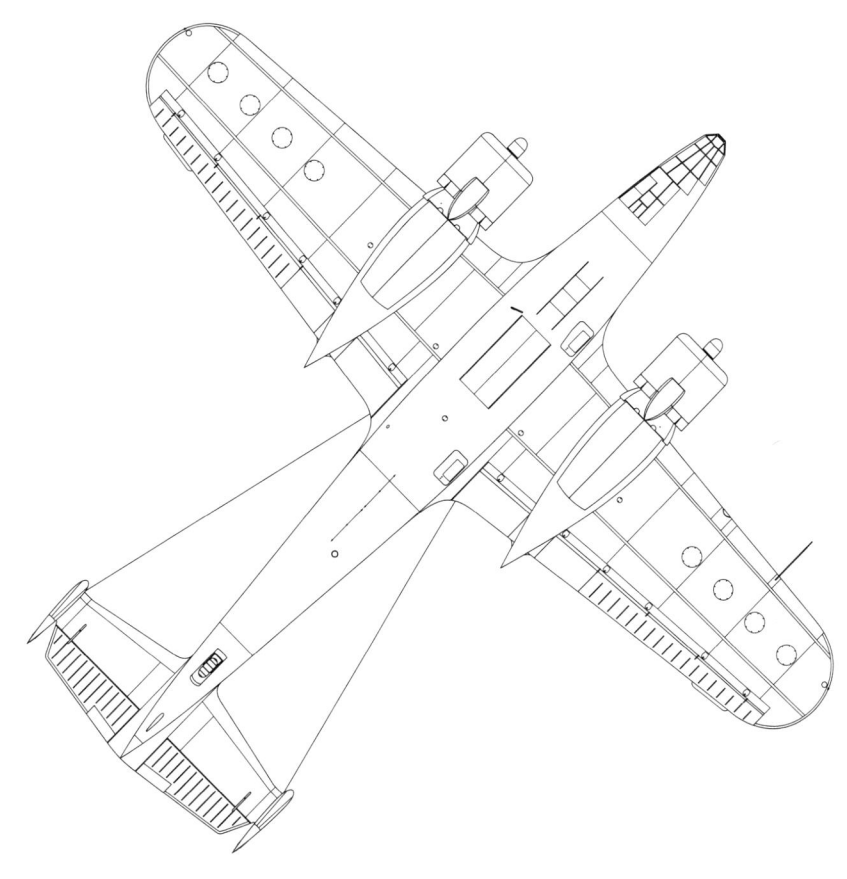

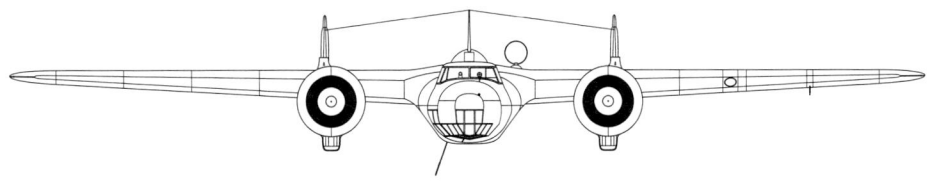

Drawings: Vojislav S. Stankov

Scale 1/144

Dornier Do 28D-2 Skyservant

Development

Development of the Skyservant began in the mid-1950s to provide the next step in a logical progression of STOL (Short Take Off and Landing) utility airplanes which had begun with the Do 25 and continued with Do 27 and finally Do 28. Do 25 was *Dornier's* first post war design built in Spain and 50 were built localy by CASA as C-127 and *Dornier* produced a total of 571 airplanes in Germany between 1956 and 1966. Do-28 was a twin engine development of Do 27 with the same fuselage with a new nose fairing, the same wing structure with extended span and engines carried on stub wings or sponsoons cantilevered off the fuselage side.

Batches of 60 of Do 28A and Do 28B were produced between 1961 and 1971. At the same time *Dornier* explored various designs which were broght together under project P350. These derivatives were entierly new design airplanes belonging to the same category but with greater capacity and improved utility. For this project, *Dornier* negotiated with the Federal German Ministry of Economics to obtain financial support for this project. One of the earlier P350 studies was an airplane with 3,800 kg gross weight, wing span of 17 m, conventionally wing mounted *Lycoming* IGSO-540 flat six engines and twin tail. This proposal was underpowered and by the late 1963 P350, which became known as Do 28, was scaled down to 3,130 kg

Dornier Do 28A D-IBOB first prototype made its first flight on 29 April 1959. [Dornier]

Legendary Claude Dornier sitting in the cockpit of Do 28A first prototype. [Dornier]

D-IBOB during one of its test flights. [Djordje Nikolić]

gross weight and 15 m wing span with new 340 hp *Lycoming* IGSO-480 engines.

By the time *Dornier* was able to launch construction of the prototype with a development loan of 5.4 Million DM (*Deutchmark*) from the *Bundesministerium für Wirtschaft und Energie* (Federal German Ministry of Economics and Energy), the design has undergone further refinement. The most important change was the decision to carry the engines on sponsons, so that the cantilever wing would remain clean. The 385 hp

Dornier Do 28D V1 first prototype during early stages of construction. This photo was taken on 4 November 1965. [Airbus Corporate Heritage]

Dornier Do 28D-2 V1 D-INTL first flight on 23 February 1966. Note the long pitot tube at the nose of the airplane and overall natural metal appearance. [Airbus Corporate Heritage]

Claudius Dornier, Claude Dornier's son and Drury Wood, Dornier Chief Test Pilot, following one of the test flights. [Airbus Corporate Heritage]

IGSO-540 engine had better STOL performance and full span fixed leading edge slats and mechanically operated double slotted flaps were fitted. The fixed mainwheel legs were attached at the ends of the sponsoons carrying the engines and a steerable, fully swivelling tailwheel was adopted in place of the previously projected nosewheel for convenience of operation from unprepared airfields. The fuselage retained a simple rectangular section with maximum headroom of 1.55 m and a minimum of 1.22 m. The width was 1.39 m and the length excluding cockpit was 3.96 m. Double doors on the port side provided easy loading of bulky freight items and Do 28D could carry up to

13 passangers and one pilot. Take off distance, in still air and from a dry runway, was estimated to be 258 m and a rate of climb of 1.27 m/s was maintainable with one engine only. The Do 28D differed from the Do 28A in that the Skyservant had little in common with the original version beyond its configuration and basic wing. The fuselage and the tail were entierly new, as were the more modern systems and equipment. The larger fuselage made it possible to offer much greater variety of special equipment for roles such as air ambulance, aerial survey, photogrammetry and patratrooper or supply droppoing as well as passager and cargo roles. The new design also envisioned the use of skis or floats.

The Do 28D-2 prototype was completed in the early 1966 and was first flown on 23 February 1966. While the tests flights were ongoing, *Dornier* produced another six airplanes which were granted formal type approval on 24 February 1967. On 6 November Do 28D-1 was type approved along with the two original D models converted to the new standard. Due to the steady backlog in orders, *Dornier* kept producing about one airplane per week at Oberpfaffenhoffen production line.

To take advantage of general product improvement programmes, *Dornier* introduced a new standard Skyservant from the 50th airplane. Designated the Do 28D-2, this has remained the standard production model for export since the completion of the *Luftwaffe* and *Bundesmarine* orders. The differences between the D-1 and D-2 versions included an increase in gross

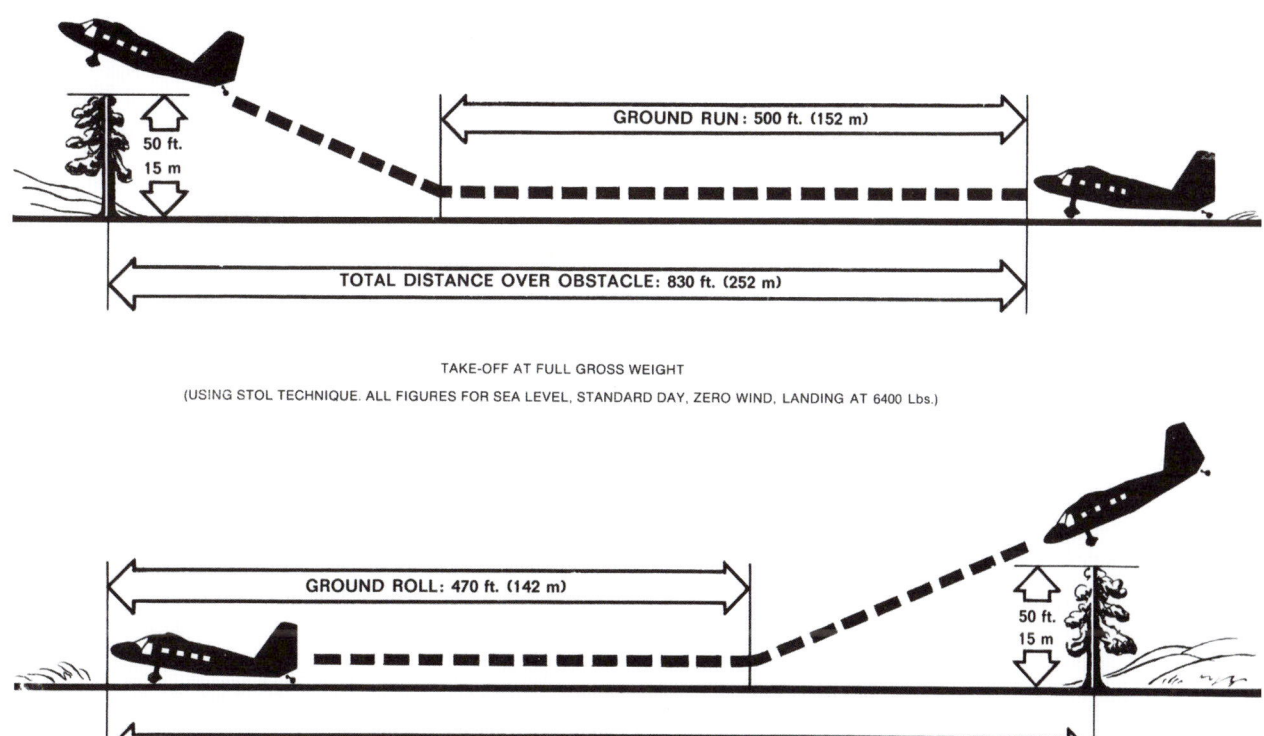

GROUND RUN: 500 ft. (152 m)

TOTAL DISTANCE OVER OBSTACLE: 830 ft. (252 m)

50 ft.
15 m

TAKE-OFF AT FULL GROSS WEIGHT
(USING STOL TECHNIQUE. ALL FIGURES FOR SEA LEVEL, STANDARD DAY, ZERO WIND, LANDING AT 6400 Lbs.)

GROUND ROLL: 470 ft. (142 m)

50 ft.
15 m

TOTAL DISTANCE FROM OBSTACLE: 770 ft. (233 m)

STOL properties of the Do 28D-2 are demonstrated in this factory graphic. [Airbus Corporate Heritage]

weight of 150 kg, an internal redesign to extend cabin length by 15 cm aft, elimination of the leading edge slats from the inboard wing and aerodynamic improvements to the flaps and ailerons, an aerodynamically improved slab tailplane, new fuel tanks in the rear of the engine nacelles with additional 79.5 liters of fuel, twin contact tail wheel, provisions for the removal of mainwheels without detaching the spats, improved heating system, new landing lamps in the wing tips and 48 additional improvements to the *Lycoming* engines. Some previous optional features were made standard on the D-2 model such as dual controls, twin 100 Amp generators and directional slaved gyros. Further developments increased the gross weight to 3,842 kg for civillian and 4,015 kg for military versions. The military versions carried auxiliary fuel tanks on hard points below each wing up to a total permitted load of 200 kg. The D-2 version had an exceptional STOL performacne with the STOL take off run of 240 m and STOL landing of 190 m.

The civilian version is equipped to seat 10 passangers in additon to two pilots. A luggage compartment is provided behind the rear seats and is accessible from the cabin. All seats are removable for modification to cargo role. The military version seats 12 on side beches. For dropping suppliers, roller track can be fitted and the door removed or replaced with an optional sliding door.

Dornier Do 28D-2 attained six world records in accordance with the *Fédération Aeronautique Internatonale* (FAI – World Air Sports Federation) on 15

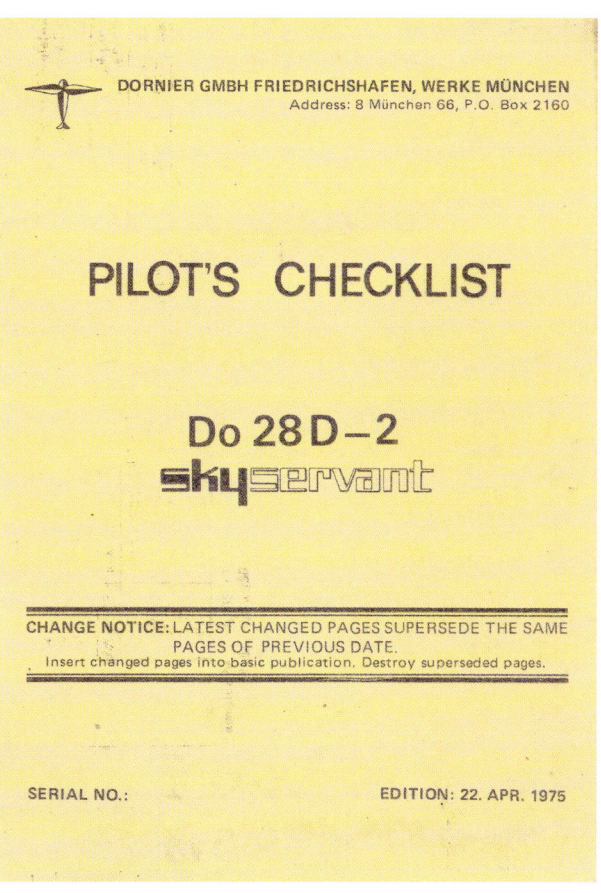

DORNIER GMBH FRIEDRICHSHAFEN, WERKE MÜNCHEN
Address: 8 München 66, P.O. Box 2160

PILOT'S CHECKLIST

Do 28 D–2
sky servant

CHANGE NOTICE: LATEST CHANGED PAGES SUPERSEDE THE SAME PAGES OF PREVIOUS DATE.
Insert changed pages into basic publication. Destroy superseded pages.

SERIAL NO.:   EDITION: 22. APR. 1975

Do 28D-2 pilot's checklist cover page. [Spomenko Marković]

March 1972 at the controls of *Dornier* chief test pilot Frank Tuytjens. He flew the airplane D-IBYR from Oberpfaffenhofen attaining the following records in *FAI* Category C1E for airplanes with piston engines of 3 to 6 ton weight:

1. Climb time to 3,000 m of 6.6 minutes
2. Climb time to 6,000 m of 16.2 minutes
3. Climb time to 9,000 m of 44.4 minutes
4. Maximum altitude without cargo of 9,963 meters
5. Greatest load to an altitude of 2,000 m of 1 ton
6. Maximum altitude of 8,630 m with load of 1,000 kg

These flights were witnessed and confirmed by FAI and *Deutscher Aero Club e.V.* (Daec – German Aero Club). The attained wolrd records were confirmed by independent experts with FAI in Paris who reviewed the barogram records. Do 28D-2 D-IBYR was not modified for this occasion, it was a standard production machine built in 1968, apart from measuring devices it had no other modifications. The same airplane was used for demonstration flights in the entire Europe and Scandinavia, in the Middle East, in the Himalaya mountain area and it even crossed the south Atlantic in both directions. During the course of these flights it accumulated a total of 1,200 flight hours.

Further versions od the Do 28, the D-3 and D-4, never came to fruition but D-5 did result in a sole prototype constructed which had first flight on 9 April

First Dornier AG certificate issued to Yugoslav Air Force pilot Budimir G. Djordjević for the Command-Pilot-Rating qualification. [Djordjević Family]

The arrival of first Do-28D-2 to Batajnica airfield on 22 August 1972. Note the absence of national insignia. [Djordjević Family]

A photograph of Budimir "Buda" Djordjević from his pilot ID. [Djordjević Family]

A page from Budimir Djordjević's pilot logbook with some of the first flights with Do 28D-2 at the beginning of August while in Germany during training on this airplane type. [Djordjević Family]

Datum	Tip vazduhoplova	Broj vežbe	Kratak sadržaj zadatka	Ukupno Letova	sati	min	Od toga — Lični nalet — Vizuelno Letova	sati	min	Instrum. Letova	sati	min	Noću Letova	sati	min	Nastavnički nalet — Vizuelno Letova	sati	min	Instrum. Letova	sati	min	Noću Letova	sati	min	Zaloga po GA	Trenažer ILS	Letova	sati	min
			Preneto.	42	26	14	23	13	34	5	5	40	14	7	00														
18.2	C-47		BT-NIŠ	1	1	00	1	1	00	–	–	–																	
24.2	C-47		NIŠ-BT	1	1	00	–	–	–	1	1	00																	
23.5	G-47	2	ŠK. KRUGOVI	6	0	36	6	0	36																				
23.5	G-47	5	ZONA IFR	1	1	30				1	1	30																	
13.6	C-47	5	ZONA IFR	2	3	00				2	3	00																	
6.8	Do28		Inform. let	1	0	45	1	0	45																				
7.8	Do28		ŠK. KRUG. ZONA	1	1	12	1	1	12																				
8.8	Do28		ŠK KRUG. ZONA	6	1	15	6	1	15																				
8.8	Do28		ZONA	1	1	00	1	1	00																				
11.8	Do28		ŠK.KRUG ZONA	8	1	24	8	1	24																				
12.8	Do28		EMER KRUGOVI	7	1	50	7	1	50																				
			Za prenos:	77	40	44	54	22	34	9	11	10	14	7	00														

— 40 —

1978. The Do-28D-5 Turbo Sky version was powered by 400 hp *Avco Lycoming* LTP 101-600 turboprop, which considerably improved performance.

In total of 120 Do 28, Do 28A and Do 28B were produced, of which at least six Do-28s were fitted with EDO floats. Over 250 Dornier Do 28D and D-128, which replaced the D-2 model as of 1980, of all versions were produced and sold to more than 30 countries, of which 54 were D-1 model and 172 were the D-2 model.

Introduction into Service

Vojno geografski institut (VGI - Military Gerographical Institute) from Belgrade started conducting aerial photogrametry filming over Yugoslavia and border areas with automated *Wild* and *Zeiss Jena* cameras.

Douglas C-47, *Junkers* Ju-52, *Avro* Anson, *Ikarus* 214 and even *Petlyakov* Pe-2 were used for this role. The work of VGI stabilized during the sixties, and in the first half of the seventies aerial photogrametry flights were conducted by the C-47 as well as the Yugoslav single engine *Utva* 60AF. Both were later replaced with *Dornier* Do 28D-2, which had the fuselage belly modified to add camera openings, which were supplied by VGI.

The first Do 28D-2 with registration number 70501 was received on 22 August 1975 within 119. TRAP (*Transportni Avijacijski Puk* - Transport Aviation Regiment). Do 28 along with two C-47s (71211 for photogrametry and 71288 for radar calibration) were reassigned during mid 1977 to 675.TRAE (*Transportna Avijacijska Eskadrila* - Transport Aviation Escadrille) based at Batajnica airbase.

Ppuk Djordjević at the commands of Skyservant, Batajnica airfield. [Djordjević Family]

293

Do-28D-2 registration number 70501 at Batajnica airfield. [Djordjević Family]

The unit's primary role until that time was the transport of Commander in Chief Marshall Josip Broz Tito and other high military ranking officials. As a replacement for the obsolete C-47, on 18 July 1977 the second example of Do 28D-2, registration number 70502, was received.

The first trio which left for Oberpfaffenhofen in Germany for conversion trainining conisted of Budimir Djordjević "Buda", Mihalj Šarnjai "Šari" and Zoran

Dimitrijević. They returned with *Dornier* issued certificates, signed by the Chief of Dornier factory pilots and instructors. Accordingly they became the first three JRV (*Jugoslovensko Ratno Vazduhoplovstvo* – Yugoslav Air Force) pilots which took on training of their pilot collegues teaching them to fly the Skyservant.

According to the memories of Slavko (Alojz) Cirkvenec "Šilja", also an instructor, 70501 lacked au-

Flyers in front of 70501 at Batajnica, from left to right: Unknown, eng. Miroslav Šešević, Djokić, Budimir Djordjević and flight mechanic Stojković. [Djordjević Family]

In front of Skyservant at Batajnica, from left to right: Spomenko Marković, Budimir Djordjević and Aleksandar Aleksić. [Spomenko Marković]

Standing in front of Do 28D-2 from left to right are: cameraman Brana Pejanović, Budimir Djordjević, pilot Josip Djetvaj and cameraman Nikolić. Kneeling are: flight mechanics Dragan Stojković and Starčević. [Djordjević Family]

A group photograph of Do 28D-2 pilots at Batajnica airbase airfield in 2001. Standing from the left are: pilot Marton Lečei, flight mechanic Nenad Arizanović (†), pilot Ištvan Viši, flight mechanic Dragan Grbić "Grba", pilot Saša Pejčić, pilot Aleksandar Aleksić "Mrčo" and flight mechanic Dragan Stojković "Soća". Kneeling fromt the left are: pilot Spomenko Marković "Moki", unknown flight mechanic and pilot Josip Djetvaj. [Aleksandar Aleksić]

topilot initially and received it finally ten years later in 1985 during an overhaul. Slavko also remembers that there was no special selection to train new Dornier pilots, hence they arrived from various branches of military aviation, including jets.

In accordance with the annual plans Do 28D-2s flew on average 100 hours to satisfy the needs of the military and additonal 200 hours for civliand needs and terrain photography. Crews spent significant time at various airfields throughout the country, depending where the terrain phtography was taking place. 675. TRAE crews were always joined by the members of aerial photography department from VGI who were trained to use cameras. These missions were of national inerest. For a long time VGI was the only institution authorized to conduct aerial photography for use with survey and land research. One of the special tasks was aerial photography with thermal scanner, in cooperation with VGI and *Industroprojekt* (later *INA-projekt*) above both the continental and sea areas of the country.

The military conducted thermal photography for INA's needs in Kvarner gulf. Above Istria and towards the international waters, photographic flights were conducted to survey locations for sea drilling oil platforms between Pula towards the Italian coast. All flights were conducted during the day, with caution that wind never exceeds two beauforts. All records were stored on a magnetoscope tape. Some of INA technicians which took part in these flights were prof. ing. Viktor Knap, ing. Matija Denih and ing. Vilibald Kidrić.

Skyservants of course conducted tasks for the needs of the Yugoslav Army and Navy. They mostly flew above the Adriatic Sea, during day time for photography of submarine trails from Rogoznica towards Biševo. During night time they used scanners. During these missions, they were temporarily based at Split airfield. According to the recollections of those which took part, similar tasks were conducted for the needs of JRM (*Jugoslovenska Ratna Mornarica* – Yugoslav Navy) combat ships.

Not all was routine, and there were missions which stood out and which are of course part of pilot profession. There were several situatons, some more and some less serious as well real anegdotes. One was flying with

Dornier Do 28D-2 pilot, officer Aleksandar Ž. Aleksić, known amongst his collegues by the nickname "Mrčo" from 27. class ŠROA (*Škola Rezervnih Oficira Avijacije* – Reserve Aviation Officer School). [Aleksandar Aleksić]

a rather pretty lady colleague from Skoplje air traffic control, when the airplane entered according to the memory of those on board into the worst hail storm cloud ever, which led to her horrified conclusion that she will never step aboard an airplane again. To make the entire event more interesting, the anti-hail defense fired rockets in the direction of the Dornier's flight path. Fortunately, without consequence for the crew and the passangers.

During the routine flight from Brnik (Ljubljana) to Batajnica (Belgrade) airbase, the meteorologist warned a pilot in training to pay attention to the concentration of cumulonimbus en route. Just as they gathered altitude, they entered in the mentioned cloud with horrific rain and wind gusts. There was thunder and lightning all around. A seasoned instructor told the young pilot: „*Keep your hand on the throttle and hold it by the horns*", implying the control column. After some time of very difficult and demanding flying, the pilot in training who was dripping with sweat said: "*Where the hell is that cumulonimbus!!!*".

During another routine flight towards Zadar, the Skyservant crew requested weather information en route

from the meteorologist on duty. "*When are you leaving?*", asked the meteorologist. "*In two hours*", answered the lead pilot. "*All right, let me know tomorrow how the weather was during your flight!*", said the meteorologist calmly as he walked away from the confused pilots at Batajnica airfield which stared at him in return.

War In The Former Yugoslavia

Routine missions for VGI were interupted by the start of the civil war in Yugoslavia in 1991. It is interesting to mention that when fighting began in Croatia, on the same day one of the Skyservants flew over Croatia photographing the terain for a Croatian civilian company. During the afternoon, following landing at a civilian airfield in Croatia, the crew went to the hotel for some coffee and drinks. Some time later, armed Croatian paramilitary unit raided the hotel looking for "Chetnik" airplane's crew which allegeldly bombed them that day?!? Owing to the clever hotel staff, the civilian clothes worn by the crew and cold blodedness, airmen which were still members of JNA (*Jugoslovenska Narodna Armija* – Yugoslav National Army) remained undiscovered avoiding a serious incident.

As the hostilities intensified both Dorniers were used for psychological and propaganda missions. Deep inside the teritories of the secessionists, they dropped leaflets prepared by an unknown state service. One of the men who took part in those flights, a non-comissioned officer at the start of his carreer, recalls that the only safety precatuion taken during flights above the territory of the breakaway republic of Croatia were parachutes which they carried in case of getting hit and being forced to bail out. The crew members never complained about the missions, and the airplanes never changed the white colour scheme which was highly visible.

Flights which originated from Batajnica airbase in the autumn of 1991 were conducted by two pilots and two flight mechanics, with the airplane filled with leaflets weighing around 600 kg in boxes. Due to the slow rate of climb, they climed en route to 7,000 m altitude in complete radio silence and with all navigation instruments switched off, safe from the Croatian anti-aircraft defenses. During one of these "deaf and blind" flights, the crew reached as far as Klagenfurt in Austria! As Dorniers lacked pressurized cockpit, they used oxygen masks. Above a designated region, they dropped the leaflets through the camera hatch in the fuselage. The boxes were only cut open and thrown

677.TRAE flight suit patch. Polar goose with old fashioned pilot hat and goggles. Patch was worn on the chest on the right hand side. [Aleksandar Ognjević]

One of the destroyed hangars at Batajnica airfield following a NATO attack in March 1999. [Spomenko Marković]

Strike by NATO forces against the infrastructure of Batajnica airbase was fierce. Still, most of the equipment was dislocated to safety in a timely manner. Heavily damaged Super Galeb G-4 jet can be seen in the background. [Spomenko Marković]

Dornier Do 28D-2 pilot kap Spomenko Marković "Moki" in the cockpit at Batajnica airbase. Together with kap Ištvan Viši and mechanic Dragan Grbić he conducted the first war time flight in a Skyservant. [Spomenko Marković]

out of the airplane where the air current dispersed the contents. Special care was made to ensure the area and the accuracy of the drop. The participant in those flights indicated it was physically demanding to conduct those flights due to a large number of leaflets.

At the beginning of 1992, Dorniers flew over Sarajevo, dropping leaflets in vain and attempting to call for peace in the country which sank deeper into war.

Do-28 returned to routine misions in the summer of 1992. Due to the reorganization, they were reassigned from 675.TRAE in 1997 and joined 677.TRAE and following the military reorganization in 2006, they were reassigned again to 138.MTRAE (*Mešovita Transportna Avijacijska Eskadrila* - Mixed Transport Aviation Escadrille). In the meantime due to expired resources and the lack of ability to conduct an overhaul (the last overhaul was conducted in Germany in 1985 and 1986), 70502 was grounded first. Until December 1993, it conducted a total of 3,114 flights lasting a total

of 3,032.15 flight hours. 70501 continued to fly until 5 March 2007 and it accumulated a total of 4,143 flight hours duirng 5,634 flights. Due to the need for aerial photography under the patronage of VGI, the superiors had to make a decision if the Do 28D-2 will receive and overhaul or if a new airplane will get purchased. The cost was a deciding factor as an overhaul of an old airplane costs practically 70% of a new airplane with bettter performance. As a result, Do 28D-2 were declared surplus and will likely be struck off charge from Serbian Military V and PVO (*Vazduhoplovstvo i protiv vazdušna odbrana* – Air Defence).

1999 NATO Aggression

The war situation during the first half of the 1990s in the already broken up SFRJ (*Socijalistička Federativna Republika Jugoslavija* – Socialist Federative Republic of Yugoslavia) inevitably spilled over to the southern Serbian province of Kosovo and Metohija. There was no true peace there since the 1980s. Clashes of small and large intensity between the Albanian terrorists and the Serbian Police became a full military confrontation with all elements of a war. The Albanian terrorism and separatism was fully supported and financed by the

Kap lk Saša Pejčić was at the controls of Dornier during the second flight during the war, together with kap Spomenko Marković and mechanic Dragan Grbić when they flew over from Trstenik sports airfield to Belgrade-Surčin. [Aleksandar Aleksić]

Another view of the devastated hangar at Batajnica airfield, with clear evidence of strikes with cruise missiles and bombs. In the background is a destroyed twin seat Super Galeb G-4 jet airplane. [Spomenko Marković]

West, which along with numerous poor decisions made by the Serbian politicians, slowly but surely pushed the country to an uncertain and unpredictable future. Short-sightedness in anticipating the goals of global world politics further worsened the already bad situation.

The Albanian terrorism intensified during 1998 and in the spring of 1999 causing exactly that which the Albanian separatists wanted. The joint operation by the Serbian Police and Army was successful, while the terrorists during their withdrawal forced with them a large number of Albanian population. The international community was "horrified" by such events, which itself was responsible for, and the threats towards Serbia and Montenegro started to arrive on a daily basis. Of course, no one mentioned the suffering of Serbian and non-Albanian population, daily kidnappings, torture and killing of Serbian civilians, police officers and soldiers. As a consequence, negotiations began at Rambouillet, France, in the spring of 1999 followed by an unprecedented blackmail of a sovereign European country. What requirements were laid in front of SRJ (*Savezna Republika Jugoslavija* – Federal Republic of Yugoslavia) is best described in the statement by an unnamed American diplomat: "*No normal person could accept such conditions*". Yugoslavia was literally supposed to sign its own occupation with direct presence of NATO troops on its territory. This was of course out of the question.

During this time Do 28D2s were stationed at Batajnica airbase in the vicinity of Belgrade. 70502 was at the time unserviceable while 70501 in the afternoon hours around 16:00 to 16:30 at the day of the first NATO attack on 24 March 1999 conducted a technical check flight. This check lasted some fifteen minutes, flying around the perimeter of Batajnica airfield. Skyservant

was piloted by kap Spomenko Marković "Moki", while besides him sat another pilot, kap Ištvan Viši. The third crew member was a mechanic Dragan Grbić "Grba". The crew was aware that NATO attack was inevitable, the question was when? Dark suspicions were supported by the lack of any radio communication in the air, which was unusual. This all pointed towards the fact that the attack could begin at any moment. At that point kap Ištvan Viši broke the deafening silence by saying:

Moki take a good look at the airfield, perhaps we are seeing it for one last time!!!

The crew landed at Batajnica and several hours later air attack alert siren was sounded. Bombs rained down on Batajnica airfield. It was struck by cruise missiles, iron and cluster bombs of all available types and calibers. Tomorrow morning the men returned to the airfield between 08:00 and 09:00 hrs. They witnessed immense destruction of the infrastructure and equipment. The unserviceable 70502 was struck by one small piece of shrapnel which pierced its wing skin causing minimal damage, while with a lot of luck, 70501 remained undamaged. The order was issued for 677.TRAE to immediately fly over to Surčin civilian airfield in the vicinity. Kap Marković recalls the apocalyptic scene from that first morning during the war:

Everything at Batajnica was in a state of chaos, airfield was struck with missiles and cluster bombs. All around was debris, shattered concrete, plowed over soil and smoke… Hangars were destroyed, equipment was burning. It was a horrible scene! I started our Dornier and began taxiing under power across the runway which was still not cleared. All around us was smoke, scattered equipment and ruins…

Five or six escadrille airplanes, amongst which was Do 28, successfully flew at tree top level to Surčin.

A group of aviators in front of Do 28D-2 at Batajnica airfield. [Aleksandar Aleksić]

Prior to the "swan's flight", the pilot jargon for the last flight in an airplane. From left to right: Pilot Saša Pejčić, pilot Aleksandar Aleksić, flight mechanic Dragan Stojković, pilot Josip Djetvaj and pilot Ištvan Viši. Visible on the right is VIP transport airplane Jak-40. [Aleksandar Aleksić]

A group of aviators under the port wing. [Spomenko Marković]

Dornier crew remained the same as during the technical trial the prior day.

Already tomorrow, on 26 March, a new order was received to dislocate. This time the order was to fly to Trstenik sports airfield which was some 140km away from Surčin. The arrival of four transport *Antonov* An 26 from Niš airfield was also expected. The flight was conducted at lowest possible altitude, following the terrain. During this flight, a large scale NATO attack was in full force. All airplanes successfully reached their new destination.

In accordance with the order, they were supposed to return to Surčin not long thereafter, but the flight was cancelled due to a large scale attack against Belgrade. Tomorrow they finally flew over to Surčin. In spite of the weak batteries in Dornier, crew managed to start the engines, flying at minimal altitude following the terrain, "hopping" over small hills and forests managed

Preparations for pre-flight check of Do 28-D2 at Batajnica airfield. Port engine is started. [Aleksandar Aleksić]

A head on view of Skyservant during the pre-flight check, Batajnica airfield. [Aleksandar Aleksić]

to complete their task. Crew consisted of kap Ik Saša Pejčić, kap Spomenko Marković and mechanic Dragan Grbić. Dornier was immediately hidden in one of JAT's (*Jugoslovenski Aero Transport* - Yugoslav Air Transport) hangars. It remained there until the end of the war, as well as some time after, until Batajnica airbase was put back into service.

Retirement

It should be noted that Skyservant's successor visited Belgrade towards the end of the 1980s. Pilot and instructor Buda Djordjević had the opportunity to fly the Dorner Do 228. He did not hide his amasement with the Do 28's younger sibiling. Unfortunately, less competent people at the time were involved in decision making, and despite the availability funds the purchase od Do 228 never materialized.

677.TRAE flight mechanic Nenad Arizanović (†) in Do 28-D2 cockpit at Batajnica airfield. Arizanović had an untimely death due to health complications on 22 January 2010. [Predrag Grandić]

The Ministry of Defence of the Republic of Serbia finally decided to purchase a new airplane, *Piper* Seneca V, for aerial photogrametry. The majority of pilots which

Last checks before a technical control flight of Dornier from Batajnica airfield. [Aleksandar Aleksić]

Skyservant taxiing towards the take off runway at Batajnica. On the left is a row of Antonov An-26 twin engine military transporters. [Aleksandar Aleksić]

Pilot Aleksandar Aleksić "Mrčo" during the last visit to his Dornier at Batajnica airfield. [Aleksandar Aleksić]

The last good bye and a "kiss on the nose" for his favorite and reliable Dornier following the retirement. Pilot Aleksandar Aleksić at Batajnica airfield March/April 2001. [Aleksandar Aleksić]

Pilots Saša Pejčić and Aleksandar Aleksić in front of the memorial made from a tail of an An-26 at Batajnica airbase airfield in 2001. It is interesting that the goose, the escadrille mascot, is pointing in the direction opposite to that on the badge. [Aleksandar Aleksić]

flew Dorniers for decades had minimal or no affinity whatsoever towards the Senecca, since they considered it inferior to the Skyservant. The arrival of this new airplane marks the end of service of two *Dornier* Do 28D-2, which have for a a long time remained grounded at Batajnica airfield. Both airplanes remained in the open with 70502 having its propellers removed.

Their crews have for a long time said that Skyservants were the only airplanes in the Air Force to brought profit to the military, while all the others were just wasting money.

They covered every square meter of the SFRJ as they photographed the terrain.

70501 and 70502 at Batajnica airfield after being struck off charge. Note that the 70502 in the background lacks the propellers. [Predrag Stamenković]

Construction Features

The airplane cabin was designed so that the airline type seats or folding bench seats may be installed for passenger or troop carrying missions. There are two pilot doors available, one on each side of the cockpit. A large cargo door allows for easy loading and unloading and it consists of two equal sections. The nose compartment is used for installation of the electronic equipment. The fiberglass nose section may be swung to the right, allowing easy access to the equipment. Both cabin and cockpit are soundproof and heat insulated with linings made of fire-resistant materials. The instrument panel consists of three single panels which are hinged to provide easy access to the instruments for maintenance.

The engines and fuel tanks form nacelles which are attached to the ends of the support running through the bottom of the fuselage. The fixed landing gear is also attached to the end of this support. The flight control surfaces are metal structures covered with nylon fabric to reduce airplane mass and to provide durability and strength. The outboard of each wing is equipped with a fixed slap. The flaps are double slotted and the ailerons are of the single slat type. The entire horizontal tailplane is used as an elevator with flap-actuated trim tabs which also, through a designated lever, may be hand adjusted.

Do 28D-2 was powered by two air cooled and fuel injected *Lycoming* IGSO-540-A1E l six cylinder opposed engines each rated at 385 hp for take off and 364 hp max continuous power at sea level. The engines obtain cooling air through the intake in the cowling, where the cooling air flows through the upper compartment of the engine through the cylinder fins to the lower compartment. The exhaust pipes protrude into

Do-28D-2 cockpit showing the instrument panel and pilot and co-pilot control columns. [Predrag Grandić]

Pilot Spomenko Marković "Moki" in the cockpit of his Dornier at Batajnica airfield. Pilot is wearing a camouflaged flight suit. [Spomenko Marković]

Entire four man Skyservant crew immediately before take off from Batajnica airfield. Pilot Marković is at the controls on the left. [Spomenko Marković]

Seen here is the pilot's cockpit as well as the aft area with the camera and operator seats. [Predrag Grandić]

Double doors on the port fuselage side swing wide open to allow for bulky cargo loading. [Predrag Grandić]

385 hp Lycoming IGSO-540 installation in the Do-28D-2 V1 prototype. [Airbus Corporate Heritage]

Detail of the starboard engine pylon as seen from the cockpit. [Hans Ulrich Willbold]

Wild RC-10 camera installation in one of the Skyservants. Note the operator seat immediately behind. [Predrag Grandić]

A set of special lenses for Wild RC-10 camera. [via Aleksandar Aleksić]

Camera operators in the cockpit of Do 28-D2 during a terrain photography mission at an unknown location. [Spomenko Marković]

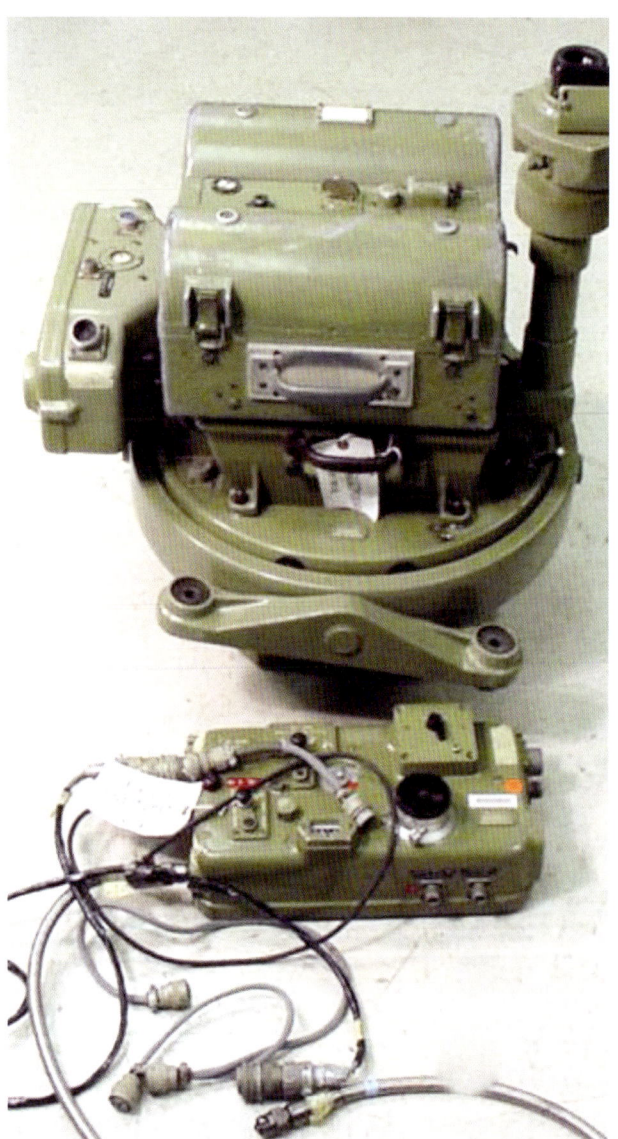

The older type of camera, Wild RC-8, was also sometimes used, but the newer RC-10 proved to be superior. [via Aleksandar Aleksić]

the augmentator tubes which assist in evacuating the hot air from the lower engine compartment. The engine lubrication system is a full pressure, dry sump type. The required pressure is provided by a gear type pump on the accessory housing. The engine is equipped with a non-regulated supercharger mounted at the rear section of the engine and gear-driven by the crankshaft with the gear ratio of 11.27:1. Dual ignition is provided by two *Scintilla* S6RN-1208 (left) and S6RN-1209 (right) magnetos, where both magnetos are of the high tension

Belgrade from the air, photographed from Dornier's cockpit. [Spomenko Marković]

Famous football stadium, "Marakana", the home of Belgrade's Red Star club, photographed here from Skyservant's cockpit. [Spomenko Marković]

type and are mounted on each side of the reduction gear housing. Electrically driven starters are actuated by means of the combined ignition/starter switches.

The engines drive three blade *Hartzell* variable pitch, constant speed full feathering propellers with clockwise rotation in flight direction. Propellers have 2.36 m diameter. Control of the propellers is accomplished by two propeller controls and the engine driven propeller governors. Propeller pitch is varied by controlling the oil pressure acting on the propeller servo cylinder. Each propeller blade is provided with a de-icer fitted with an inboard and outboard heating element, each with a heat output of approximately 126 W.

The fuel supply system consists of two metal tanks, each with 452 liter capacity and located in the rear of each engine nacelle, two auxiliary fuel pumps, an engine driven fuel pump for each engine, two fuel quantity transmitters, one dual fuel quality gauge, two fuel filters, three fuel selector levers and associated vales and fuel lines. The two main tanks are interconnected by means of a cross feed line. There are provisions for one 250 liter drop tank under each wing.

The airplane electrical system is primarily 28 VDC. An inverted is provided to supply 115 VAC, 400 cycle current. Each engine drivers a 30V 300A generator providing approximately 9 kW power. The voltage of each generator is controlled by a voltage regulator to an almost constant value of 28 VDC. Two 12 V 33Ah batteries are located in the cockpit. They are connected in series to function as one battery.

The AN/ARC-34 C/G UHF radio set is comprised of a transceiver located in the electronics compartment, the UHF antenna located under the nose section, microphone buttons and UHF switches. The radio set provides two way amplitude modulated voice communication across 3,500 channels in the frequency range between 225,00 to 399,95 MHz. The 618M-3A radio set is comprised of a transceiver located in the electronics department, the combined VHF/UHF antenna below the nose and the VHF switch on the right hand side of each interphone unit. The radio provides two way amplitude modulated voice communication in the frequency range between 116,00 to 151,975 MHz.

Two camera doors are located in the bottom of the fuselage. The doors are opened individually with a crank. Two small holes are provided in the floor just forward of the front vertical camera position to insert the crank.

The dual flight control system is a conventional one. The linkage between the control column and the control surfaces is mechanical and consists of cables and

Front view showing the landing gear and engine arrangement. Note the landing light in the nose. Batajnica airbase, 2007. [Djordje Nikolić]

pushrods. Cables are used for rudder control and trim tab actuation. Longitudinal control is accomplished with the control column and the elevator. Provided trim tabs can be adjusted by means of a trim control wheel on the cockpit ceiling and by flap actuation. Lateral control is effected by the control wheel and the ailerons which are also provided with trim tabs. Directional control is achieved by means of rudder pedals and rudder. The rudder is equipped with an adjustable trim tab and the corresponding control crank is located on the cockpit ceiling. The double slotted flaps are actuated by an electric geared actuator through control rods and bell cranks. There are three positions: 0º, 20º and 52º. The inboard sections extend to 52º while the outboard ones to 34º. The wing aspect ratio is 8.3:1 and dihedral is 1.5º.

The airplane is equipped with a conventional non-retractable landing gear. The main gear is an oleo-pneumatic shock absorber strut type, fitted with a fork mounted low pressure tire. It is streamlined by a fiberglass fairing. The tail gear assembly consists of a laminated fiberglass spring and 360º swiveling, self centering pneumatic tire. The twin contact tail wheel assists in preventing tail wheel flutter. The undercarriage track is 3.52 m and the wheelbase is 8.63 m.

The emergency equipment consists of engine fire detection system, engine fire extinguishing system, a hand fire extinguisher, crash axe, four first aid kits and an emergency exit.

Technical Specifications

Technical specifications Dornier Do 28D-2 in Yugoslav service	
Quantity used:	2
Crew:	2 + 2 (photo version)
Years of Service:	1975-2007
Span:	15.5 m
Length:	11.3 m
Height:	3.9 m
Wing area:	29.0 m²
Engine:	Two 385 hp Lycoming IGSO-540-A1E
Empty weight:	2,190 kg
Maximum weight:	3,700 kg
Maximum speed:	320 km/h
Cruise speed:	272 km/h at 3,050 m with 65% power
Service ceiling:	7,750 m
Maximum range:	850 km (with maximum payload)
Armament:	None

Camouflage and Markings

Both airplanes were delivered from the factory in overall White livery which they retained throughout their service. A Yugoslav flag band consisting of Blue, White and Red fields wrapped around the entire fuselage, from the nose to the tail. The wing leading edge was painted Black to assist with de-icing (the so called de-icing boots)

70501 at an airshow showing its overall white lively which was retained throughout their service. A Yugoslav flag band consisting of blue, white and red fields is wrapped around the entire fuselage, from the nose to the tail. [Predrag Stamenković]

Note the detail of the Yugoslav flag band around the fuselage, red band around the engine pylon as well as Hartzell propeller logos. [Djordjević Family]

and the wing tips, the vertical stabilizer and horizontal stabilizer tips were painted in Red colour. The engine nacelles had a Red band around the entire length.

During the service in the SFRJ, standard Yugoslav insignia was applied on both sides of the fuselage (inter-rupting the flag band around) along with the topside of the port wing and underside of the starboard wing. The rudder had a Yugoslav flag band applied the entire length and the registration number was painted on the vertical stabilizer. The airplane registration number,

Following the break up of SFRJ, a small Yugoslav flag was applied on the vertical stabilizer. The code "501" and "502" as well as a SFRY roundel were deleted from the fuselage. [Aleksandar Aleksić]

To commemorate 100 Years of Serbian Air Force, 70501 had the official logo commemorating this occasion applied on the engine pylons as well as the new Serbian insignia on the fuselage and flag on the vertical stabilizer. [Predrag Stamenković]

The wing and horizontal leading edges were painted black to assist with de-icing. The wing tips as well as the vertical and horizontal stabilizer tips were painted red. [Predrag Stamenković]

501 or 502, was applied on the fuselage next to the insignia as well as on the underside of the port wing and topside of the starboard wing.

Following the break-up of the country in 1991, a new style insignia, the so called "Pepsi-Cola" roundel, was applied in place of the previous markings. The insignia and the registration number were removed from the fuselage and the Yugoslav flag band was painted continuous, which indicates that the entire airplane likely received a new coat of paint due to their pristine appearance. The large flag band from the rudder was removed and its place a smaller size SRJ flag was applied on the vertical stabilizer. The registration numbers were removed from the wings and the fuselage.

During the celebrations at Batajnica airfield commemorating the 100th Anniversary of the Serbian Air Force, 70501 received the new Serbian roundel with "Kosovo cross" on the fuselage, along with a Serbian flag on the vertical stabilizer. Additionally, a logo commemorating the anniversary was applied on the outboard side of the engine pylon. Interestingly, the Yugoslav flag band around the fuselage remained in place.

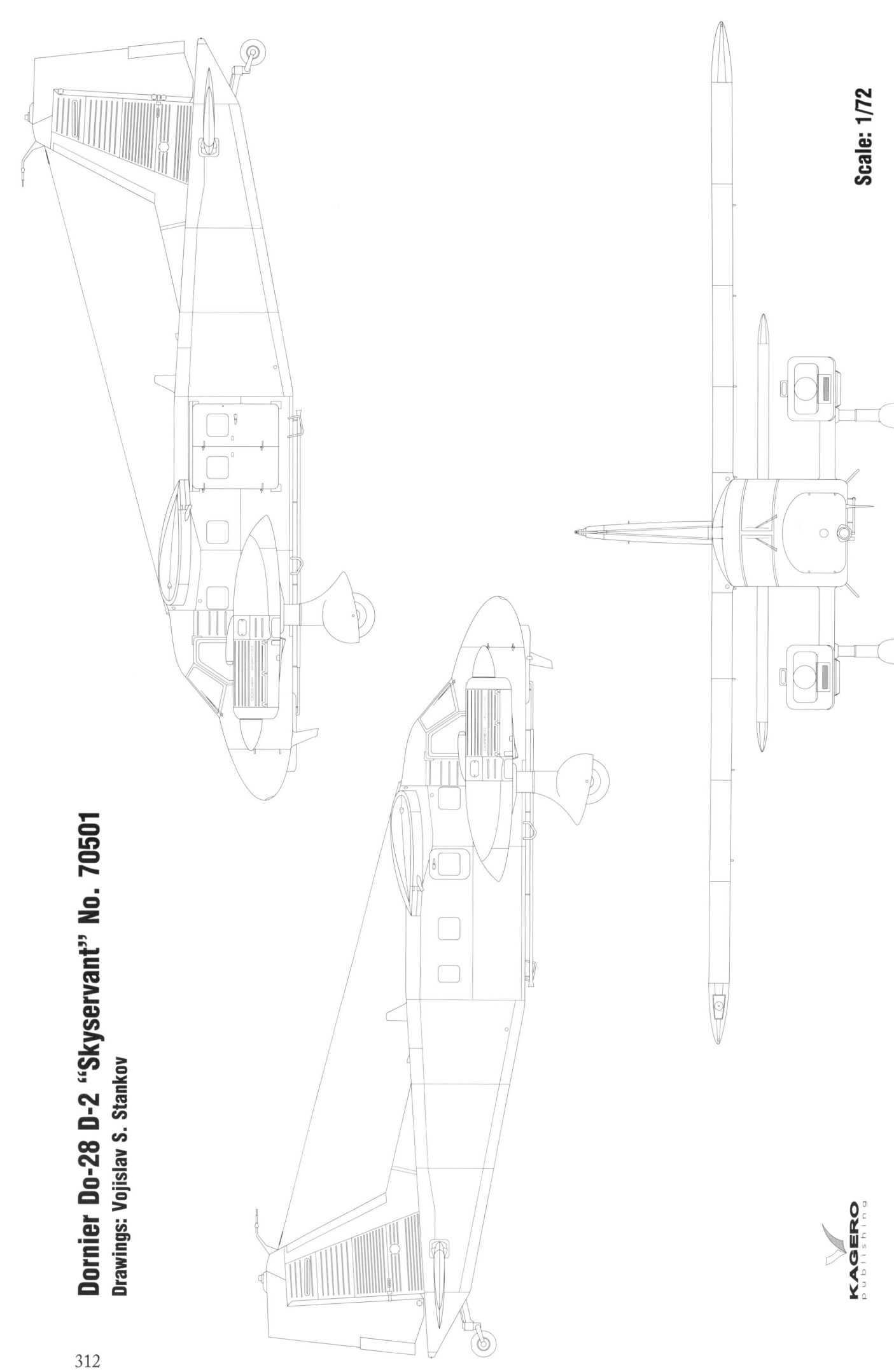

Dornier Do-28 D-2 "Skyservant" No. 70501
Drawings: Vojislav S. Stankov

Scale: 1/72

KAGERO
publishing

Dornier Do-28 D-2 "Skyservant" No. 70501

Drawings: Vojislav S. Stankov

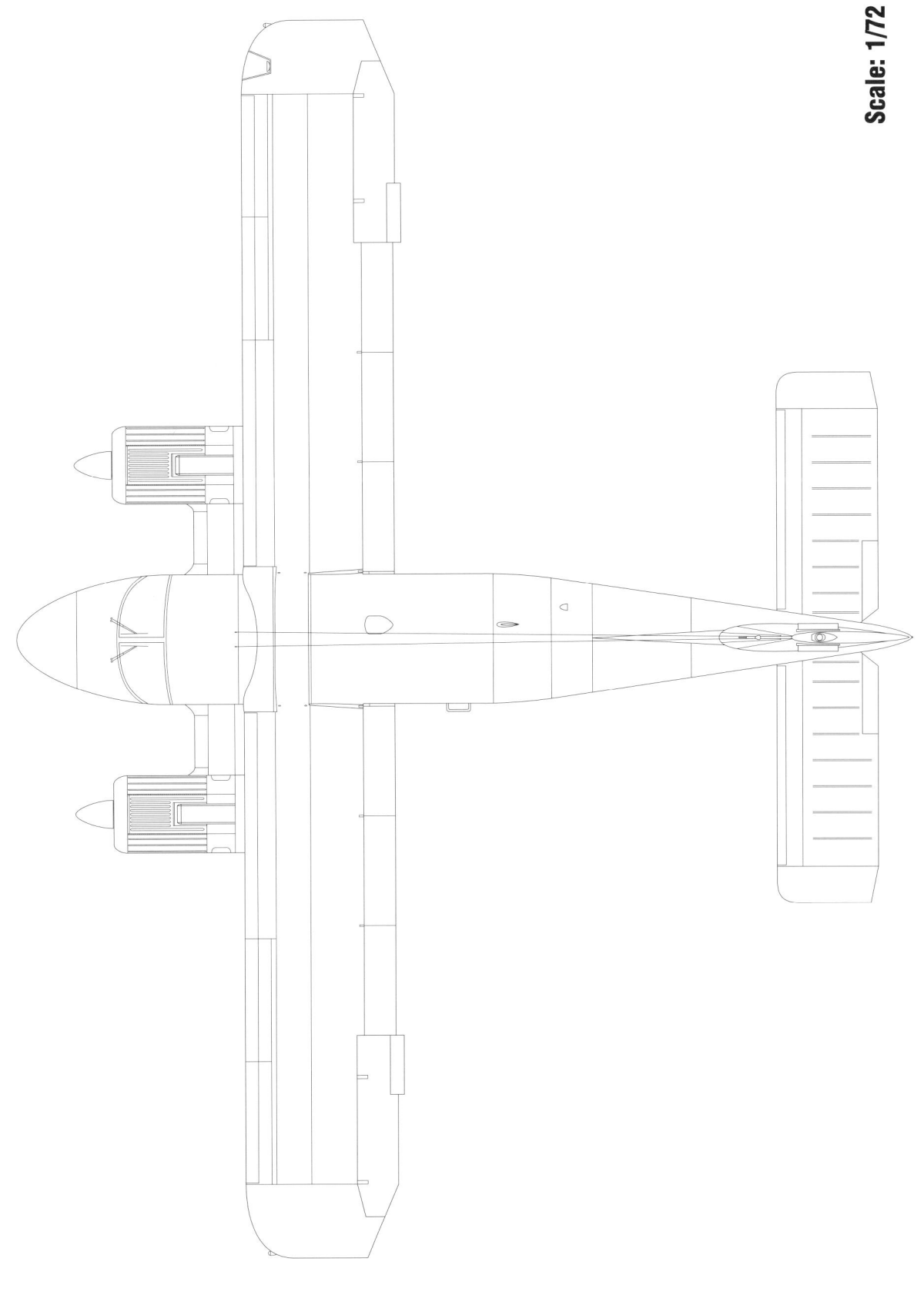

KAGERO
publishing

Dornier Do-28D "Skyservant"   No. 50701

Drawings: © Vojislav S. Stankov

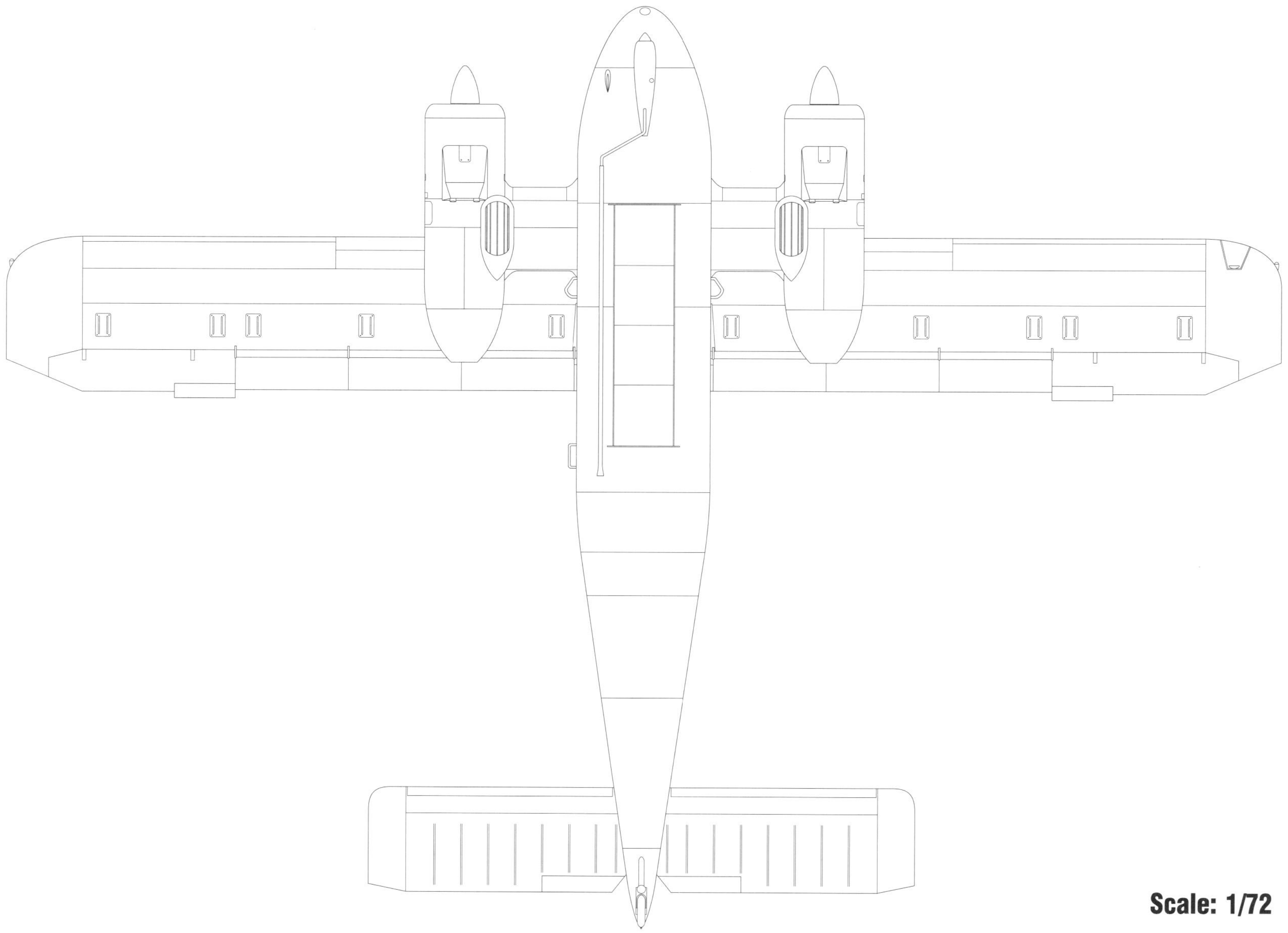

314

Dornier Do-28 D-2 "Skyservant" No. 70501

Drawings: Vojislav S. Stankov

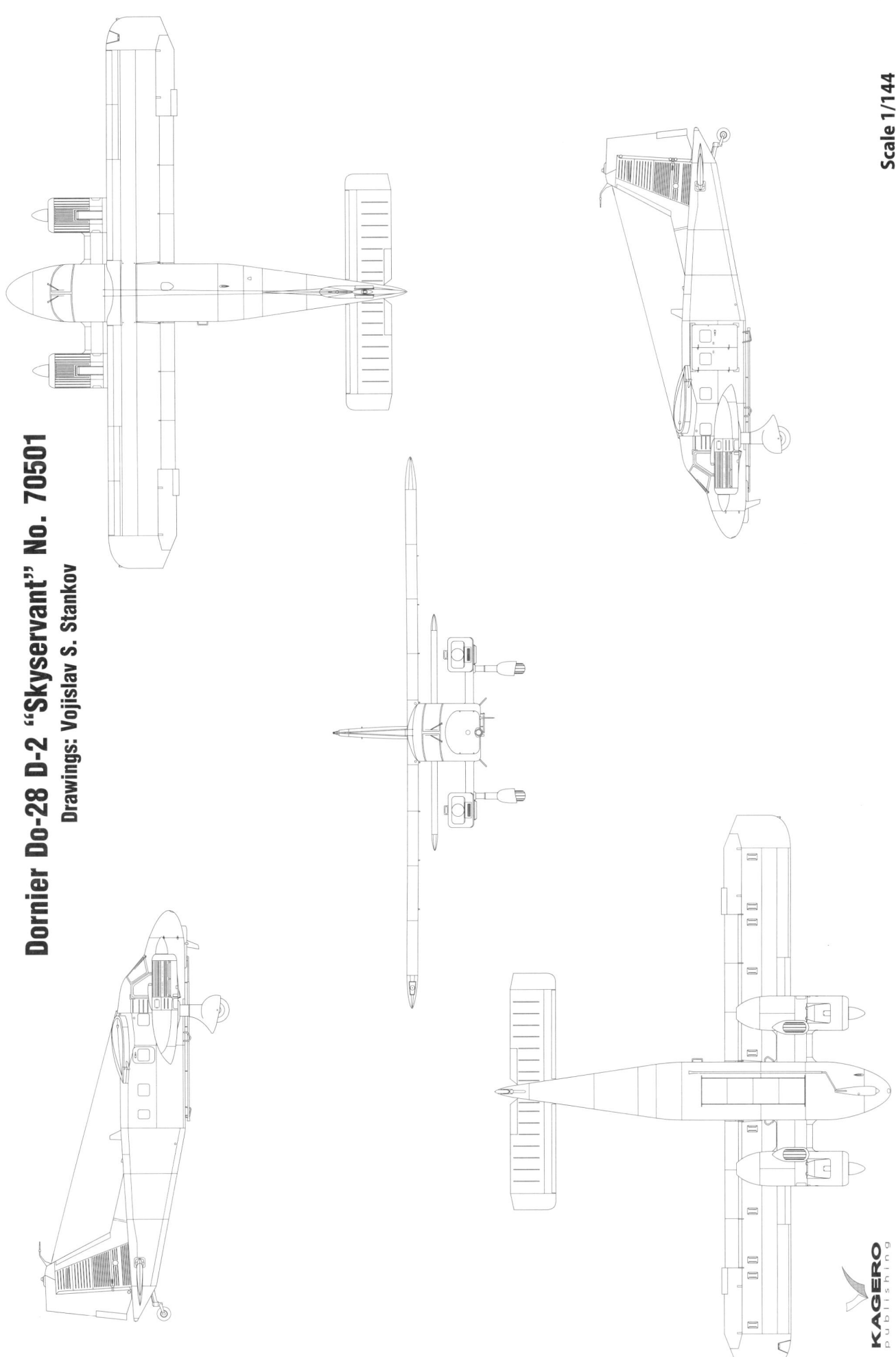

Scale 1/144

KAGERO
publishing

Appendices

Appendix 1. Maps of Kingdom of Yugoslavia

KJ Map 1941 [Aleksandar Ognjevic]

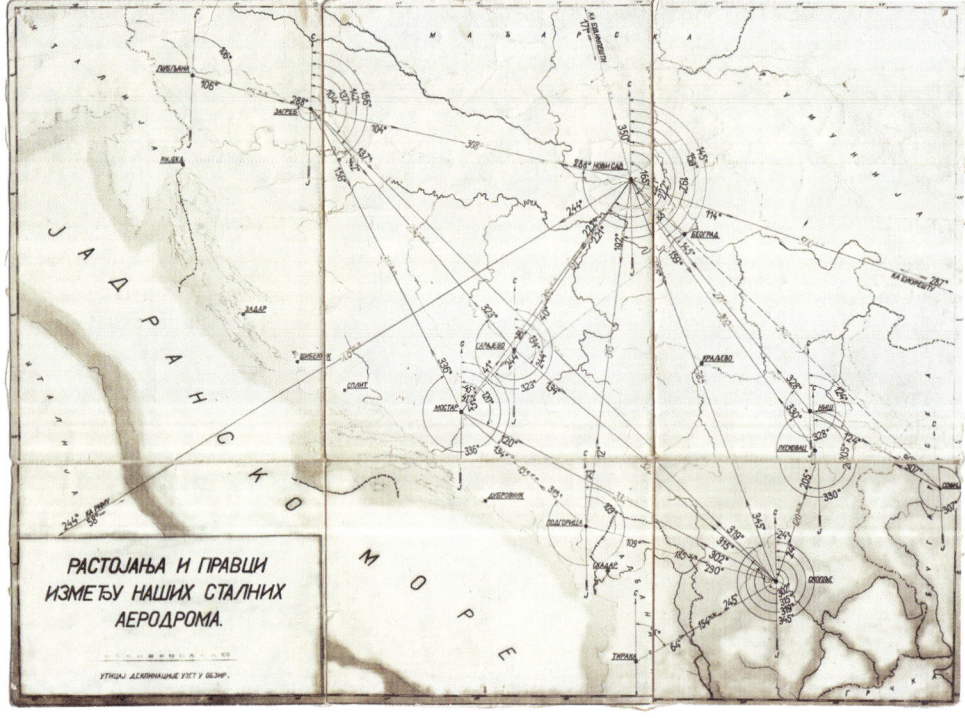

Pilot Map - Kingdom of Yugoslavia [Dragan Savić]

KJ Coastline 1941 [Aleksandar Ognjević]

Appendix 2. Air Force and Navy Rank Comparison

VVKJ		Luftwaffe		Royal Air Force	
Rank	abb.	Rank	abb.	Rank	abb.
armijski đeneral	arm đen	General Oberst	GenOb.	Air Chief Marshall	ACM
divizijski đeneral	div đen	General Leutnant	GenLt.	Air Marshall	AM
vazduhoplovni brigadni đeneral	vazd brig đen	General Major	GenMaj.	Air Vice Marshall	AVM
pukovnik	puk	Oberst	Oberst	Group Captain	G/Cpt
potpukovnik	ppuk	Oberstleutnant	OberstLt.	Wing Commander	W/Cdr
major	maj	Major	Maj.	Squadron Leader	S/Ldr
kapetan I klase	kap Ik	Hauptmann	Hptm.	Flight Lieutenant 1st class	F/Lt
kapetan II klase	kap IIk	-	-	-	-
poručnik	por	Oberleutnant	Oblt.	Flying-Officer	F/O
potporučnik	ppor	Leutnant	Lt.	Pilot-Officer	P/O
narednik-vodnik I klase	nv Ik	Stabsfeldwebel	Stfw.	Flight Sergeant	F/sgt
narednik-vodnik II klase	nv IIk	Oberfähnrich	Obfhr.	-	-
narednik-vodnik III klase	nv IIIk	Oberfeldwebel	Obfw.	-	-
narednik I klase	nar Ik	Feldwebel	Fw.	Sergeant	Sgt
narednik II klase	nar IIk	Unterfeldwebel	Ufw.	-	-
podnarednik I klase	pnar Ik	Unteroffizier	Uffz.	Corporal	Cpl
podnarednik II klase	pnar IIk	Stabsgefreiter	Stgefr.	Senior Aircraftman	SAC
kaplar	kpl	Gefreiter	Gefr.	Airman 1st Class	A1C
redov	red	Flieger	Flg.	Airman 2nd Class	A2C

According to VVKJ service rule, non-commissioned officers which met the requirements for officer rank without completing the Military Academy were entitled rank of *vojno-tehnički činovnik* of various classes which were equivalent to specific officer ranks:

viši vojnotehnički činovnik I klase	vvtč Ik	brigadni đeneral
viši vojnotehnički činovnik II klase	vvtč IIk	pukovnik
viši vojnotehnički činovnik III klase	vvtč IIIk	potpukovnik
viši vojnotehnički činovnik IV klase	vvtč IVk	major
niži vojnotehnički činovnik I klase	nvtč Ik	kapetan I klase
niži vojnotehnički činovnik II klase	nvtč IIk	kapetan II klase
niži vojnotehnički činovnik III klase	nvtč IIIk	poručnik
niži vojnotehnički činovnik IV klase	nvtč IVk	potporučnik

PVKJ		Royal Navy
Rank	abb.	Rank
Admiral	-	Admiral
Vice-Admiral	-	Vice Admiral
Kontra-Admiral	-	Rear admiral
No equivalent	-	Commodore
Kapetan bojnog broda	kbb	Captain
Kapetan fregate	kf	Commander
Kapetan korvete	kk	Lieutenant-Commander
Poručnik bojnog broda	pbb	Lieutenant
Poručnik fregate	pf	Sub-Lieutenant
Poručnik korvete	pk	Mid-shipman
Narednik-vodnik	nv	Warrant-Officer
Narednik	nar	Petty Officer
Podnarednik	pnar	Able body
Kaplar	kap	Seamen
Mornar	-	Sailor

Appendix 3. VVKJ Unit Structures

brigada (Brigade)	Comprised of two to three Bomber or Fighter Regiments
puk (Regiment)	Composed of two to three fighter or two bomber Aviation Groups
grupa (Group)	A tactical unit consisting of two to three Escadrilles
eskadrila (Escadrille)	A unit consisting of nine to 12 airplanes

Appendix 4. Abbreviations

ausländish	a	Foreign
-	AME	Air Ministry Experimental
Aeroplanska radionica	AR	Airplane Workshop
Aerodromska Četa	AČ	Airfield Company
Bombardirovičen orlijak	BO	Bomber Group
Broj	Br	Number
Bombarderski puk	BP	Bomber Regiment
Construcciones Aeronáuticas SA	CASA	Aeronautical Constructions SA
Costruzioni Meccaniche Aeronautiche S.A.	CMASA	Mechanical and Aeronautical Constructions S.A.
-	CO	Commanding officer
Deutscher Aero Club e.V.	Daec	German Aero Club
Državna Fabrika Aviona	DFA	State Airplane Factory
Dr. Kurt Herberts	DKH	-
Deutchmark	DM	German post war currency
Deutscher Luftsportverband	DVL	German Air Sports Organization
Deutsche Verkehrsfliegerschule	DVS	German commercial pilot school
Vojno evidencijski broj	EvBr.	Army evidence number
Fédération Aeronautique Internatonale	FAI	World Air Sports Federation
Fabrique Nationale Herstal	FN	National Factory Herstal
Federativna Narodna Republika Jugoslavija	FNRJ	Federative Peoples Republic of Yugoslavia
Hidro Eskadrila	HE	Hydroescadrille
Her Majesty's Ship	HMS	-
Hidro Komanda	HK	Hydro Command
-	HQ	Headquarters

-	hp	Horse power
-	IAACC	Inter-Allied Aeronautical Commission of Control
Industrija aeroplanskih motora	IAM	Aero Engines Industry
Internationella Luftfartsutställningen i Stockholm	ILIS	International Aviation Exhibit in Stockholm
Jugoslawien	j	Yugoslavia
Jugoslovenski Aero Transport	JAT	Yugoslav Air Transport
Jugoslovenska Narodna Armija	JNA	Yugoslav National Army
Jugoslovenska Ratna Mornarica	JRM	Yugoslav Navy
Jugoslovensko Ratno Vazduhoplovstvo	JRV	Yugoslav Air Force
Kampf	K	Combat
Kraljevski Brod	KB	Royal ship
Kraljevina Jugoslavija	KJ	Kingdom of Yugoslavia
Kraljevska Mornarica	KM	Royal Yugoslav Navy
Komanda Mornarice	KM	Naval Command
Kraljevina Srba, Hrvata i Slovenaca	KSHS	Kingdom of Serbs, Croats and Slovenes
Komunistička Partija Jugoslavije	KPJ	Yugoslav Communist Party
Komanda vazduhoplovstva	KV	Air Force Command
Lovački puk	LP	Fighter Regiment
Mešovita Transportna Avijacijska Eskadrila	MTRAE	Mixed Transport Aviation Escadrille
Ministarstvo Vojske i Mornanrice	MViM	Ministry of Army and Navy
-	NACA	National Advisory Committee for Aeronautics
Nezavisna Država Hrvatska	NDH	Independent State of Croatia
Pilotska škola	PS	Pilot School
Pomorsko Vazduhoplovstvo Kraljevine Jugoslavije	PVKJ	Royal Yugoslav Naval Air Service
Protiv vazdušna odbrana	PVO	Air Defence
Royal Air Force	RAF	Air force of the United Kingdom
Reichverband der Deutschen Luftfahrtindustrie	RDL	Reich Society of Airplane Manufacturers
Reichluftfahrtministerium	RLM	Reich Air Ministry
Reichsmark	RM	German currency between 1924 and 1948
-	RPM	Revolutions per minute
Reichsverkehrsministerium	RVM	German Ministry of Transport
S.A. Italiana Costruzioni Meccaniche	SAICM	S.A. Italian Mechanical Constructions
Sociedad Colombo Alemana de Transportes Aéreos	SCADTA	Colombian-German Air Transport Company
Socijalistička Federativna Republika Jugoslavija	SFRJ	Socialist Federative Republic of Yugoslavia
Sivo-maslinasta Boja	SMB	Gray-Olive color
Savezna Republika Jugoslavija	SRJ	Federal Republic of Yugoslavia
-	STOL	Short Take Off and Landing
Škola Rezervnih Oficira Avijacije	ŠROA	Reserve Aviation Officer School
Transportna Avijacijska Eskadrila	TRAE	Transport Aviation Escadrille
Transportni Avijacijski Puk	TRAP	Transport Aviation Regiment
-	UK	United Kingdom
-	USAF	United States Air Force
Versuch	V	Trial
Vazduhoplovna Baza	VB	Aviation Base
Vereinigte Deutche Metallwerke	VDM	United German Metal works
Vazduhoplovna grupa	VG	Aviation Group
Vojno geografski institut	VGI	Military Gerographical Institute
Vazduhoplovni Puk	VP	Aviation Regiment
Vazduhoplovno-tehnicki Zavod	VTZ	Aviation Technical Depot
Vazduhoplovstvo Vojske Kraljevine Jugoslavije	VVKJ	Royal Yugoslav Army Air Force
Vazduhoplovna škola bombardovanja	VŠB	Air bombardment school
Werke Nummer	W.Nr.	Construction number
Zrakoplovstvo Nezavisne Države Hrvatske	ZNDH	Air Force of the Independent State of Croatia

Selected Bibliography

A Dornier freight carrier, Flight, November 23 1933

Air Enthusiast nö.30: *The Anals of the "Pencil",* England 1986

Miloš Aćin: *Spomenica palih Srba vazduhoplovaca 1941 – 1945,* Srpska krila, Washington D.C., USA 1975

Ulf Balke: *Der Luftkrieg in Europa 1939 – 1941,* Bechtermünz Verlag, Germany 1997

Walter Barth: *Bemerkenswerte Untersuchungsergebnisse aus der Arbeit des Windkanals der Dornier-Werke 1935-1945,* Germany 1954

Phil Butler: *War Prizes,* UK 1994

Sven Carlsen, Michael Meyer: *Die Flugzeugführer-Ausbildung der Deutschen Luftwaffe 1935-1945 – Band I,* Zweibrücken, Germany 1998

John Carr: *On Spartan Wings,* Barnsley, Great Britain 2012

Boris Ciglić, Dragan Savić: *Dornier Do 17 The Yugoslav Story,* Beograd, Serbia 2007

Boris Ciglić: *Seaplanes of Bocche,* Belgrade, Serbia 2014

Dušan Ćirović: *Vazduhoplovne žrtve 1913 – 1945,* private edition, Zemun, Yugoslavia 1970

CMASA: *Instruzioni per il montaggio e la regolazione dell' idrovolante Dornier-Wal Cabina*

Peter W. Cohausz, Flugzeugclassic 11/2010: *Seeflugzeug Dornier Do 22,* Germany 2010

Bojan Dimitrijević, Milan Micevski, Predrag Miladinović: *Kraljevsko Vazduhoplovstvo 1918-1944,* Beograd, Serbia 2012

Bojan Dimitrijević, Milan Micevski, Predrag Miladinović: *Kraljevsko Vazduhoplovstvo 1912-1945,* Beograd, Serbia 2016

Bojan Dimitrijević, Milan Micevski, Predrag Miladinović: *Jugoslovensko Pomorsko Vazduhoplovstvo,* Beograd, Serbia 2014

Bojan Dimitrijević: *Istorija 20. Veka 1/2014: The Yugoslav Floatplane Squadron in Aboukir 1941-1942,* Beograd, Serbia 2014

Brigitte Kazenwadel-Drews: *Claude Dornier – Pionier der Luftfahrt,* Bielefeld, Germany 2007

Miroslav Filipović: *Kraljevski avioni,* Kraljevo, Serbia 1995

Der Flieger, Heft 6, Juni 1940, Germany 1940

Zvonimir Freivogel, Achille Rastelli: *Adriatic Naval War 1940-1945,* Zagreb, Croatia 2015

Danijel Frka, Šime Oštrić: *Aeromagazin: Hidroavion Dornier Wal,* Portorož, Yugoslavia 1990

Daniel Frka, Bojan Dimitrijević: *The Naval Aviation in the Adriatic 1918-1991,* Zagreb, Croatia 2016

Günter Frost, Karl Kössler, Volker Koss: *Dornier von den Anfangen bis 1945,* Königswinter, Germany 2010

Günther Frost (ADL): *Das Flugboot Dornier "Wal" (Do J) Teil 1, 2, 3,* Online PDF edition

Günther Frost (ADL): *Torpedoflugzeug und Seeaufklärer Dornier Do D,* Online PDF edition

Nebojša D. Djokić, Radovan M. Radovanović: *Purchase of weapons from Weimar Germany for the Army of Kingdom of SCS/Yugoslavia,* (IZVOR)

Dornier GmbH: *A Brief History of the Company,* Friedrichshafen, Germany 1983

Dornier Werke GmbH: *25 Jahre Dornier – Ein Vierteljahrhundert Pionierarbeit fur die Luftfahrt,* Friedricshafen, Germany, 1939

Dornier-Metall-Flugzeuge: *Grossflugboot Dornier-Wal*

Dornier PR Abteilung: *Dornier Typenblatt: Aufklärer Do-D,* München, Germany

Dornier PR Abteilung: *Dornier Typenblatt: Do-Wal-Flugbook,* München, Germany

Dornier PR Abteilung: *Dornier Information 72D8: Skyservant,* München, Germany

Dornier-Werke GmbH Friedrichshafen: *Typenblatter der Dornier-Baumuster 1939*

Die Dornier Post Nr.14, Dez/. Jan. 1937/38

Die Dornier Post, Februar/März 1938

Die Dornier Post, Juni/Juli 1938

Die Dornier Post, Juni/Juli 1937

Die Dornier Post, August/September 1937

Dornier Post 2-72: Dornier Skyservant, München, Germany

Dornier Do.C4, Flight, May 24 1934

Dornier's versatile Skyservant, Air International January 1979

Fashionably Old-fashioned – Dornier's Versatile Skyservant, Air International/January 1979

Giancarlo Garello: *Prede Di Guerra,* Torino, Italy 2007

Zeitengenossisches Archiv St. Gallen: *Dornier,* St. Gallen, Switzerland 1934

Manfred Griehl: *Dornier Flugzeuge seit 1915,* Bobingen, Germany 2009

Manfred Griehl: *Dornier Wal,* Vicenza, Italy 2012

Manfred Griehl: Dornier Bombers and Reconnaissance Aircraft 1925 – 1945, London, UK 1990

Homologierungs-Akte Flugzeugmuster Do 22, A.-G. für Dornier – Flugzeuge, Altenrhein, Eidgenössische Luftamt, 7 December 1934

Vladimir Isaić, Danijel Frka: Naval *Aviation At the Eastern Coast of the Adriatic Sea 1918-1941 (Volume 1),* Zagreb, Croatia 2010

Vladimir Isaić: *Mornarički Glasnik Br. 5 – Pomorsko Vazduhoplovstvo Jugoslavije u ratu 1941,* Beograd, Yugoslavia 1970

Milorad Janković: *Aprilski rat,* Politika, April - July 1973, Beograd, Yugoslavia

Zoran Jerin: *Avions No. 104, Les Dornier 22 Yougoslaves,* Boulogne sur Mer, France 2001

Toma Ješanovic: *Politika, Jedan nepoznati izveštaj,* Beograd, Yugoslavia 1998

Edvard Kocent-Zieliński: *Dornier Do-17 cz.I,,* Lublin, Poland 1997

Aleksandar Kolo: *Aerosvet 12/1988 - U potrazi za slobodom,* Novi Sad, Yugoslavia 1988

Aleksandar Kolo: *Aeromagazin specijalno izdanje – Dejstva Pomorskog Vazduhoplovstva Jugoslovenske Mornarice,* Portorož, Yugoslavia 1991

Aleksandar Kolo: *Vesnik vojnog muzeja broj 40, Pomorsko vazduhoplovstvo Jugoslovenske kraljevske mornarice u Aprilskom ratu,* Beograd, Serbia 2013

Karl Kössler (ADL): *Dornier Do 17 – Dichtung und Wahrheit,* Online PDF edition

Michel Ledet: *Avions No 101: Le Dornier Do 22,* France 2001

Michel Ledet: *Avions No 102: Le Dornier Do 22 deuxième partie,* France 2001

Les Ailes, No 791, 13-8-1936, France 1936

Franjo Lolić: *Front 39/1977, Prkosna i hrabra eskadrila br. 1000,* Beograd, Yugoslavia 1977

Franjo Lolić: *Front 10/1978, Hidroavion Do-H br. 1023,* Beograd, Yugoslavia 1978

Franjo Lolić: *Glasnik RV i PVO Br.1 - Dejstva Jugoslovenske hidroplanske eskadrile u Drugom Svetskom ratu u sastavu RAF-a,* Beograd, Yugoslavia 1976

Siniša Luković: *Heroj Aprilskog rata 1941 – Poručnik bojnog broda I klase Vladeta Petrović,* Boka Kotorska, Montenegro 2011

Sinisa Luković: *Ratni dnevnik 20. hidroeskadrile (Prepis operacijskog dnevnika 20. HE),* Boka Kotorska, Montenegro 2011

Jože Maden: *Teleks, Vojna vihra nad Boko,* Ljubljana, Yugoslavia 1987

Materialamt der Luftwaffe: *Flight Manual Do 28,* 1 November 1985

Kenneth A. Merrick: Thomas H. Hitchcock, *The official monogram painting guide to German aircraft 1935-1945,* Boylston, USA 1980

Ministero dell' aeronautica: *Catalogo nomenclatore per idrovolante wal Miltaire*

Dimitar Nedialkov: *Air Power of the Kingdom of Bulgaria Part IV,* Sofia, Bulgaria 2001

Heinz Nowarra: *Die Fliegenden Bleistifte Dornier Do 17 und Do 215,* Friedberg, Germany 1978

Andrija Pavlović: *Dejstvo 3. vazduhoplovne mešovite brigade u Aprilskom ratu* (unpublished manuscript)

Rafael Perhauc: *Delo, Letalci niso priznali poraza,* Ljubljana, Yugoslavia 1967

Šime Oštrić: *Avions No. 123 Juin 2003: Laid et lourd: le Dornier DoY en Yougoslavie,* Boulogne sur Mer, France 2003

Šime Oštrić: *Avions No. 125 Août 2003: Laid et lourd: le Dornier DoY en Yougoslavie,* Boulogne sur Mer, France 2003

Ognjan M. Petrović: *Dornier Do H* (Serbia, not published)

Ognjan M. Petrović: Nabavke Nemačkih hidroplana za PV, Let 2/2000, Beograd, Serbia 2000

Ognjan Petrović: *Vesnik Vojnog Muzeja 38, Vojni aeroplani Kraljevine Jugoslavije, modernizacija- Nemački tipovi, 2. nastavak,* Serbia

Evgenije Radanov: *Uspomene jednog letača,* Beograd, Yugoslavia 1995

Aleksandar Radić: *Do kraja veran geodetima,* Arsenal Br.48

Karl Reis: *Recherchen zur Deutschen Luftfahrzeugrolle Teil 1 1919-1934*, Mainz, Germany 1977

Dipl.-Ing Helmut Schneider: *Flugzeug-Typenbuch 1936*, Leipzig, Germany 1936

Dipl.-Ing Helmut Schneider: *Flugzeug-Typenbuch 1941 Gekürzte Ausgabe*, Leipzig, Germany 1941

Christopher Shores, Brian Cull, Nicola Malizia: *Air war over Yugoslavia, Greece and Crete 1940-1941*, London, Great Britain 1987

Službene novine, London, Great Britain 1941

Tehnički uput za avion Do 17Ka-2 Gnome-Rhône K-14NO, Tipografska radionica Štaba vazduhoplovstva vojske, Zemun,1938

The Flying Pencil, Flight, April 14 1938

Janez Žerovac: *Zapisani nebu*, Radovljica, Slovenia 1991

Oskar Ursinus: *Flugsport XXVIII Jahrgang*, Germany 1936

Unsere Chronik: Dornier-Wal, Dornier PR Abteilung, München, Germany

Vojna enciklopedija, Vojnoizdavački zavod, Belgrade, Yugoslavia 1978

War in the air, Flight, May 29 1941

Internet resources
www.paluba.info
www.adl-luftfahrthistorik.de/
www.forum.12oclockhigh.net/
www.luftwaffe-research-group.org
www.swissair00.ch

Archival Funds
Airbus Group Archives, Friedrichshafen, Germany
Luftarchiv Hafner, Ludwigsburg, Germany
Dornier Museum, Friedrichshafen, Germany
Militararhiv, Bundesarchiv, Freiburg, Germany
Vojni Arhiv, Belgrade, Serbia
Muzej vazduhoplovstva - Beograd, Belgrade, Serbia
Arhiv Vojnoistorijskog Instituta, Belgrade, Serbia
Istorijski Institut JNA – Belgrade, Serbia
Arhiv Jugoslavije, Belgrade, Serbia
The Museum of Flight, Seattle WA, USA
USAF Museum, Dayton OH, USA
National Air and Space Museum, Washington DC, USA
Imperial War Museum, London, UK
RAF Museum, London, UK
The National Archives, Kew, UK
230 Squadron Archives, Benson, UK
Stato Maggiore Aeronautica, Roma, Italy
Staatsarchiv St. Gallen, St. Gallen, Switzerland
Hrvatski Pomorski Muzej, Split, Croatia
Muzej Revolucije Naroda Hrvatske/Hrvatski Povijesni muzej, Zagreb, Croatia

Selected Archival Funds

Airbus Group Archives, Friedrichshafen, Germany
Dornier-Werke GmbH, Do 22 Kg 15/3, Friedricshafen, Germany 1938
Dornier-Werke GmbH, Handbuch fur Dornier Schwimmerflugzeug Do 22 Kj, Friedrichshafen, Germany 1938
Baubeschreibung Nr. 1200 28 Juli 1935
Beureilung der Flugzeugmusters Do 22, 22. November 1934
Bericht uber das Nachfliegen des Flugzeugmusters Dornier Do 22 in Friedrichshafen am 20.11.1934.
Bericht uber das Nachfliegen des Flugzeugmusters Dornier Do 22 in Friedrichshafen am 22.11.1934.
Geschichtliches über die Schwimmerflugzeuge Do C3 und Do C2A
Dornier Werke GmbH, Mehrzwecke-Flugzeug Dornier Do 22, Ein Beispiel für folgerichtige Konstruktionsarbeit
Dornier Werke GmbH, Flugzeugstückzahlen 1934-1945, Landflugzeuge (Serienproduktionsaufträge)
Dornier Werke GmbH, Ausgabe Nr. 12, Dornier-Flugzeugmuster die 1934/1935 gebaut wurden
Dornier Werke GmbH, Flugzeug Zusammensellung Bl. Nr. 8, 9
Dornier Werke GmbH, Seeflugzeug Dornier Do 22
Trimmflug Do C3
Trimmflug Do C2A
Tabelle über die Rüstgewichte der Type Do 22, Bl. 1, 2, 3, 4, 5
Dornier-Werke GmbH, Typenblatt Mehrzweck-Land- und Seeflugzeug Do 22

Flugzeugstuckzahlen 1934-1943
Fabrikations – Bestand am 27. September 1924, Brochüre 28
Flugzeug-Zusammenstellung, Bl.5/Forts.6, Bl.6/Forts.7, Bl.7/Forts.8, Bl.8, Bl.Nr.9
Baubeschreibung Do D
Baubeschreibung Do Y
Baubeschreibung Nr. 1203
Jahresbericht 1930
Dornier Typenblatt: Do D, Dornier-Werksnachichten 4/1974
Do17Ka-Jugoslawien 8/3, Baujahr 1937/38
8-17 K,00-100 Übersicht
8-17 K,00-101 Längsschnitt
Staatliche Flugzeugwerke Kraljevo (Jugoslawien) – Ausführungen in Zusammenhand mit der Weideraufnahme der Arbeiten in den Flugzeugwerken Kraljevo
Flugzeuge für die Kroatische Luftwaffe
Beircht über eine Reise nach Jugoslawien betroffend Lizenzbau Do.17 K v.19 – 30.4.37.

Luftarchiv Hafner, Ludwigsburg, Germany
Do 22 Kl Handbuch fur Dornier-Wasser-u. Landflugzeug Do 22 Kl
Do 22 Kl – Vorschriften für Hispano-Motor 12 Ydrs

Vojni Arhiv, Belgrade, Serbia
Popisnik 17, kutija 390, fascikla 3, dokument 8
Popisnik 17, kutija 393, fascikla 11, dokument 12
Popisnik 17, kutija 396, fascikla 1, document 5
Popisnik 17, kutija 1071, fascikla 1, document 6
Popisnik 17, kutija 1103, fascikla 1, dokument 2, 9, 11, 23, 25, 26, 36. 39, 44
Popisnik 17, kutija 1104, fascikla 1, dokument 49
Popisnik 17, kutija 1105, fascikla 1, dokument 46, 54, 56
Popisnik 17, kutija 1105, fascikla 2, dokument 21, 31, 32 i 34
Arhiv Jugoslavije, Belgrade, Serbia
Fond 54, fascikla 340, arhivska jedinica 504
Fond 65, fascikla 907, arhivska jedinica 1693
Fond 65, fascikla 907, arhivska jedinica 1695

RAF Museum. London, UK
AM (Air Movement) Form 59: YU 302, 306, 307, 308, 309, 311, 312, 313
AM (Air Movement) Form 59: YU 3348, 3363

The National Archives, Kew, UK
AIR 27/1422, AIR 27/1425, AIR 27/30-1, AIR 27/30-2, AIR 27/30-3, AIR 27/30-4

Muzej vazduhoplovstva - Beograd, Belgrade, Serbia
Fond Kraljevine Jugoslavije, kutija 44

Istorijski Institut JNA- Beograd, Belgrade, Serbia
Popisnik 10, kutija 376, fascikla 1, dokument 2
Popisnik 36, kutija 376, fascikla 12, dokument 3
Popisnik 47, kutija 376, fascikla 3, dokument 4
Popisnik 14, kutija 376, fascikla 2, dokument 2

Personal Collections:
Serbia: Aleksandar Ognjević, Dejan Milojević, Aleksandar Smiljanić, Aleksandar Kolo, Ognjan Petrović, Mario Hrelja, Milan Micevski, Bojan Dimitrijević, Nenad Jovanović, Nebojša Milovanović, Miloš Milosavljević, Momir Milinović, Aleksandar Milošević, Vladeta Vojinović, Mileta Protić, Igor Černiševski, Marko Babić, Petar Bosnić, Željko Marković, Vojislav Stankov, Spomenko Marković, Djordjević Family, Aleksandar Aleksić, Predrag Grandić, Boris Ciglić, Šime Oštrić, Predrag Stamenković, Dragan Drašković
Croatia: Josip Novak, Robert Čopec, Mario Raguž, Tomislav Aralica, Danijel Frka, Dinko Predoević
Slovenia: Tomaž Perme, Dalibor Jovanović, Marko Ličina
Italy: Giancarlo Garrelo, Gregory Alegi, Roberto Gentilli
Germany: Günter Frost, Oliver Fisher, Volker Koss, Peter Petrick
France: Michel Ledet
Switzerland: Herald Schiess, Peter Simeon
Netherlands: Jan van den Heuvel
Greece: Andrew Stamatopoulos
Romania: Denés Bernád
Hungary: Gyorgy Punka
USA: Djordje Nikolić, Tod Rathbone
UK: Ian Carter, Andrew Crawford, Andrew Thomas

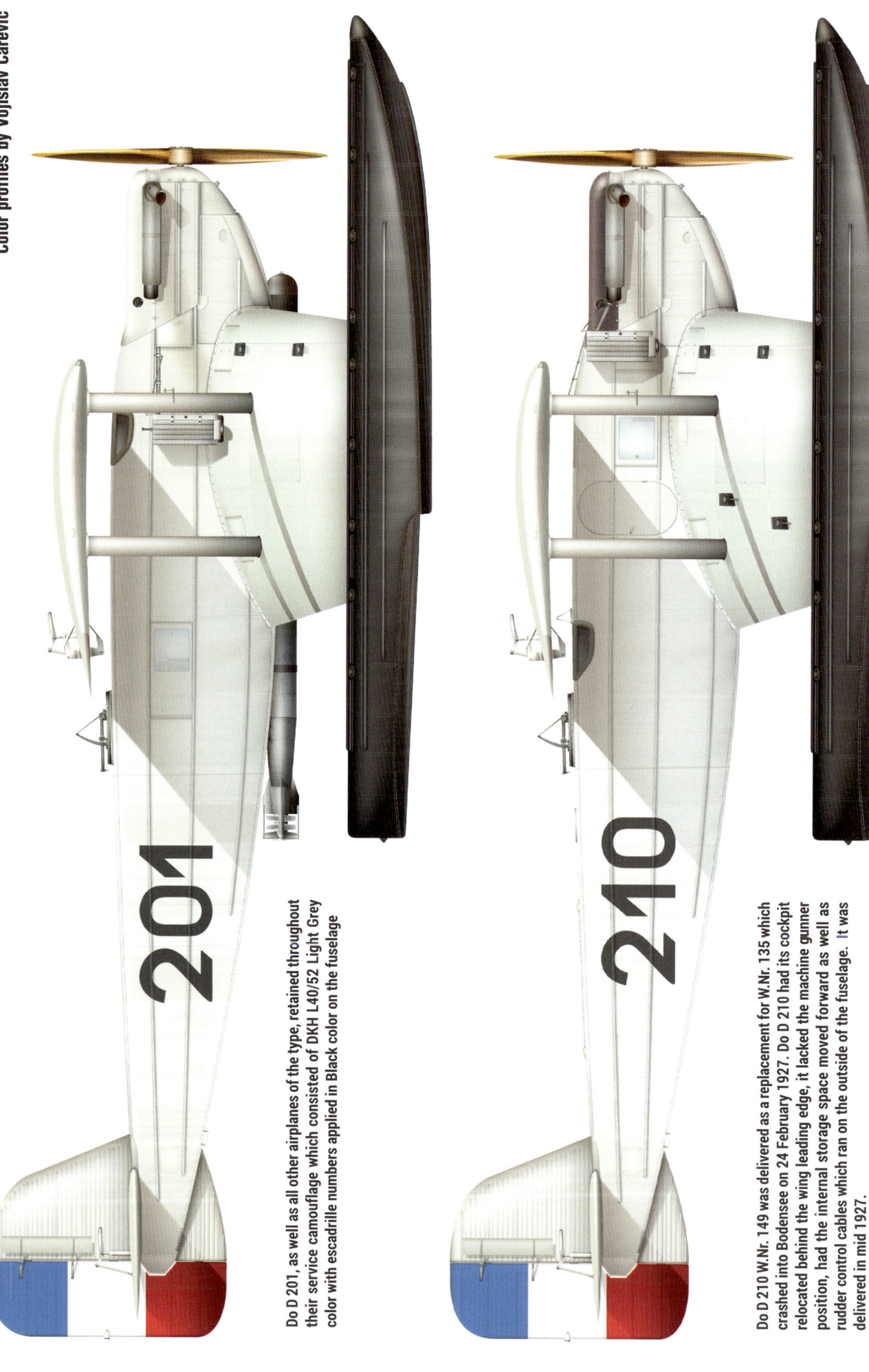

Do D 201, as well as all other airplanes of the type, retained throughout their service camouflage which consisted of DKH L40/52 Light Grey color with escadrille numbers applied in Black color on the fuselage

Do D 210 W.Nr. 149 was delivered as a replacement for W.Nr. 135 which crashed into Bodensee on 24 February 1927. Do D 210 had its cockpit relocated behind the wing leading edge, it lacked the machine gunner position, had the internal storage space moved forward as well as rudder control cables which ran on the outside of the fuselage. It was delivered in mid 1927.

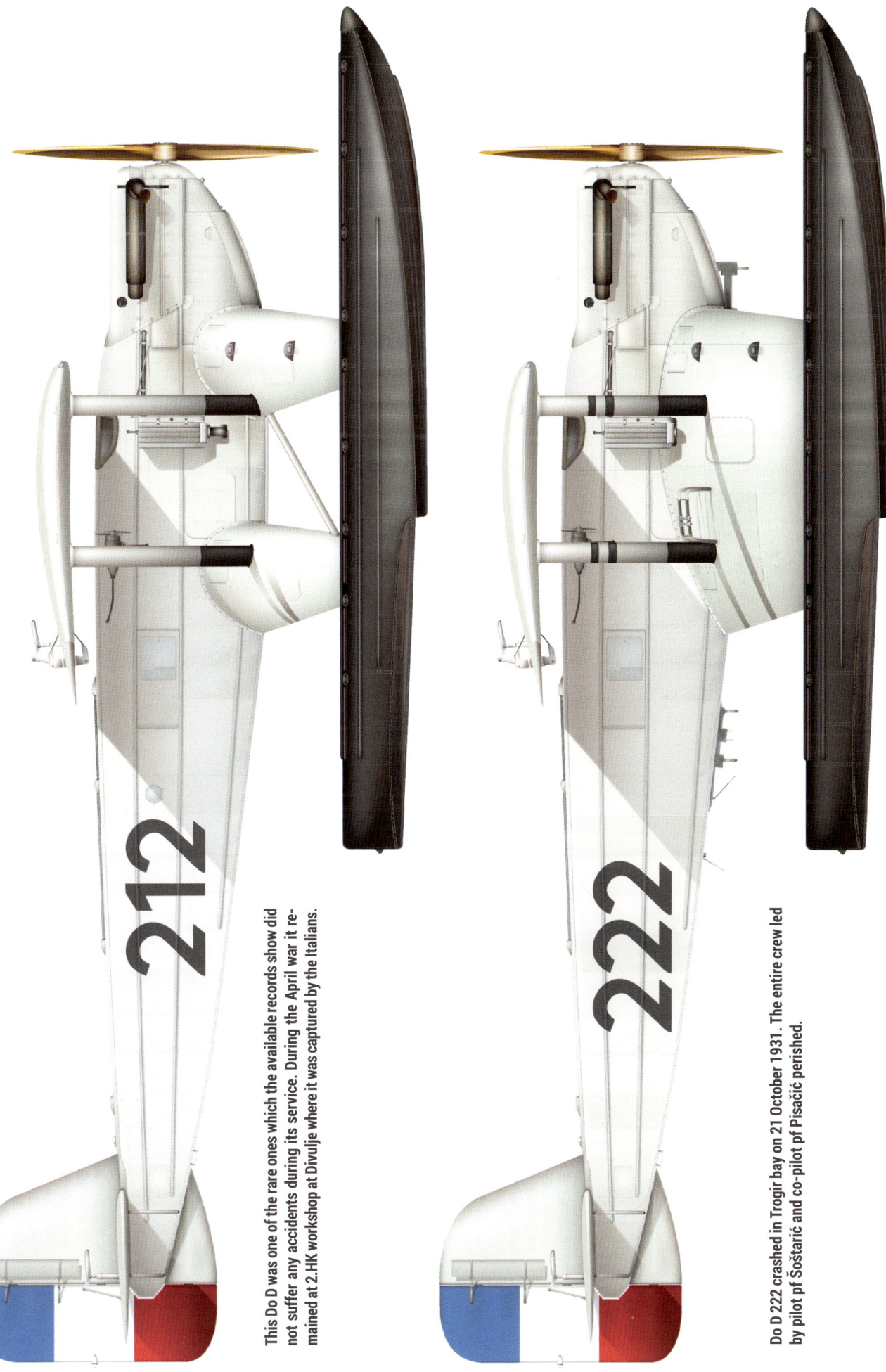

This Do D was one of the rare ones which the available records show did not suffer any accidents during its service. During the April war it remained at 2.HK workshop at Divulje where it was captured by the Italians.

Do D 222 crashed in Trogir bay on 21 October 1931. The entire crew led by pilot pf Šoštarić and co-pilot pf Pisačić perished.

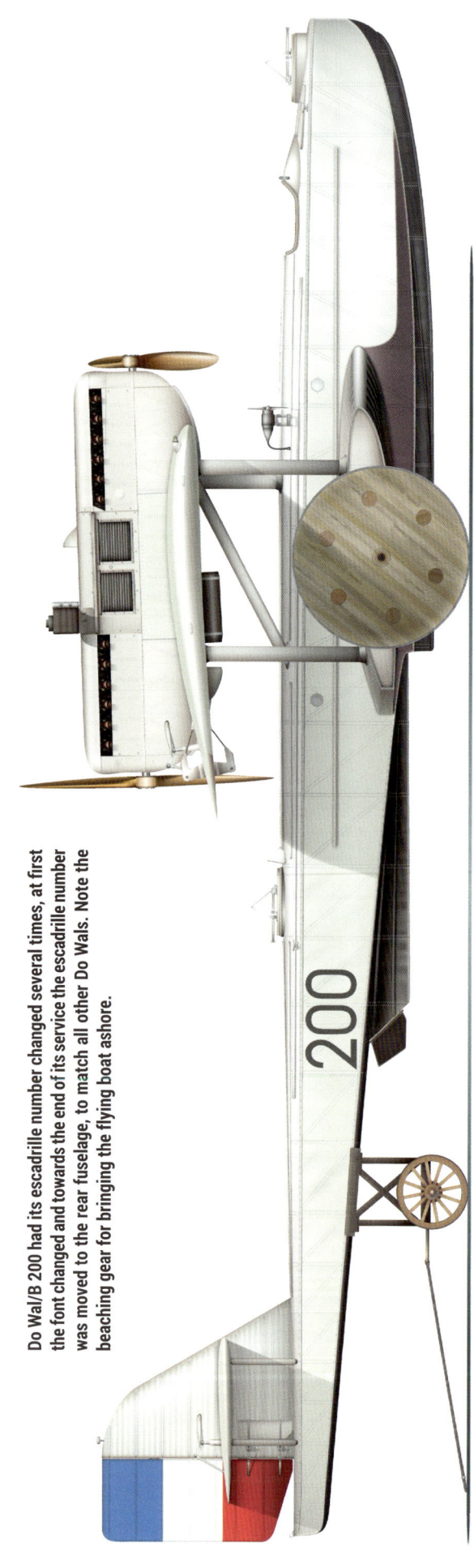

While in service Do Wal/B 200 was painted several times with a Light Grey color which was most likely the DKH L40/52.

Do Wal/B 200 had its escadrille number changed several times, at first the font changed and towards the end of its service the escadrille number was moved to the rear fuselage, to match all other Do Wals. Note the beaching gear for bringing the flying boat ashore.

Do Wal/J 256 was one of several Wals which had an interesting yellow triangle with a red border and large black number 6 inside applied on the vertical stabilizer. The purpose of this triangle remains unknown.

The second BMW VI powered machine was Do Wal/B 257 which was delivered in 1933. During the April war it was damaged on 12 April when it ran aground.

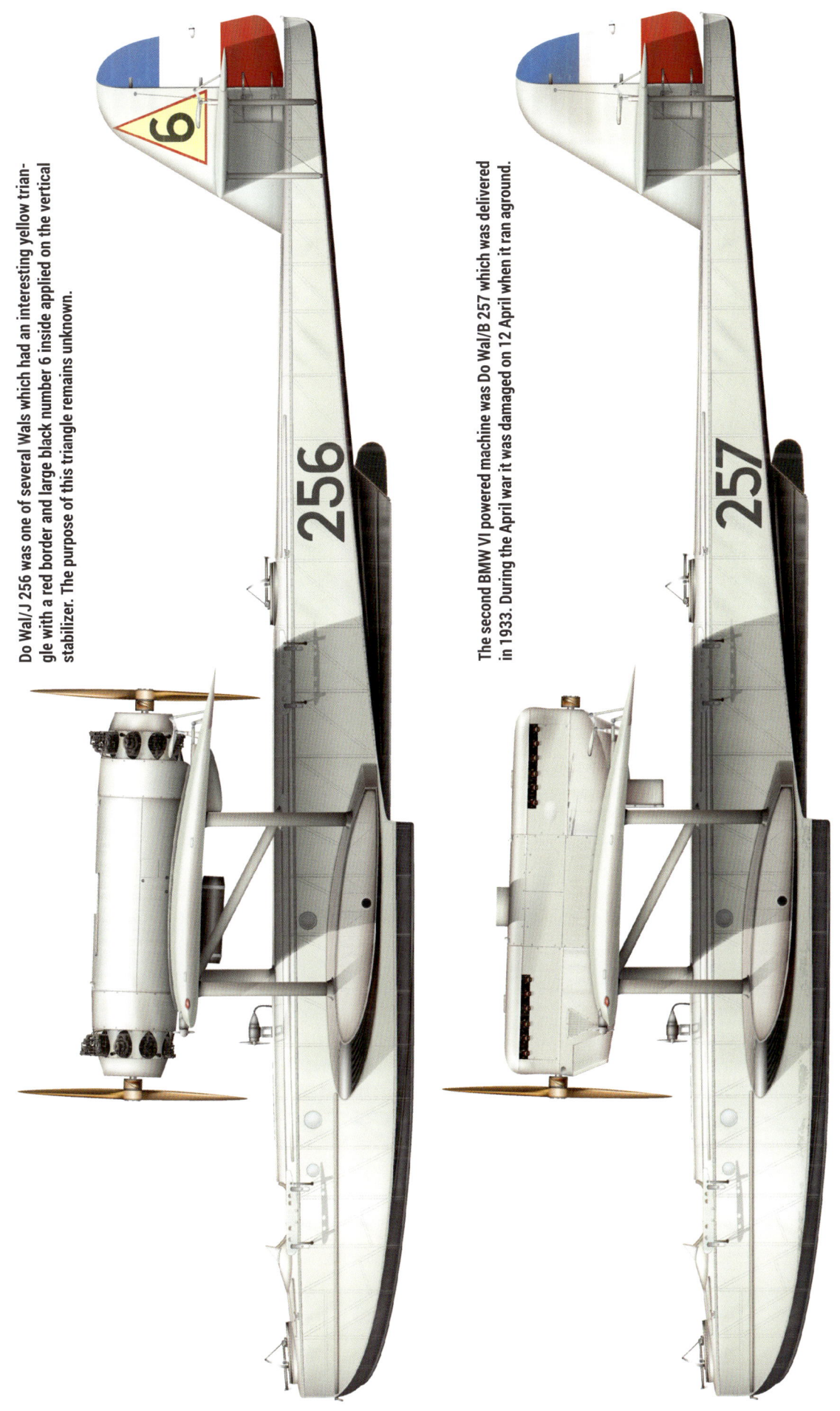

Color profiles by Vojislav Carević

Do Wal/H 260 was one of three Wals powered by a Hispano-Suiza Ydrs engine. It carried an interesting hastily applied green paint on wing top surfaces as well as part of the fuselage topside behind the wings.

Wings of Do Wal/H260 before and after the ad hoc camouflage application during April war. Note that the "Kosovo cross" insignia was covered in the process.

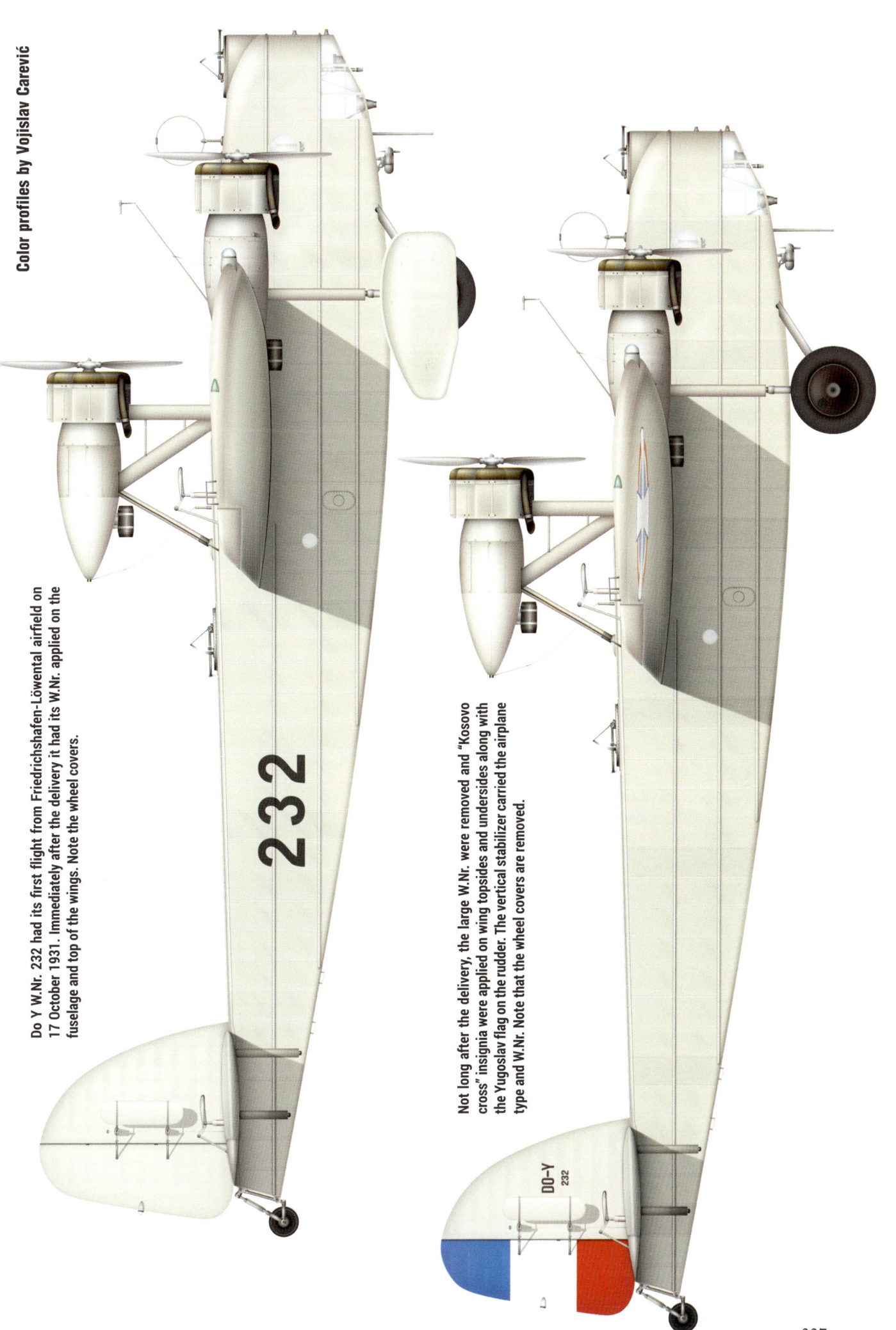

Do Y W.Nr. 232 had its first flight from Friedrichshafen-Löwental airfield on 17 October 1931. Immediately after the delivery it had its W.Nr. applied on the fuselage and top of the wings. Note the wheel covers.

232

Not long after the delivery, the large W.Nr. were removed and "Kosovo cross" insignia were applied on wing topsides and undersides along with the Yugoslav flag on the rudder. The vertical stabilizer carried the airplane type and W.Nr. Note that the wheel covers are removed.

DO-Y
232

Color profiles by Vojislav Carević

W.Nr. 232 was painted in overall *Sivo-mas/inasta boja* (SMB – Gray-Olive color) and had White escadrille number 173 applied on the fuselage. Note the Cyrillic letter "ħ" in White color on the vertical stabilizer.

Do Y W.Nr. 233 at the latter part of its active service. Note that the airplane type and W.Nr. were moved to the White field of the Yugoslav flag on the rudder.

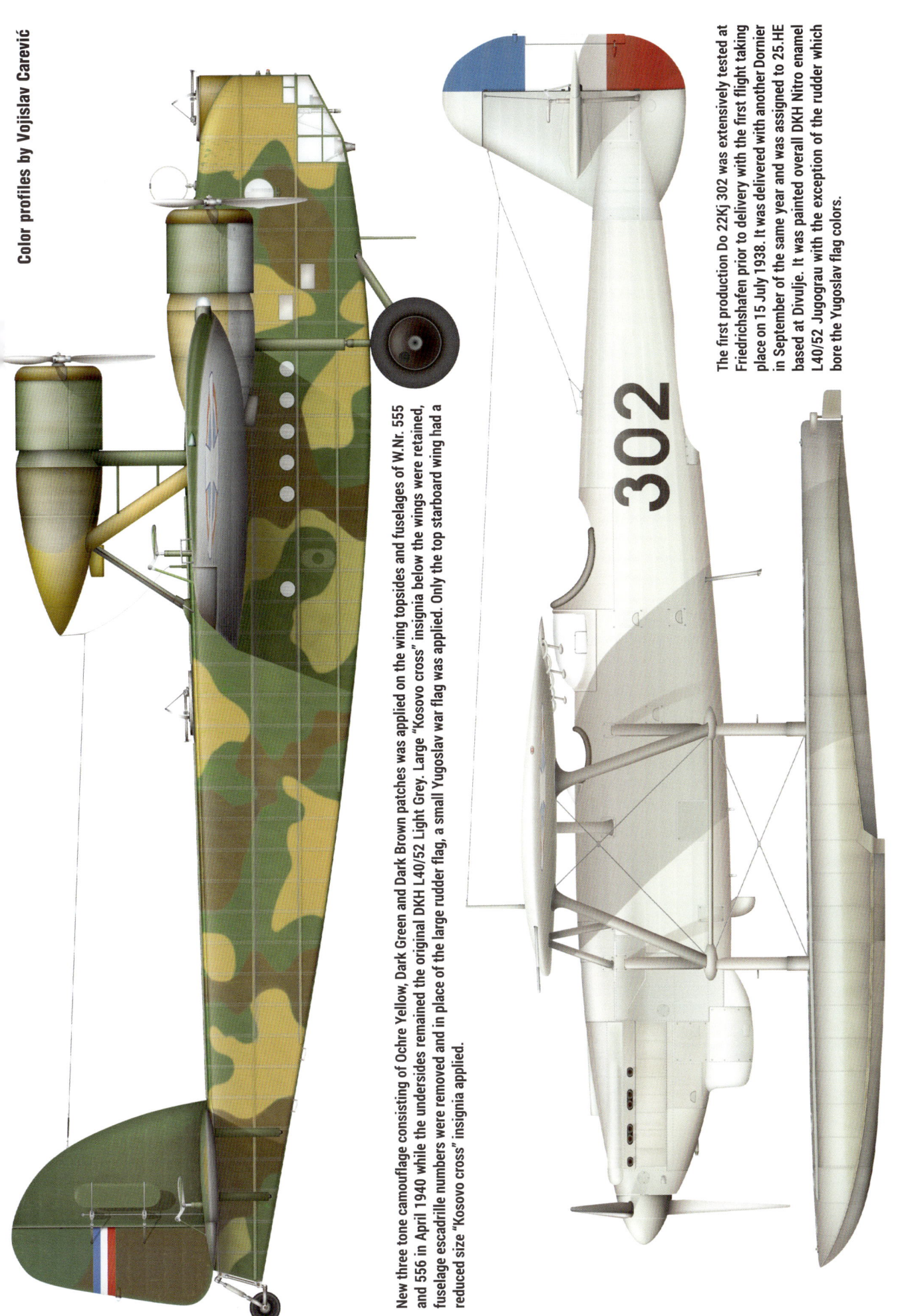

Color profiles by Vojislav Carević

New three tone camouflage consisting of Ochre Yellow, Dark Green and Dark Brown patches was applied on the wing topsides and fuselages of W.Nr. 555 and 556 in April 1940 while the undersides remained the original DKH L40/52 Light Grey. Large "Kosovo cross" insignia below the wings were retained, fuselage escadrille numbers were removed and in place of the large rudder flag, a small Yugoslav war flag was applied. Only the top starboard wing had a reduced size "Kosovo cross" insignia applied.

The first production Do 22Kj 302 was extensively tested at Friedrichshafen prior to delivery with the first flight taking place on 15 July 1938. It was delivered with another Dornier in September of the same year and was assigned to 25.HE based at Divulje. It was painted overall DKH Nitro enamel L40/52 Jugograu with the exception of the rudder which bore the Yugoslav flag colors.

329

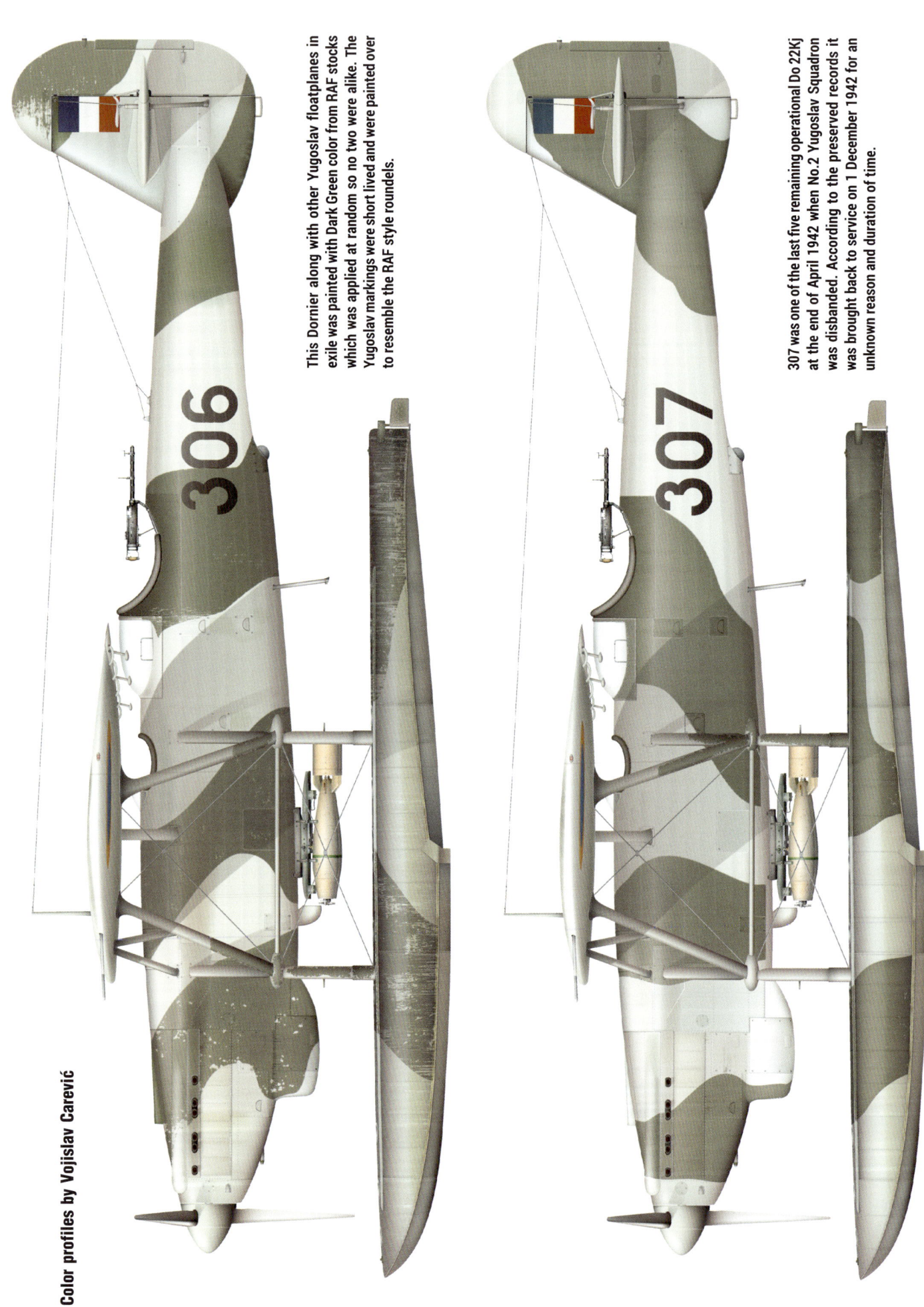

Color profiles by Vojislav Carević

This Dornier along with other Yugoslav floatplanes in exile was painted with Dark Green color from RAF stocks which was applied at random so no two were alike. The Yugoslav markings were short lived and were painted over to resemble the RAF style roundels.

307 was one of the last five remaining operational Do 22Kj at the end of April 1942 when No.2 Yugoslav Squadron was disbanded. According to the preserved records it was brought back to service on 1 December 1942 for an unknown reason and duration of time.

Color profiles by Vojislav Carević

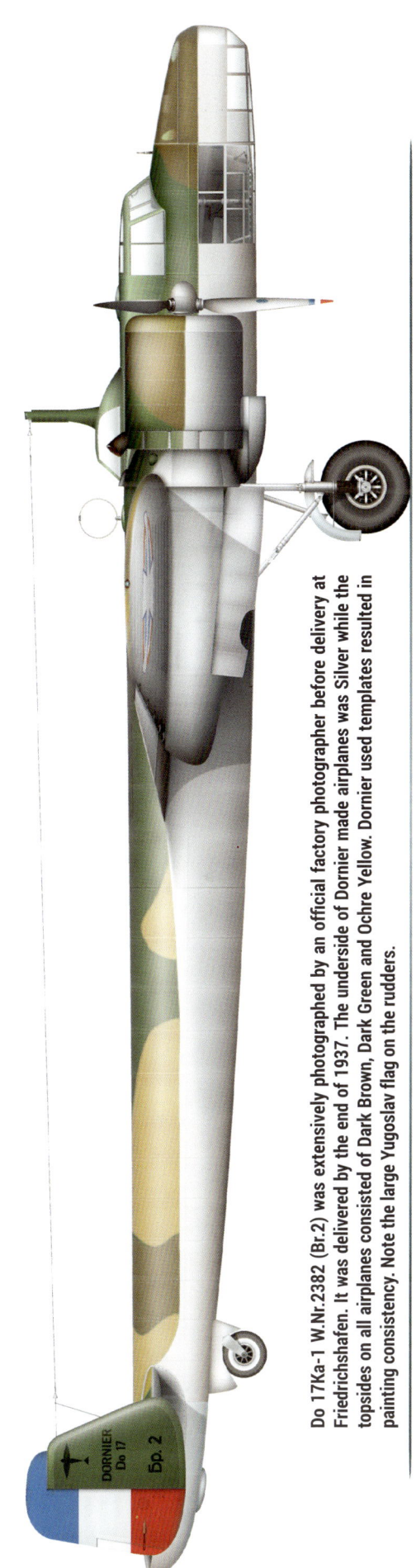

This Do 22Kj is so far the only known example which had Red, White and Blue colored roundel applied on the wing undersides. The intent of the new style markings was to avoid friendly fire due to the Yugoslav "Kosovo cross" similarity to German insignia.

Do 17Ka-1 W.Nr.2382 (Br.2) was extensively photographed by an official factory photographer before delivery at Friedrichshafen. It was delivered by the end of 1937. The underside of Dornier made airplanes was Silver while the topsides on all airplanes consisted of Dark Brown, Dark Green and Ochre Yellow. Dornier used templates resulted in painting consistency. Note the large Yugoslav flag on the rudders.

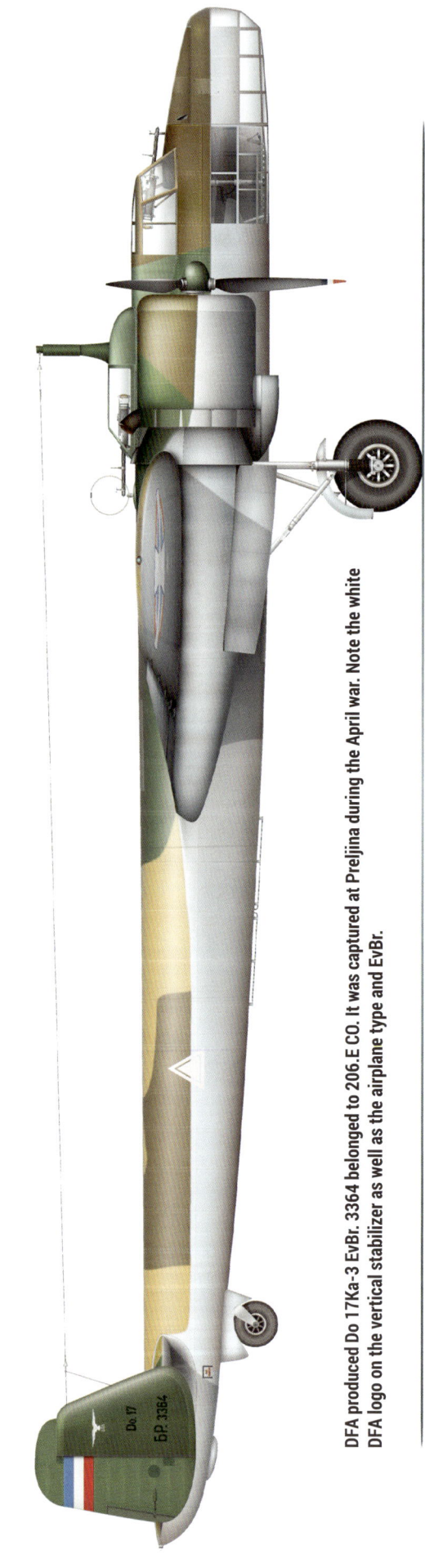

Do 17Ka-2 EvBr. 3333 (W.Nr. 2473) belonged to the second series Dornier produced airplanes. During the April war while attempting to fly over to the Soviet Union it force landed at around 15:30 3 km from Nagyskomkút and ended on its nose. This machine belonged to 210.E C0 as evidenced by the twin red circles on the fuselage.

DFA produced Do 17Ka-3 EvBr. 3364 belonged to 206.E C0. It was captured at Preljina during the April war. Note the white DFA logo on the vertical stabilizer as well as the airplane type and EvBr.

Color profiles by Vojislav Carević

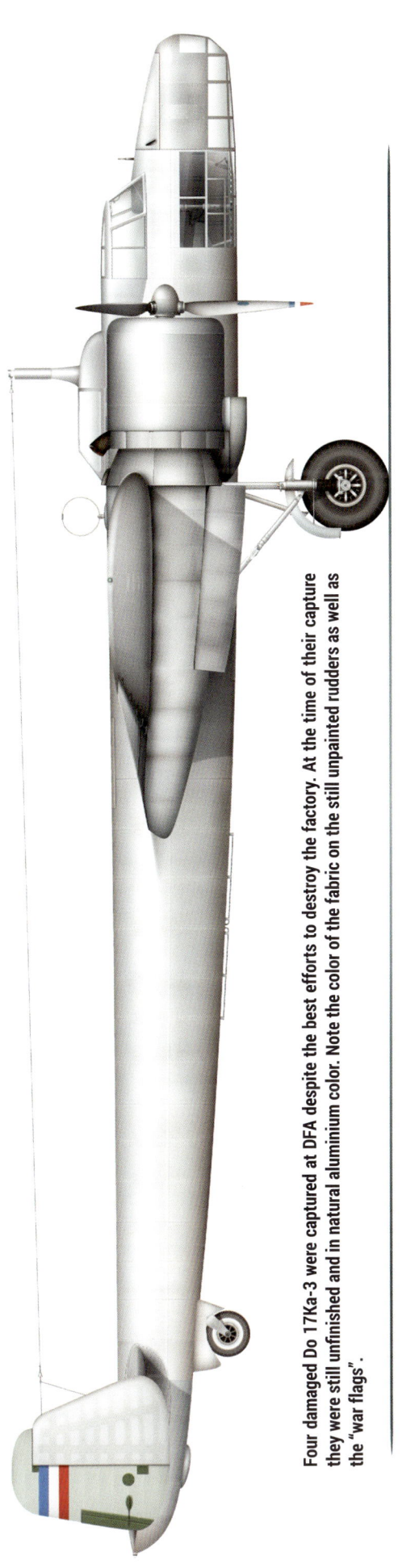

Four damaged Do 17Ka-3 were captured at DFA despite the best efforts to destroy the factory. At the time of their capture they were still unfinished and in natural aluminium color. Note the color of the fabric on the still unpainted rudders as well as the "war flags".

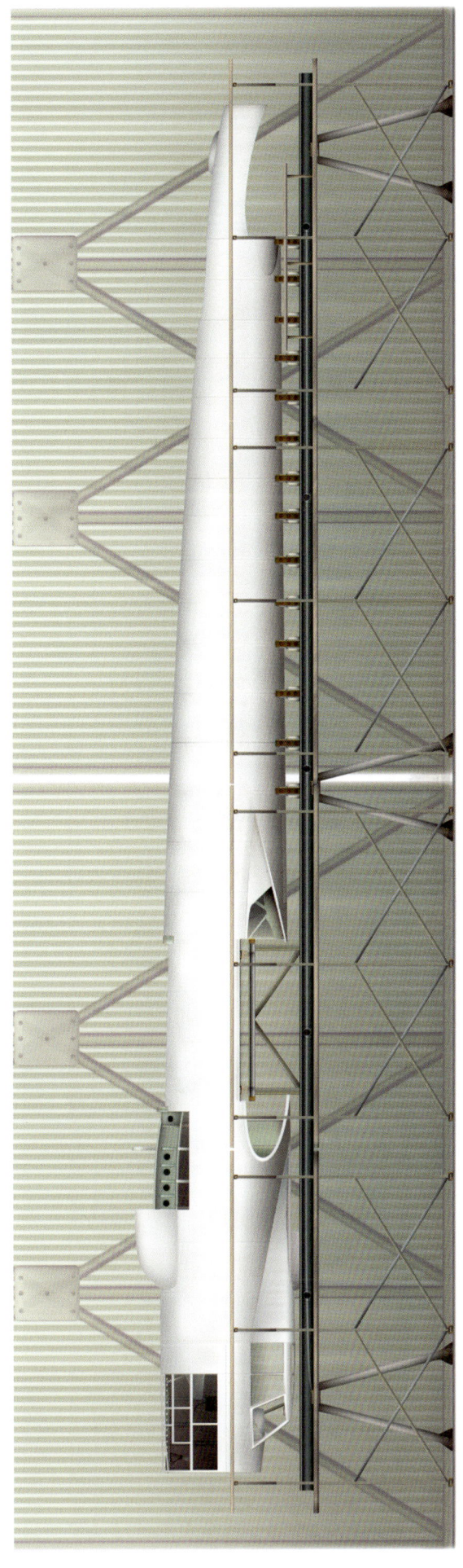

An unknown an incomplete Do 17 fuselage with a belly turret, similar to that on Do 215, was captured at DFA. This could have been an entirely ingenious design or could have been replicated from drawings provided by Dornier.

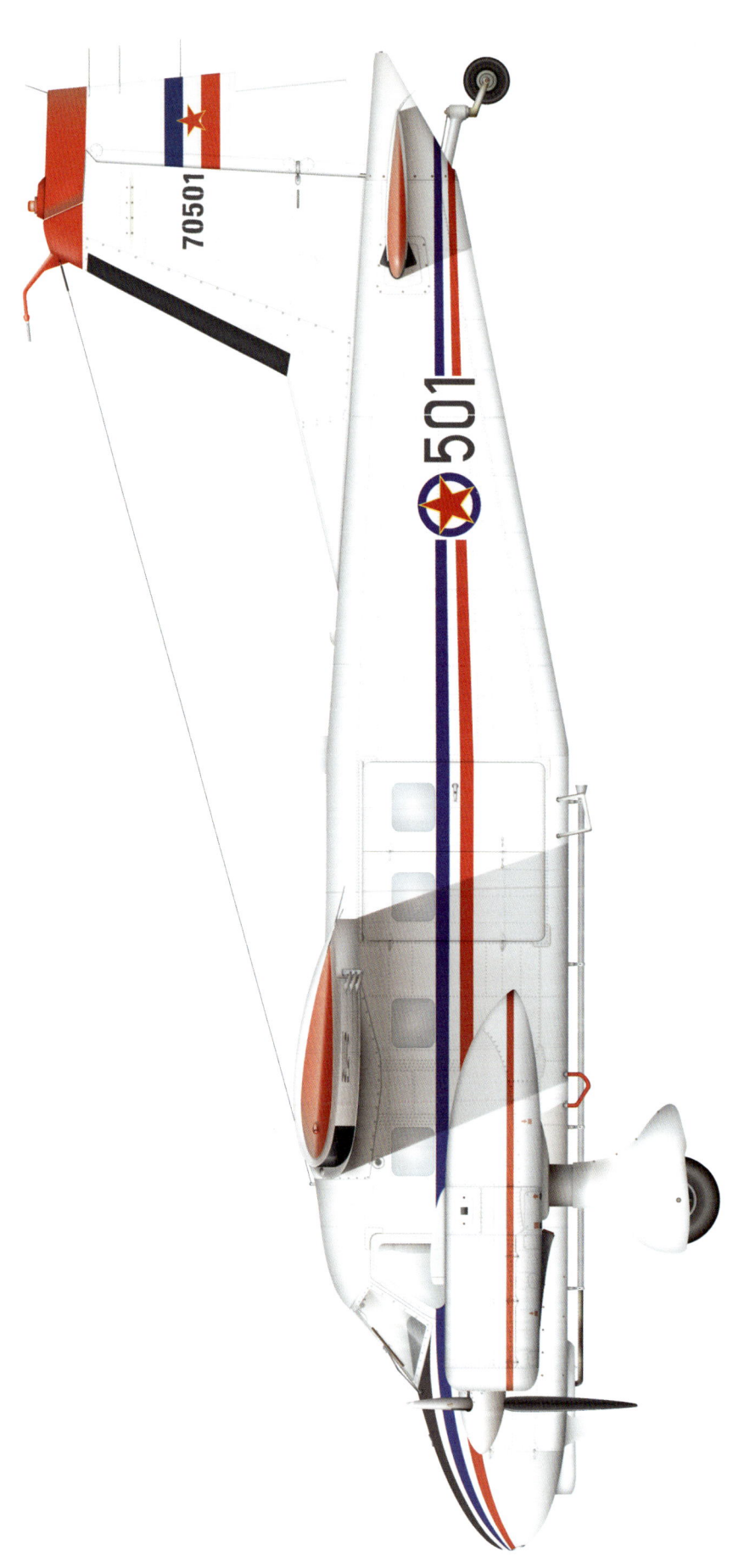

Do 28 70501 was one of the two airplanes in service. Both were delivered from the factory in overall White livery which they retained throughout their service. A Yugoslav flag band consisting of Blue, White and Red fields wrapped around the entire fuselage, from the nose to the tail. The wing leading edge was painted Black to assist with de-icing (the so called de-icing boots) and the wing tips, the vertical stabilizer and horizontal stabilizer tips were painted in Red colour. The engine nacelles had a Red band around the entire length.

70502

Following the break-up of the SFRJ in 1991, a new style insignia, the so called "Pepsi-Cola" roundel, was applied in place of the previous markings. The insignia and the registration number were removed from the fuselage. The large flag band from the rudder was removed and its place a smaller size SRJ flag was applied on the vertical stabilizer. The registration numbers were removed from the wings as well.

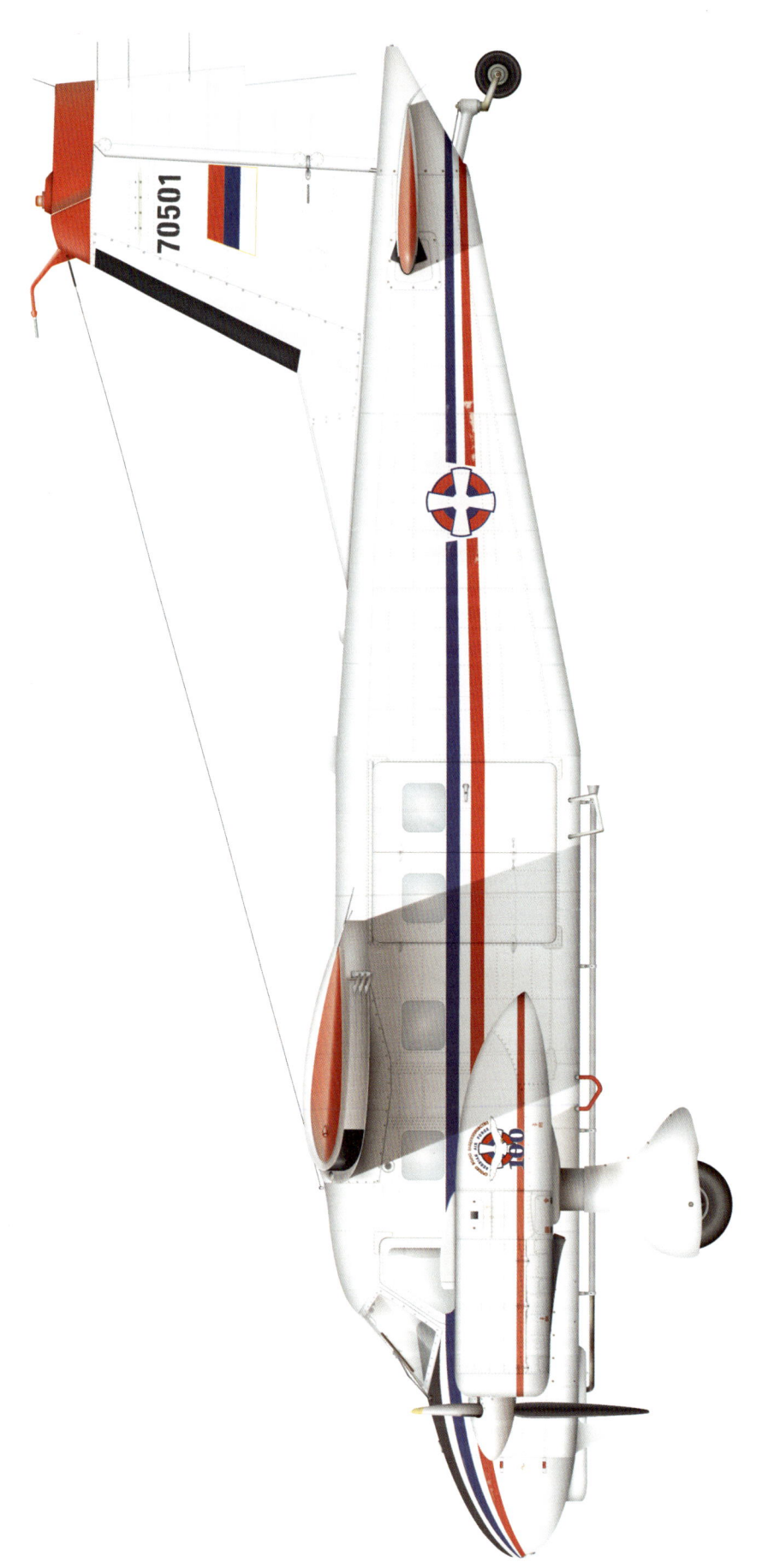

70501

During the celebrations at Batajnica airfield commemorating the 100th Anniversary of the Serbian Air Force, 70501 received the new Serbian roundel with "Kosovo cross" on the fuselage, along with a Serbian flag on the vertical stabilizer and an appropriate logo to mark the occasion on the outboard sides of its engine nacelles.